Fish Management and Aquatic Environment

Fish Management and Aquatic Environment

Editors
Professor (Dr.) Arvind Kumar

&

Dr. Pushpalata Dubey

2017
Daya Publishing House®
A Division of
Astral International (P) Ltd
New Delhi 110 002

First Published, 2006
Reprinted, 2017

ISBN: 978-81-7035-429-1(Hardbound)

Published by : **Daya Publishing House®**
A Division of
Astral International Pvt. Ltd.
– ISO 9001:2008 Certified Company –
4760-61/23, Ansari Road, Darya Ganj
New Delhi-110 002
Ph. 011-43549197, 23278134
E-mail: info@astralint.com
Website: www.astralint.com

Printed at : **Sanat Printers**

Preface

Fish is the most efficient among farm animals in converting feed into nutritious proteinous food. Fish enjoys a very special consideration and place in human civilization from early period. It was regarded as one of the treasures of water. The nutritive and medicinal value of fish has been recognized from time immemorial. Fresh fish flesh is an excellent source of protein for human diet. Traditionally the fish is considered as the "protein for poor". Some fishes are traditionally used for curing a few human diseases. Fishes are used as tool for biomedical research and are also a bio-control agent of vector and pests of medical and agricultural importance. Therefore, the fish must be considered as integral part of overall development of human health and wealth and its importance for human beings can't be neglected.

Population pressure, pollution, over exploitation, destruction of habitat, urbanization, siltation and introduction of exotic species are some of the important activities that lead to loss of fish biodiversity. The prime reason for the loss of fish genetic diversity is water pollution. Fish is the bio-indicator of aquatic ecosystem health. If the fishes can survive and reproduce in a particular aquatic ecosystem, that ecosystem is considered to be a healthy one. Thus, fish is extremely valuable for human beings. There is an urgent need for the development of fish farming in India to supply sufficient amount of fish to the poor people. So, the aim of this book is to provide a wide-ranging and authoritative coverage of topics of fish management for sustainable development. The all research articles are written by authors who are internationally acknowledged experts in the different fields of aquaculture with the intention to provide a treasure in the form of book to the students, researchers and academicians as well.

My special thanks and appreciation go to the scientists whose contributions have enriched this volume. I wish to express my sincere gratitude to Dr. P.C. Hembram Hon'ble Vice Chancellor, S.K.M. University, Dumka who has been a source of constant inspiration. I owe my special thanks to Professor M. C. Dash, Hon'ble Vice Chancellor of Sambalpur University, Professor H.R. Singh, Hon'ble Vice Chancellor of Allahabad University, Professor K.C. Pandey, Former Vice Chancellor of Meerut University, Professor N.C. Datta of Calcutta University, Dr. S.N. Kaul of NEERI, Professor U.S. Bagde of Mumbai University, Professor S.K. Konar of Kalyani University, Professor D.K. Belsare of Bhopal University, Professor U.C. Goswami of Gauhati University, Professor P.C. Mishra of Sambalpur University, Professor

P. Natarajan of Kerala University, Professor P.S. Murthy of Bangalore University, Professor Ajit Varma of J.N.U., New Delhi, Professor P.P. Sood of Saurashtra University, Professor A.L. Bhatia of Jaipur University, Professor A.K. Mittal of B.H.U., Professor S.P. Hosmani of Mysore University, Professor K. Kapoor of Udaipur University, Professor K.B. Reddy of Nagarjuna University, Professor M. Vikram Reddy of Pondicherry University, Professor B.K. Tiwari of NEHU, Professor G. Tripathi of Jodhpur University, Professor R. Ramalingam of Annamalai University, Professor Sharif U. Ahmad of Nagaland University, Professor G.K. Kulkarni of Aurangabad University, Professor S.U. Mehram of Nagpur University, Professor S. K. Battish of PAU, Professor B. Manihar Sharma of Manipur University, Professor B.D. Joshi of Hardwar University, Professor G.C. Pandey of Faizabad University, Professor K.C. Sharma of Ajmer University, Professor A.K. Dobryal of Garhwal University, Professor Kaushal Kumar of Gorakhpur University, Professor M. Raziuddin of Hazaribag University, Professor Gurdeep Singh of I.S.M., Dhanbad, Dr. P.K. Goel of Karad, Dr. Ram Kumar of Delhi University, Professor S.P. Roy and Shri Tribhuwan Poddar of Bhagalpur University, Bhagalpur as well as Dr. Chandan Bohra of B.S.K. College, Barharwa (Jharkhand).

I also express my deep sense of gratitude to my parents whose blessings have always prompted me to pursue academic activities deeply. I am also thankful to my sweet wife, *Kumari Bimla* and my two lovely sons, *Kumar Pallav Shivshankaran* and *Kumar Prasun Ramakrishnan* whose natural smiles extended to me relief all through this tiresome endeavour.

Last but not the least, I am also thankful to Mr. Anil Mittal, Proprietor, Daya Publishing House, Delhi for taking keen interest in bringing out of this book. Finally, I will always remain a debtor to all my well-wishers for their blessings, without which this book would not have come into existence.

Dr. Arvind Kumar

Dr. Pushpalata Dubey

Contents

Section II: Aquatic Environment

Section I

FISH MANAGEMENT

Chapter 1

Transgenic Fish: Possibilities and Threats

Partha Bandyopadhyay* and Arvind Kumar**

**Latika Sea Food Private Limited, Jalalkhanbard, Contai –721 401, Purba Medinipur, West Bengal, India*
***Environmental Science Research Unit, Post Graduate Department of Zoology, S.K.M. University, Dumka –814 101, India*

ABSTRACT

The potential benefits of animal biotechnology will become a reality only if the regulatory system can review these products in a comprehensive manner that merits public confidence and gives product developers a clear pathway to the marketplace. Issues in Science and Regulation of Transgenic Fish describes some of the products that could be created through the application of genetic engineering to aquaculture as well as the potential environmental and food safety issues. But, as demonstrated by the current plan to regulate transgenic fish, it is unclear if regulators have the tools they need to adequately evaluate these new products, and there is reason to wonder if innovation is getting ahead of our ability to manage it. Without a clearly articulated and transparent road map to guide agency review and approval, it will remain difficult for developers to bring products to market and even more difficult to assure the public that the next generation of agricultural biotechnology products are undergoing a regulatory review process that includes a careful, thorough, and rational consideration of risks and benefits.

Keywords: *Transgenic fish, Animal biotechnology, Ecological risk.*

Introduction

Over the past decade, genetic engineering has emerged as one of the most powerful transforming technologies known to this generation. Scientists can now transfer the beneficial traits of a particular gene from one organism to another in far less time than needed in traditional breeding and with more precision. The application of genetic engineering to agriculture already makes it easier for farmers to grow certain crops and reduces the use of certain pesticides. But this technology has also triggered debate among scientists, philosophers, environmental advocates, public health officials, business leaders, and regulators over a range of issues–from environmental safety and ecological impacts to the ethics of altering a genome.

The potential benefits of animal biotechnology will become a reality only if the regulatory system can review these products in a comprehensive manner that merits public confidence and gives product developers a clear pathway to the marketplace. In the mid-1980s, federal policy declared that biotechnology products would be evaluated under the same laws and regulatory authorities used to review comparable products produced without biotechnology. However, as scientists develop new varieties of organisms through genetic engineering that confound conventional categorization, regulators are increasingly pressed to develop creative interpretations of existing laws to address the unique issues raised by these products. For example, the Food and Drug Administration (FDA) has indicated its intention to use its authority to approve "new animal drugs" under the Federal Food, Drug, and Cosmetic Act (FFDCA) as the basis for regulating transgenic fish.

Issues in Science and Regulation of Transgenic Fish describes some of the products that could be created through the application of genetic engineering to aquaculture as well as the potential environmental and food safety issues associated with such products. The report also analyzes the process through which regulators plan to evaluate transgenic fish. By consolidating both the scientific and regulatory issues concerning genetically modified fish, this report hopes to facilitate a robust discussion about genetically modified fish and the adequacy of the regulatory process through which they may be brought to market.

Development and Economic Implications

The first reports of the application of genetic engineering to animals appeared in the 1980s. Since then, there has been a burst of genetic modification activity in aquaculture research and development. Fish make attractive candidates for genetic modification for several reasons. One is that they produce eggs in large quantities and those eggs develop outside the fish's body. A second, equally compelling reason is that, aquaculture is one of the fastest growing food-producing sectors globally.

Given such demand, it is not surprising that the majority of aquaculture biotechnology research and development efforts to date have focused on improving production. Scientists have modified at least 14 fish species–including varieties of carp, trout, salmon and channel catfish–so they will grow 2 to 11 times faster than their non modified counterparts. Current research efforts are not, however, limited to improving fish farm productivity. In an attempt to identify new sources for the production of pharmaceutical substances, scientists have engineered tilapia with a human gene that makes the fish capable of producing a compound essential to the production for Factor VII, a substance vital to the clotting of human blood. Scientists have also tried to improve the cold tolerance of goldfish, so that this popular ornamental fish could thrive in different climates than its non-modified counterpart. And scientists have altered some marine microorganisms to reduce their dependence on light so they might live in unique environments. While improving aquaculture efficiency could lower costs and generate economic benefits to producers and consumers, global supply and demand combined with consumer acceptance will largely determine whether these benefits will be realized.

Transgenic Fish Research in India

India has developed experimental transgenics of rohu fish (*Labeo rohita*), zebra fish, cat fish and singhi fish (*Heteropneustas fossilis*).Genes, promoters and vectors of indigenous origin are now available for only two species namely rohu and singhi for engineering growth. Transgenic rohu recently produced from indigenous construct at Madurai Kamaraj University has proved to be eight times larger than the control siblings. This transgenic rohu attains 46 to 49 gram body weight within 36 weeks of its birth. In India, research in transgenic fish was initiated in Madurai Kamaraj University (MKU), Centre for Cellular and Molecular Biology (CCMB), Hyderabad and National Matha College, Kollam with borrowed constructs from foreign scientists (Sharma, 2004).

Taking the research further to promote the transgenic fish programme, the Indian Council of Agricultural Research (ICAR) plans to develop autotransgenesis in commercially important fish species with growth hormone gene. ICAR plans also include production of pharmaceutical and other industrial products from piscine origin, development of transgenic native glow fish varieties, fish biosensors for monitoring aquatic pollution, isolation of genes, promoters and synthesis of effective gene constructs, researches in embryonic stem cells and *in vitro* embryo production (Sharma, 2004). Indian scientists are concentrating on developing transgenic fish through autotransgenesis which involves just increasing the copies of growth hormone genes present in a fish as opposed to allotransgenesis which amounts to transfer of genes from different species. The increase in growth hormone genes leads to an increase in flesh content. Indian scientists feel that autotransgenesis is safer and less controversial.

Environmental Issues

Conventional aquaculture practices have raised a number of environmental issues including, but not limited to pollution and the impacts on wild fish populations should fish escape from fish farms. The application of genetic engineering to aquaculture adds significant new dimensions to these existing concerns, especially the multiple possibilities related to gene flow and the ecological disruption associated with escapes.

Gene Flow

One of the larger environmental concerns raised by transgenic fish is the possibility that a transgenic species raised in open water pens will escape and spread novel traits into the ecosystem by breeding with wild relatives, a biological process known as "gene flow." Gene flow between transgenic or conventionally bred fish and wild populations is an environmental concern because it may present a threat to natural biodiversity.

According to one scientific model, if a transgenic fish escaped and mated with wild fish, gene flow could follow one of three scenarios:

Purge Scenario

When the net fitness of a transgenic fish is lower than that of its wild relatives, natural selection will quickly purge from the wild population any novel gene(s) introduced by the transgenic fish. In theory, evidence of the novel trait will disappear from subsequent generations.

Spread Scenario

When the net fitness of a transgenic fish is equal to or higher than the net fitness of a wild mate, gene flow is likely to occur and the genes of the transgenic fish will spread through the wild population. This means evidence of the transgenic genome would persist in subsequent generations.

Trojan Gene Scenario

When the net fitness of a transgenic fish is altered such that the fish has enhanced mating success but reduced adult viability (*i.e.*, chances of surviving long enough to mate), introduction of that fish into the wild population could result in a rapid decline of the wild population. Essentially mating success would ensure the spread of the novel gene throughout the population, but the inability to survive would reduce the population size of subsequent generations and potentially lead to extinction.

Lastly, genetic changes in a species may impact the resilience of fish communities (the interconnected groups of fish species and other aquatic organisms living in the same environment). Although scientific understanding about what drives the resilience of a fish community is evolving, the introduction of transgenic fish could lead to a resilience reduction that may be irreversible or difficult and expensive to undo.

Transgenic Fish as Invasive Species

Even if they don't breed with wild relatives, transgenic fish that escape into natural ecosystems could be an environmental nuisance by becoming an invasive species. This danger mainly arises for those transgenic fish endowed with new genes that improve such fitness traits as breeding capabilities and the ability to withstand harsh conditions.

Risk Mitigation

It is important to note that developers of transgenic fish are attempting to reduce- or eliminate both gene flow and invasive species risks by sterilizing transgenic fish. It is also noted that an escaped, sterile fish might still engage in courtship and spawning behavior, disrupting breeding in wild populations. Waves of escaped sterile fish could also create ecological disruptions as each group is replaced by another equally strong group of transgenic sterile fish.

Ecological Risk Assessment of Growth-Hormone Transgenic Fish

GH transgenic coho salmon have the potential to grow more than twice the rate of normal fish (Devlin *et al.*, 1994, 2004). In the hatchery where food is abundant, transgenic fish can quickly grow large and have the potential to out-compete normal conspecifics (Devlin *et al.*, 1999). This suggests that transgenic fish may displace normal salmon in the wild. On the other hand, to sustain their growth, transgenic fish are more willing to take risks when feeding (Sundström *et al.*, 2003). Under semi-natural conditions, this has lead to higher mortality, particularly at lower food abundance (Sundström *et al.*, 2004). Since release of transgenic fish to the wild is prohibited, we cannot study directly how transgenic fish trade-off feeding for survival in the natural environment.

Instead, we simulate the situation with elevated growth hormone levels using implants in brown trout (*Salmo trutta*) in small Swedish streams. We found that elevated hormone levels cause an increase in growth, but did not affect the survival of manipulated individuals (Johnsson and Björnsson, 2001). This indicates that GH transgenic fish may survive and compete with conspecifics in the wild. However, details on juvenile survival and adult reproductive success will also play important roles for the overall success of transgenic fish in the natural environment.

Transgenic Fish Could Threaten Wild Populations?

Purdue University researchers have found that releasing a transgenic fish to the wild could damage native populations even to the point of extinction. The Purdue research was part of an effort by Purdue and the U.S. Department of Agriculture to assess the risks and benefits of biotechnology and its products, such as genetically modified fish. Atlantic salmon was introduced to a population of unmodified fish. The research was conducted in banks of aquariums in a laboratory setting. The results warn that transgenic fish could present a significant threat to native wildlife. This experiment showed that the genetic change also reduces the offspring's ability to survive and a transgenic animal could bring a wild population to extinction in 40 generations (Muir and Howard, 2000). This extinction results from a phenomenon that called the "Trojan gene hypothesis."

Transgenic Fish Pose Extinction Threat

New evidence shows genetically modified fish released into the environment could drive wild fish populations to extinction. The finding promises to reopen the debate about how these transgenic animals should be regulated. Fish that are genetically modified to be larger or reach adult size more rapidly are sought after by commercial fish producers. None of the species has been approved by the Food and Drug Administration for human consumption but an aquarium fish, known as the Glofish, that has been genetically modified to fluoresce reddish-pink, went on the market and is available at pet stores for about $5. Watchdog groups are concerned about the impact transgenic fish could have on both humans and the environment. The new study by Purdue University researchers, which will appear in the proceedings of the National Academy of Sciences Journal, suggests those concerns might be warranted (Mitchell, 2004). This study showed for the first time that genetically modified fish could have better mating success than wild types. Salmon gene for growth hormone into male medaka fish resulted in animals that were 83 per cent larger than their wild counterparts (Mitchell, 2004). Further research showed the larger transgenic fish out-competed the wild males for females and were responsible for 75 per cent of all matings. The resultant offspring from the transgenic fish matings, however, had a reduced viability.

Food Safety Issues

As with the environmental issues raised by transgenic fish, genetic engineering adds a layer of complexity to food safety issues associated with conventional fish, both farmed and wild. It is important to note, however, that genetic engineering could potentially reduce certain fish-specific food safety dangers.

Allergies to conventional fish and shellfish are common in the general population. Some scientists have expressed concern that the genetic engineering process could increase the allergic potential of fish, particularly through the introduction of novel proteins that never before existed in the food chain. In time, researchers may determine that genetic engineering similarly allows them to address toxins and other food safety concerns unique to conventional fish.

Restaurants and Grocers Asked to Avoid Transgenic Fish

Three environmental groups are asking seafood retailers across the United States to pledge not to sell genetically engineered fish and to oppose their commercialization. Citing potential negative human health effects and a threat to the genetic purity of wild salmon, Friends of the Earth, the Center for Food Safety, and Clean Water Action announced a new campaign on October 18, 2001 that aims to prevent the commercialization of genetically engineered fish (Anon., 2001). Transgenic fish of various species of salmon, tilapia, channel catfish and others are being actively investigated worldwide as possible new food producing varieties. The groups cite a pending application to the U.S. Food and Drug Administration (FDA) for market approval of an experimental salmon developed by Aqua Bounty Farms, also known as A/F Protein. The company describes itself as a development-stage biotechnology company with offices in the United States and Canada. A/F Protein developed the transgenic fish technology out of its research on fish species living in the high north that have a compound in their blood that lowers the freezing point of the whole fish so that they are safe from freezing through the frigid Arctic winters (Anon, 2001). The Food and Drug Administration could approve the application at any time, the groups fear, putting the first engineered fish on American dinner plates, grocery shelves and in restaurants across the country.

Regulating Transgenic Fish

In 1986, federal regulators adopted the Coordinated Framework for Regulation of Biotechnology, which reflected the policy decision that no new laws were needed to regulate the then anticipated biotechnology products. Instead, the framework at this time, agency officials have stated that the biological material used to transform a genetically modified fish, and the product expressed by the fish's transformed genetic construct, both fall under the legal definition of a "new animal drug." Therefore, the Food and Drug Administration (FDA) has indicated it will use the same process to regulate transgenic fish that it currently uses to review a new animal drug. For example, the agency already has the authority to review a conventional drug that would increase growth hormone levels in fish. Therefore, one could perceive the FDA's regulatory approach as simply applying the same regulatory scrutiny to another method–genetic engineering–designed to achieve the same end result: a faster growing fish.

One way to evaluate the FDA's new animal drug application process is to examine how it measures against five elements often thought to be part of an adequate regulatory review system:

Legal Authority

It is unclear if the FDA has the comprehensive legal authority needed to address all the food safety and environmental issues associated with transgenic fish.

Adequacy of Risk Management Tools

The new animal drug approach gives the FDA a great deal of authority before and after product approval. The developer must prove the safety of a product before it is allowed on the market and the FDA can restrict the use of the product with labels, conditions of use, and post-approval monitoring.

Transparency, Clarity and Public Participation

These three elements are the backbone of public confidence in the safety of an approved product. Together, they mean the public has access to, and understands, the information used to determine if a product is fit for approval. It also means that before an agency makes an approval decision, the public can provide input.

Resources and Expertise

Even if one could stretch the FDA's new animal drug application authority to include the necessary environmental assessments, it is unclear if the agency has the staff and resources needed to do so.

Efficiency and Coordination

The expertise to deal with food safety and environmental risks associated with transgenic fish is spread over several agencies. In 2001, the Council on Environmental Quality and the Office of Science and Technology Policy published case studies that provide informal guidance as to how these agencies might work together. But no other official or formal explanation of the agencies' approach to these issues has been made public. In the absence of a clear statement of how these authorities and agencies will work together, it is difficult to judge whether the regulatory system for reviewing transgenic animals is efficient, sufficient and well coordinated.

Conclusion

Triploids have been created in a wide variety of fish. Sterile triploids have been suggested as a means to contain transgenes in transgenic stocks released to the natural environment, as if the method were "fool proof". Peer reviewed studies on transgenic fish indicates that the technology is fraught with uncertainties and problems that have not yet been resolved. But, as demonstrated by the current plan to regulate transgenic fish, it is unclear if regulators have the tools they need to adequately evaluate these new products, and there is reason to wonder if innovation is getting ahead of our ability to manage it. Without a clearly articulated and transparent road map to guide agency review and approval, it will remain difficult for developers to bring products to market and even more difficult to assure the public that the next generation of agricultural biotechnology products are undergoing a regulatory review process that includes a careful, thorough, and rational consideration of risks and benefits.

References

Anon. 2001. Restaurants and grocers asked to avoid transgenic fish. *Environmental News Network*, (Internet).

Devlin, R.H., Biagi, C.A. and Yesaki, T.Y. 2004. Growth, viability and genetic characteristics of GH transgenic coho salmon strains. *Aquaculture*, 236: 607–632.

Devlin, R.H., Johnsson, J.I., Smailus, D.E., Biagi, C.A., Jonsson, E. and Björnsson, B.T. 1999. Increased ability to compete for food by growth hormone-transgenic coho salmon *Oncorhynchus kisutch* (Walbaum). *Aquaculture Research*, 30: 479–482.

Devlin, R.H., Yesaki, T.Y., Biagi, C.A., Donaldson, E.M., Swanson, P. and Chan, W.K. 1994. Extraordinary salmon growth. *Nature*, 371: 209–210.

Johnsson, J.I. and Björnsson, B.T. 2001. Growth-enhanced fish can be competitive in the wild. *Functional Ecology*, 15: 654–659.

Mitchell, S. 2004. *Transgenic Fish Pose Extinction Threat*. United Press International, Florida, USA.

Muir, W.M. and Howard, R.D. 2000. Transgenic fish could threaten wild populations. *Purdue News Service*, USA.

Sharma, A.B. 2004. Scientists Develop Transgenic Fish Species. *The Financial Express*, India.

Sundström, L.F., Devlin, R.H., Johnsson, J.I. and Biagi, C.A. 2003. Vertical position reflects increased feeding motivation in growth hormone transgenic coho salmon (*Oncorhynchus kisutch*). *Ethology*, 109: 701–712.

Sundström, L.F., Lõhmus, M., Johnsson, J.I. and Devlin, R.H. 2004. Growth hormone transgenic salmon pay for growth potential with increased predation mortality. *Proceeding of the Royal Society of London*, B271(55): 350–352.

Chapter 2

Studies on Fisheries of Tilapia-Dominated Perennial Tank

A. Madhusudhan Rao and Ravi Shankar Piska

Department of Zoology, University College of Science, Osmania University, Hyderabad – 500 007, India

ABSTRACT

The present work deals with the studies on fisheries of a Tilapia-dominated perennial tank, Julur, Nalgonda District, Andhra Pradesh during three years from 2000–01 to 2002–03. The water-spread area of the tank is 14.06 ha and water is useful for irrigation and fisheries. The tank is in eutrophic nature. The fishes grouped into major carps, Tilapia, Murrels, catfishes and other food fishes, out of which Tilapia dominated with 61.04 per cent and followed by major carps (36.39 per cent). The mean fish production was 737.50 kg/ha/yr ($p>0.05$). Tilapia production reduced from 596.44 to 432.13 and further decreased to 387.85 kg/ha/yr during the above period. The major carp production increased from 215.46 to 251.76 and further to 318.99 kg/ha/yr (mean production 262.03 kg/ha/yr) during the above period ($p > 0.05$). Only major carp seed (25–30 mm total length) stocked every year (4267, 7113 and 8534 fry/ha) and other fishes depend on self-recruitment. The major carp production increased with increase of stocking densities. The number of fry for production of 1 kg fish was 25.33. The fish production per day and per fisherman were 106 and 53.34 kg respectively. The growth rate of major carps reduced significantly and attained 1 kg only during third year. Tilapia was responsible for decreased growth rate and major carp yield, due to the competition with other fishes because of voracious feeding and prolific breeding.

Keywords: *Fisheries, Tilapia, Major carps, Fish production, Growth, Perennial tank.*

Introduction

Reservoir constitute the single largest inland fishery resource, both in terms of size and fish production potential. These man made ecosystems offer enough scope for stock manipulation through ecological maneuvering, paving the way for production hikes at relatively low capital investment. Reservoir fisheries development is labour productive and ensure employment for weaker section of our society. 3.15 million ha of water spread area is available in India for reservoir fisheries. Experience has revealed that these water bodies of immense potential for fish husbandry through extensive aquaculture techniques. Reservoirs in India are considered the growing resource with enormous potential for fish production augmentation through capture fisheries. Reservoir fisheries is essentially a stocking-cum-capture system.

The exotic fish, Tilapia, *Oreochromis mosambicus* is common fish in Indian waters and is uneconomical and has less demand in Indian markets (Mitra, *et al.*, 1989; Piska and Rao, 2002; Piska, 2005; Narayana and Kohli, 2004 and Sakhare, 2005). It is dominant in some of the water bodies and effecting fisherman economy. The main objective of present work is to observe the impact of Tilapia-dominance on fish growth and production in a perennial tank.

Materials and Methods

The present work conducted in a perennial tank, Julur, Nalgonda District, Andhra Pradesh during three years period from 2000–01 to 2002–03. The water spread area is 14.06 ha and water is useful for agriculture and fisheries.

The fishes caught with the help of gill nets, cast nets, drag nets, etc. in holidays during monsoon and post-monsoon months and other than holidays during pre-monsoon months. Since Tilapia population was more, its population reduced by repeated netting. The carnivores and trash fishes also reduced by repeated netting.

Major carps seed (25–30 mm total length) stocked every year during monsoon months, except common carp seed during first year. The over all stocking of major carps was 60,000, 100,000 and 120,000 during 2000–01, 2001–02 and 2002–03 respectively. The stocking rates were 4267, 7113, and 8534 no/ha with the mean of 6639 no/ha. The stocking percentages of the catla, rohu and common carp were 42.86 per cent, 42.86 per cent, 14.29 per cent. The growth rate of fishes analysed every fortnight, then calculated for every season. Number of fry for the production of 1 kg fish calculated with the help of stocking densities and fish production. ANOVA test used according to Bailey (1959).

Results

The mean fish production was 737.50 kg/ha/yr (F 0.9619; df 2, 4; $p > 0.05$). The fish production decreased from 11829 kg (841.32 kg/ha/yr) to 9910 kg (704.84 kg/ha/yr) and further to 9375 kg (666.79 kg/ha/yr) during 2000–01, 2001–02 and 2002–03 respectively (Table 2.1).

Tilapia dominated with a mean of 61.04 per cent. Tilapia production declined from 596.44 to 337.84 kg/ha/yr during 2000–01 to 2002–03 with mean production of 444.19 kg/ha/yr (Table 2.1). The landing size of Tilapia was 100–150 g. The ratio between major carps and Tilapia production was 1 : 1.75 and the ratio between other fishes and Tilapia was 1 : 23.27. The dominance ratio of Tilapia declined from 1 : 2.78 to 1 : 1.07 with the major carps during first year to third year. Major carps occupied the second place in the fish landings with a mean of 262.03 kg/ha/yr (F 0.9619; df 2, 4; $p > 0.05$). The major carps productivity increased from 215.46 to 318.99 kg/ha/yr (3030 to 4485 kg).

Cat fishes occupied the third place in the fish landings with 10.86 kg/ha/yr. Murrels occupied the fourth place with 6.78 kg/ha/yr. The contribution of the other food fishes landings was least with 2.39 kg/ha/yr].

Table 2.1: Total Fish Landing in Julur Tank During 2000–01 to 2002–03

Sl.No.	Type of Fish	2000–2001			2001–2002			2002–2003			Total Fishes Landing (kg/ha/yr)
		Total Quantity (kg)	Fish Landing (kg/ha/yr)	Percent-age (%)	Total Quantity (kg)	Fish Landing (kg/ha/yr)	Percent-age (%)	Total Quantity (kg)	Fish Landing (kg/ha/yr)	Percent-age (%)	
1.	Major carps	3030	215.46	25.61	3541	251.76	35.73	4485	318.99	47.84	2000–01:841.32
2	Murrels	149	10.60	1.26	96	6.83	0.97	41	2.92	0.44	2001–02:704.84
3.	Tilapia	8386	596.44	70.9	6077	432.13	61.32	4750	337.84	50.67	2002–03:666.79
	Catfishes	210	14.94	1.78	166	11.81	1.68	82	5.83	0.87	Total landings 31114
	Other fishes	54	3.84	0.45	30	2.13	0.3	17	1.21	0.18	Mean:737.50 (2000-01 to 2002-03)
	Total	**11829**	**841.32**	–	**9910**	**704.84**	–	**9375**	**666.79**	–	

Catla dominated among the major carp landing with the production of 382.22 kg/ha/yr (Table 2.2). Next dominated fish after catla among major carps was rohu with 286.20 kg/ha/yr. Least dominated fish in major carp landings was common carp with 117.90 kg/ha/yr. Catla landing slightly decreased from 1769 to 1581 kg during 2000–01 to 2001–02, in the following year catla landings increased to 2024 kg (Table 2.2). Similar trend observed in case rohu. The major carp landings increased from 2000–01 to 2002–03 (3030 to 4485 kgs).

Table 2.2: Major Carp Landings in Julur Tank During 2000–01 to 2002–03

Sl.No.	Type of Fish	2000–2001			2001–2002			2002–2003			Total Fishes Fish	
		Total Fish Catchs (kg)	Fish Catch (kg/ha/yr)	Percent-age (%)	Total Fish Catchs (kg)	Fish Catch (kg/ha/yr)	Percent-age (%)	Total Fish Catchs (kg)	Fish Catch (kg/ha/yr)	Percent-age (%)	Catch (kg)	Catch (kg/ha/yr)
1.	Catla	1769	125.81	58.39	1581	112.45	44.65	2024	143.95	45.13	5374	382.22
2.	Rohu	1261	89.69	41.61	1202	85.49	33.95	1561	111.02	34.81	4024	286.20
3.	Common carp	–	–	–	758	53.91	21.4	900	64.01	20.06	1658	117.9
	Total	**3030**	**215.46**		**3541**	**251.76**		**4485**	**318.99**		**11056**	**786.34**

The average number of fishing days/year were 46.33. Maximum number of days engaged in fishing during pre-monsoon months in three years. The average number of fishermen per year were 108.66. The average catch per day per month was 106 kg. The average catch of fish/fisherman was 53.34 kg (Figure 2.1).

Only rohu and catla stocked during July 2000.The average growth rate of rohu was 19.83,46.09 and 45.23 g/month for 1st, 2nd, 3rd years respectively (Table 2.3). The average growth rate of catla per month were 23.95 g, 46.0 g and 61.90 g for above period. Rohu and catla and common carp seed

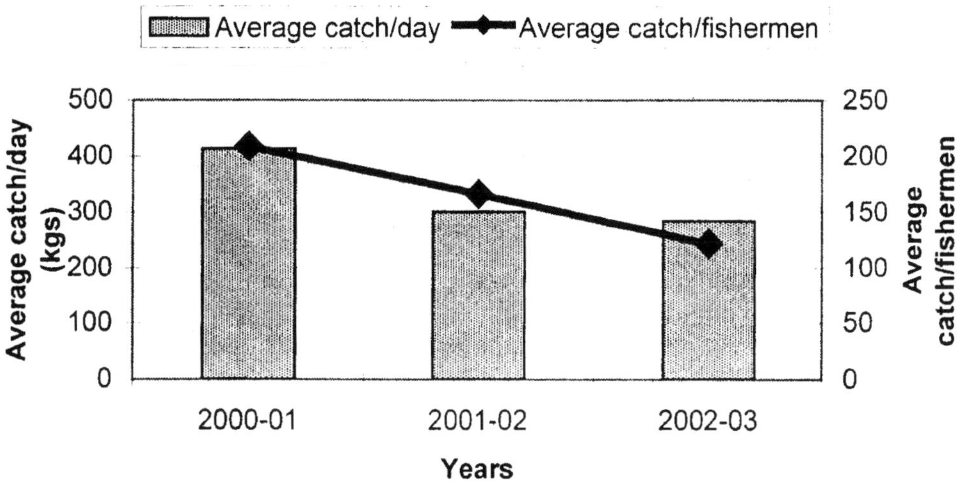

Figure 2.1: Fishing Efforts in Julur Tank During 2000–01 to 2002–03

stocked during July, 2001 and the average growth rate of rohu per month was 19.35 g and 47.35 g during first and second years respectively. The average growth rate of catla was 23.83 g/month in first and 48.28 g/month during second years respectively. The common carp growth rate per month was 25.16 g and 52.90 g during first and second years respectively. The fishes also stocked during July 2002 and the average growth rates were 21.39 g/month (rohu), 27.14 g/month (catla) and 28.46 g/month (common carp) (Table 2.3).

The number of fry for production of 1 kg fish was 19.80.28.25 and 26.75 during the above period with a mean of 25.33. During the present study 262.03, kg/ha/yr major carp production achieved with the stocking density of 6638 fry/ha (Table 2.4; Figure 2.2).

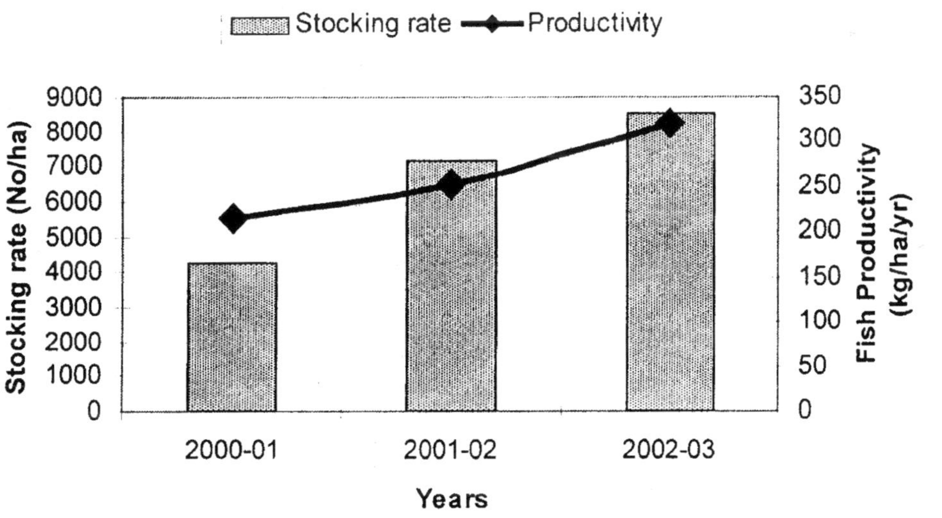

Figure 2.2: Stocking Densities and Productivity of Major Carps from Julur Tank During 2000–01 to 2002–03

Table 2.3: Growth Rate (g) of Major Carps Stocked During 2000–01 in Julur Tank

Season and Year	*2000–2001*			*2001–2002*			*2002–2003*		
	Rohu	*Catla*	*Common Carp*	*Rohu*	*Catla*	*Common Carp*	*Rohu*	*Catla*	*Common Carp*
2000–2001									
Monsoon	21.3±1.8	28.5±2.0	–	–	–	–	–	–	–
Post-monsoon	101.4±2.6	115.4±2.9	–	–	–	–	–	–	–
Pre-monsoon	252.2±3.6	304±4.1	–	–	–	–	–	–	–
2001–2002									
Monsoon	482.3±3.8	502.1±4.2	–	20.6±0.2	26.3±1.8	31±2.1	–	–	–
Post-monsoon	681.1±6.2	701.3±7.2	–	98.8±6.1	113.4±4.8	142.2±3.1	–	–	–
Pre-monsoon	805.3±8.5	856.0±8.1	–	246.2±9.8	301.4±8.6	318±3.6	–	–	–
2002–2003									
Monsoon	1108.0±20.6	1196.5±21.3	–	502.2±4.9	526.1±5.3	551.3±8.4	24.8±4.8	31.3±4.2	37.1±5.1
Post-monsoon	1236.1±22.4	1371.1±27.4	–	698.1±9.2	722.4±8.4	794.1±6.8	121.2±12.3	135.5±8.1	161.4±9.3
Pre-monsoon	1348.1±28.2	1598.8±28.4	–	821.4±14.1	880.3±12.5	952.8±12.1	271.5±13.4	342.8±14.5	358.6±14.2

Table 2.4: Stocking Density and Production of Major Carps from Julur Tank During 2000–01 to 2002–03

Sl.No.	*Year*	*Stock Rate No/ha*	*Productivity (kg/ha/yr)*	*No. of Fish for Present Production of 1 kg Fish*
1.	2000–2001	4267	215.46	19.80
2.	2001–2002	7113	251.76	28.25
3.	2002–2003	8534	318.99	26.75
	Average	6638	262.03	25.33

Discussion

The fish productivity of the Julur tank was much higher when compared to the average Indian production of 29.70 kg/ha/yr (Dehadrai, 2001). The present figures were much higher than small reservoirs of India–49.9 kg/ha/yr. (Piska, 2000). The present productivities were higher than other minor reservoirs like Baghla–106 kg/ha/yr, Bachra–139 kg/ha/yr and Gularia–100 kg/ha/yr which managed by scientific methods (Jhingran and Sugunan 1990). Devi (1997) and Piska (2000) recorded the productivity of 445 kg/ha/yr and 528 kg/ha/yr in Ibrahimbagh and Shanthamrai reservoirs Andhra Pradesh. Chary (2003) reported that productivity of Durgamcharuvu as 535.46 kg/ha/yr. Mahapatra (2003) recorded only 15.6 kg/ha/yr in Hirakud reservoir and 5–10 kg/ha/yr in other major reservoirs in Orissa. Higher productivity of fish found in Bidikiyar lake (825 kg/ha/yr) and Ambala municipal lake (500 kg/ha/yr) in Haryana (Agarwal, 1990).

Saxena (1990) reported the potential yield in Ukai (220 kg/ha/yr). Nagarjunasagar (119 kg/ha/yr). Bhavanisagar (54 kg/ha/yr). Rihand (54 kg/ha/yr) and Gobindsagar (263 kg/ha/yr) reservoirs. These figures were much less than the potential yield of present study. Burli and Sontakki (2002) reported that low fish production (4–6 kg/ha/yr) noticed in Malaprabha and Ghataprabha reservoirs of Karnataka. Tamot (2002) studied the fish production of irrigation reservoirs of Madhya Pradesh 70 kg/ha/yr. Bandyopodhyay *et al.* (2003) reported that after the introduction of Indian major carps and exotic carps in Manipur waters, dramatic change in the fish production was observed. Gowda (2002) reported the productivity of reservoirs of Karnataka as 30 kg/ha/yr with a total water spread area of 2.10 lakh ha (73 reservoirs).

Jhingran and Sugunan (1990) reported that the major carps contributed 90.54–94.76 per cent in Tillaiya and 57.44–76.61 per cent in Konar reservoirs. Devi (1997) reported that the contribution of major carps was 91.42 per cent in Ibrahimbagh and 96.40 per cent in Shathamraj reservoirs of Andhra Pradesh and Chary (2003) also reported the dominance of major carps in Durgamcheruvu of Andhra Pradesh.

In the present study Tilapia production was in declining trend, where as major carps production was in increasing trend, *i.e.*, the production levels of the above were in inversely proportion to each other. Tilapia and other fishes production levels were in decreasing trend during the above period, *i.e.*, the production levels of Tilapia and other fishes were directly proportional to each other.

Importance of Tilapia on fish production was reported by Rao and Shakuntala (1998) in Nelligudda reservoir, Karnataka and Piska and Rao (2002) in perennial tank, Nalgonda. They reported inverse relationship between major carps and Tilapia production. They also reported that Tilapia competes with major carps for natural food and space. When Tilapia was dominated there was scarcity of natural food in reservoir, which reflects the growth of carps and resulted low yields. Mitra *et al.* (1998),

Narayanan and Kohli (2004) and Sakhare (2005) reported that Tilapia was responsible for reduction of fish production in reservoirs due to its competition for food with other fishes.

This increase of major carps production was due to the stock manipulation and reduction of Tilapia and predators like murrels, cat fishes and other fishes. This indicated that the major carp percentage was dependent upon the other fishes, especially on Tilapia and predators. The major carps especially in the younger stages easily fell as pray to the predatory fishes. The present results indicated that minor carps and trash fishes never played an important role in increase or decline of major carps, due to their less number. Most of the minor carps and trash fish were herbivorous.

The average catch/fisherman/month was 36.4 kgs in Karnal lake in Haryana (Agarwal, 1990). This value was very less when compared to the values of the present study. This may be due to more number of fishermen engaged in Karnal lake. Chary (2003) reported that 3.29 kg per fishermen per day in Durgamcheruvu. Agarwal (1990), Devi (1997) and Piska *et al.* (2000) stated that the fish production increased with increase of number of fishermen engaged in fishing. They concluded that the fish production from the natural water bodies could be increased by proper utilization of manpower and material for fishing.

In the present study stocked fishes attained 1 kg only during third year. This indicated that the growth rates of major carps were very slow. Usually major carps attained 1 kg weight during first year in ponds (Piska, 1999, 2000). Where as fishes reared in natural waters fish attained 1 kg either during first year or early part of second year. (Devi, 1997; Piska, 2001 and Chary, 2003).

Piska (2000) reported that rohu grow to 401.35 g, catla to 407.32 g and common carp to 653.46 g in six months in a seasonal rain-fed tank. The growth rates in present study were less when compared earlier studies. Venugopal *et al.* (1988) reported the growth of common carp was more than 600 g and catla 400 g and rohu 300 g in 7 months rearing in a seasonal rain-fed tank. These growth rates were also more when compared to that of those in the present work.

During the present study the growth rates of rohu, catla and common carp were low. The growth rates were poor due to the competition for food and space with Tilapia. Due to the voracious feeding, omnivorous nature early maturity and prolific breeding Tilapia domination responsible for competition for food and space. Due to the presence of more number of Tilapia the natural food scarcity was found in the tank. This was indicated by analysis of plankton in the tank. During the first and second years the growth of major carps was poor when compared third year. This was due to more number of Tilapia in first two years. Tilapia population was reduced by the management and the growth of major carps was slightly improved.

Different authors suggested different stocking densities in Indian reservoirs. Mohanty (1984) suggested 1000 seed/ha in reservoirs of Orissa. Agarwal (1990) recommended the stocking rates in reservoirs of Haryana of minor reservoirs (10–100 ha) as 2000 fry/ha/yr and small reservoirs (below 10 ha) as 10000 fry/ha/yr. The present water body comes under minor reservoir category. Mathew and Mohan (1990) reported the stocking rate was 841 fingerlings/ha in Kerala and they never considered the stocking rate was low. Piska (2003) suggested the stocking density of 2000 seed/ha in reservoirs and tanks and 5000 seed/ha in the ponds in Andhra Pradesh. He also stated that the stocking density of 2000/ha was slightly high, which was due to compensate the mortality of sensitive fry in the open waters.

Sinha (1998) studied the stocking density and size of fingerlings in Indian reservoirs. Sreenivasan (1999) reported the relation to stocking, recapture of fish. Devi (1997) and Piska (2000) studied the stocking relationship with fish yield rates of Shathamraj and Ibrahimbagh reservoirs. Das *et al.* (2002),

Mahapatra *et al.* (2002) and Chary (2003) studied the stocking and fish yield of various reservoirs of India. Some of the reservoirs also stocked with advanced fry or fingerlings (Mathew and Mohan, 1990).

Different authors reported different figures for number of fry for production of 1 kg fish in different reservoirs of India–5–35 fry in reservoirs of Haryana (Agarwal, 1990), 9–12 fry in reservoirs of Andhra Pradesh (Devi, 1997; Chary, 2003).

The results indicated that the fish production can be increased by stock manipulation. The similar reported were given by Devi (1997), Velayudhan and Siddarthan (1998), Piska *et al.* (2000), Piska (2000) and Chary (2003).

Acknowledgement

We are thankful to Prof. B. Raghavendra Rao and Prof. P. Umamaheshwer Reddy for their suggestion during the present work.

References

Agarwal, S.C. 1990. *Fishery Management.* Ashish Publishing House, New Delhi.

Bailay, N.T.J. 1959. *Statistical Methods in Biology.* The ELBS and English University Press Limited, Great Britain.

Bandyopadhyay, M.K., Singh, S.K. and Sarkar, B. 2003. Possibility of raising Indian major carp seed in cases for stocking larger water Bodies. CA–1, *Nat. Sem. Fish and Shellfish Culture,* pp. 37.

Burli, A.S. and Sontakki, B.S. 2002. A comparative study of fisheries management in Malaprabha and Ghataprabha reservations of Karnataka, AR-16. *The Sixth Indian Fisheries Forum,* pp. 54.

Chary, K.D. 2003. Present status, management and economics of fisheries of a minor reservoir, Durgamcheruvu of Rangareddy District. *Ph.D. Thesis,* Osmania University.

Das, A.K., Krishna Rao, D.S., Sukumaran. P.K. and Karthikeyan, M. 2002. Strategies for management of Karnataka reservoirs. *Fishing Chimes,* 22(2): 17–21.

Dehadrai. 2001. Reservoir fisheries in India. *Proc. Nat. Sem. Riverine and Reservoir Fisheries,* p. 97–104.

Devi, B. S.1997. Present status, potentialities, management and economics of fisheries to two minor reservoir of Hyderabad. *Ph.D. Thesis,* Osmania University.

Gowda, H.S.V. 2002. Status of fisheries development in Karnataka. *Fishing Chimes,* 22(1): 86–94.

Jhingran, A.G. and Sugunan, V.V. 1990. General guidelines and planning criteria for small reservoir fisheries management. *Proc. Nat. Workshop Reservoir Fish,* p. 1–8.

Mahapatra, B.K., Vinod, K. and Mandal, B.K. 2002. Stocking of Indian major carps in Umiam reservoir of Meghalaya: An approach for culture based fisheries. *Proc. Nat. Symp. Fisheries. Enhanc. Inland Waters,* p. 43–44.

Mahapatra, D.K. 2003. Present status of fisheries of Hirakud reservoirs, Orissa. *Fishing Chimes,* 22(10 and 11): 76–79

Mathew, P.M. and Mohan, M.U. 1990. Reservoir Fisheries in Kerala status and scope for development. *Proc. Net. Workshop Reservoir Fish,* p. 111–117.

Mitra, A., T. Nayak and S.K. Sarkar. 1989. Social acceptance and economic returns of Tilapia culture in W.B. Exotic aquatic species in India sp.pub. *J. Asian Fish. Soc., Ind. Br.,* 101–104.

Mohanty, S.K. 1984. An approach for the development of reservoir fisheries in Orissa. *Proc. Freshwater Fish. Rural Devel. Sec.*, 1: 40–46.

Narayanan, A. and M.P.S Kohli. 2004. Biotype imbalances due to unscientific introduction of exotic fishes: the Indian experience. *Fishing Chimes*, 24(3): 39–41.

Piska, R.S. 1999. *Fisheries and Aquaculture.* Lahari publications, Hyderabad.

Piska, R.S. 2000. Impact of stocking densities of major can seed on fish production in a minor reservoir. *Fishing Chimes*, 20(9): 39–41.

Piska, R.S., Devi, B.S. and Chary, K.D. 2000. The present status of fisheries of Ibrahimbagh, a minor reservation of Hyderabad. *Fishing Chimes*, 20(2): 41–43.

Piska, R.S. 2001. Fisheries of minor reservoir, Shathamraj, Hyderabad. *Ind. J. Fish*, 48(2): 119–122.

Piska, R.S. and Rao, A.M. 2002. Phenomenal fish production in the Tilapia-dominated tank. *Aquacult.*, 3(2): 171–176.

Piska, R.S. 2003. Carp seed productions Scenario in Andhra Pradesh. *Fishing Chimes*, 23(1): 122–123.

Piska, R.S. 2005. Planktonic diversity in a Tilapia dominated minor reservoir. *Trends Life Sci.* (Press).

Rao, D.S.K. and K. Shakunthala. 1998. Phenomenal Fish Production in the dominated Nelligudda reservoir in Karnataka. *Fishing Chimes*, 48(4): 5–16.

Sakhare, V.B. 2005. Impact of Tilapia on indigenous fishes. *Fishing Chimes*, 24(11): 55–56.

Saxena, B.S. 1990. Potentialities production and problems of Indian reservoir fisheries. *Proc. Nat. Workshop Reservoir Fish*, p. 34– 38.

Sinha, M. 1998. Policy options for integrated development of reservoir fisheries from production to Marketing *Fishing Chimes*, 18(1): 54–59.

Sreenivasan, A. 1999. The role of small reservoir in fish production. *Fishing Chimes*, 18(10): 29–31.

Tamot, S. 2002. Fisheries Potential in Madhya Pradesh. *AR–15 The Sixth Indian Fisheries forum*, pp. 96.

Velayudhan, T.D. and Sidharthan, T. 1998. Influence of stocking on fish yield in Chittore Reservoir and its prospect on rural economy. *Fishing Chimes*, 18(4): 17–20.

Venugopal, G., H.G. Hingorani, K., Venkateshvaran and J.P. George. 1998. Culture of Carps in a seasonal tank. *Comp. Physiol., Eco.*, 13(1): 20–23.

Chapter 3

Status of Fisheries Resources in Selected Backwaters of Kerala

P.K. Sukumaran and V.K. Unnithan

Central Inland Fisheries Centre of CICFRI, Alappuzha – 688 001, Kerala

ABSTRACT

Seasonal investigations on fishery resources were conducted in ten backwaters of Kerala during 1996–1997. The study indicated that the backwaters are under deterioration and the fishery is being lost due to a variety of anthropogenic reasons. Reclamation of the backwaters coupled with the alarmingly low depth ($\bar{X}$: 2.05 m) for nine out of ten systems have reduced the water holding capacity of the backwaters to a large extent. Retting of coconut husk in the backwaters has been the most predominant activity that contributed to the organic pollution resulting in low water quality and depletion of faunal resources. The total landings from different backwaters varied from 96.8 to 2898.7 t. The study reveals that the fishery suffers from both intensive fishing and unscrupulous fishing of undersized fishes.

Keywords: *Backwater, Lagoons, Anthropogenic, Reclamation, Exploitation, Deterioration.*

Introduction

The brackish water environment is a unique ecosystem of high fertility, supporting natural fishery in the rural economy of India (Thampi, 1973). Coastal wetlands are a permanent habitat for a variety of resident and migratory aquatic fauna. The Kerala backwaters are interconnected with lagoons, bays and swamps. These backwaters (area: 2,42,600 ha) harbouring about 200 resident, migratory fish and shellfish species. The fishing activities in these backwaters support livelihood about 0.2 million fishermen and provide full-time employment to about 50,000 fishermen. The annual catch from these water bodies spread in nine coastal districts is reported to be about 33,000 t (George and Sebastian, 1970). These water bodies are served as a nursery ground for prawns *viz., Penaeus indicus, P. monodon, Metapenaeus dobsonii, M. monoceros, crabs* and fin fishes like mullets, perches and pearl spot and the breeding ground for freshwater caridian prawns (*Macrobrachium* sp.). Obviously, the lake ecosystem has a high carrying capacity and playing a crucial role in both biologically and economically. The need to protect, preserve such ecosystem is imperative. Of late, these ecosystems are reported to be under severe environmental stress arising from a variety of anthropogenic activities. Very little information is available on the backwaters (Shetty, 1965, Nair *et al.*, 1983, Kurup *et al.*, 1993). Scientific monitoring of these ecotypes aiming at its rational fishery management has been highly inadequate. Hence detailed investigations in these water bodies are envisaged to know the present status of the fishery resources.

Materials and Methods

Fishery survey was conducted quarterly in ten of the thirty backwaters along the southwest coast of India (Figure 3.1). The backwaters investigated have a waterspread area of 15102 ha, of which 22 per cent of the total backwaters available in the state. Forty-six landing centres around these backwaters examined for a period of 2–3 days in each quarter either continuously or intermittently in each quarter and the total landings were estimated based on average values for a day. Average fishing days in a month for each category gear was ascertained through repeated enquiry at each stations and the total landings were computed through extrapolation. Fish samples were collected from the major landing centres (Table 3.1). The fishes were identified with the help of Day (1967) and Thalwar and Jhingran (1991).

Table 3.1: List of Backwaters and their Area (ha)

Sl.No.	*Name of the Backwater*	*District*	*Area (ha)*
1.	Kadinamkulam (KDK)	Trivandrum	374
2.	Anchuthengu (ANG)	Trivandrum	552
3.	Ashtamudi (AST)	Kollam	6424
4.	Kayamkulam (KYK)	Kollam/Alappuzha	1652
5.	Azhikode (AZK)	Thrissur	696
6.	Chettuva (CTV)	Thrissur	714
7.	Ponnani (PNI)	Thrissur	908
8.	Mahe (MHE)	Kannur	88
9.	Valapattanam (VPM)	Kannur	3074
10.	Noolcowaram (NEL)	Kasargod	825

Results and Discussion

In Kerala, the backwaters support a rich fishery. Their relative abundance was recorded for the various systems explored. The common species that regularly contributing to the fishery is listed in Table 3.2. Of the 94 species listed, different workers, thereby establishing a close relationship of the backwater fishery with that of marine, have recorded 63 from the marine waters. In Ashtamudi backwater alone contributed 97 species belonging to 39 families (Nair *et al.*, 1983). Backwater fauna is

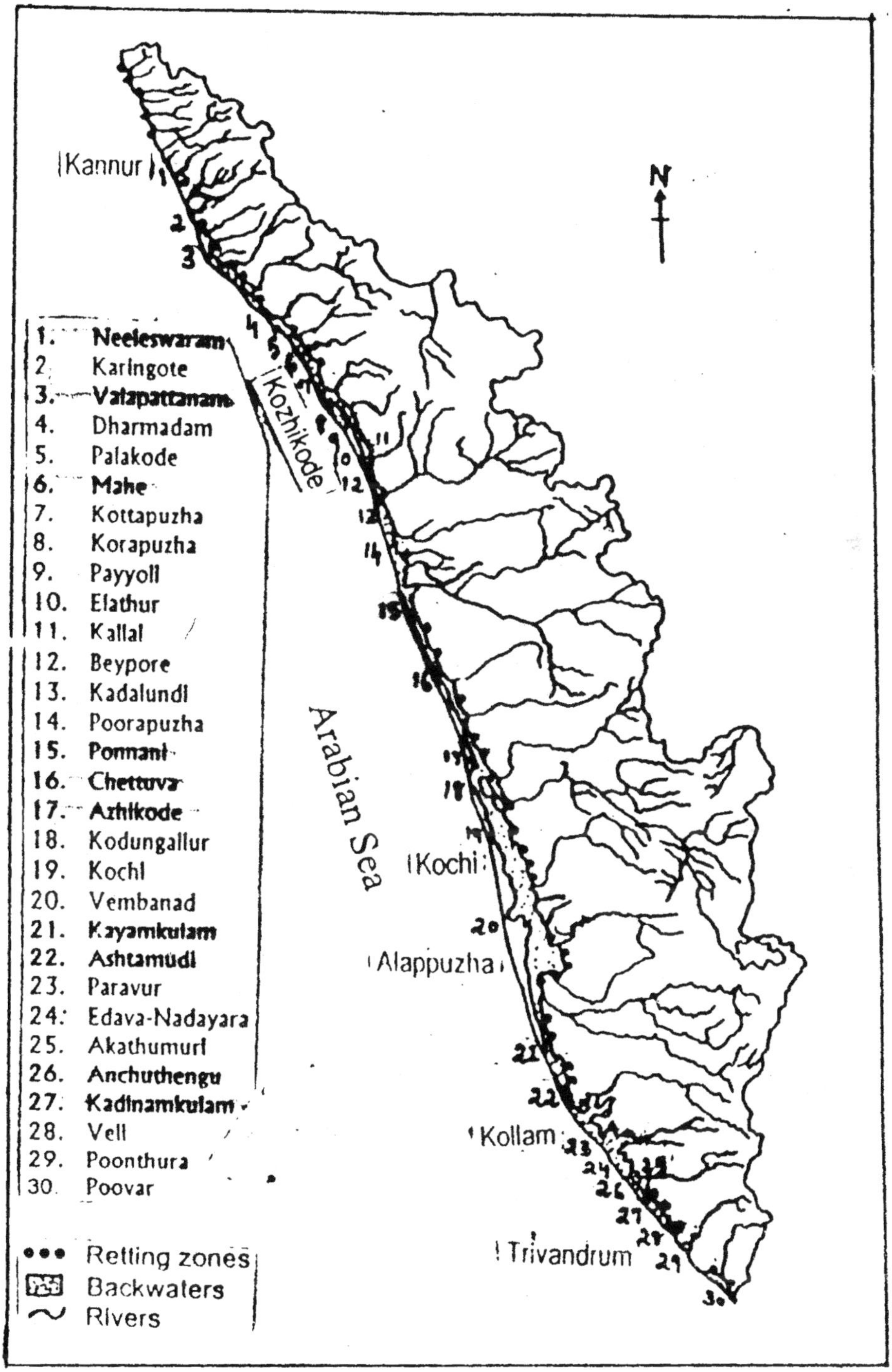

Figure 3.1: Map of Kerala Showing Important Backwaters

Table 3.2: Total Landings (t), Gear and Fisherman at Different Backwaters During 1996–97

Backwaters	*KDK*	*ANG*	*AST*	*KYK*	*AZK*	*CTV*	*PNI*	*MHE*	*VPM*	*NEL*	*Aver-age*	*Aver-age*
Area (ha)	*375*	*552*	*6424*	*1652*	*696*	*714*	*908*	*88*	*3074*	*825*	*(kg ha^{-1} yr^{-1})*	*(%)*
Major Species/Group												
Fishes												
Acanthus sp.	0.7	0.1	3	3.9	5	5.3	2.9	4.8	8.9	10.5	2.9	0.5
Gerres sp.	7.9	5.5	30.2	29.8	65.3	18.3	12.1	2.7	36.8	54.4	17.1	2.7
Platycephalus indicus	0.6	0.2	32.3	17.1	7	6.4	8.4	1.3	3	27.8	6.8	1.1
Lieoganthus spp.	1.2	2.6	9.6	29.8	6.7	12.4	12	5.7	6	18.7	6.8	1.1
Etroplus sp.	17.4	29.7	168	128.6	38	27	30.3	8.9	35.7	65.8	35.8	5.7
Megalops cyprinoids	6.1	12.4	11.8	15.9	1.5	0	2.1	3.6	2.7	1.7	3.8	0.6
Arius sp.	6.7	22.2	83.8	45.6	37.3	18.6	20.6	7.7	24	42.9	20.2	3.2
Ambassis sp.	13.8	5.4	75.4	141	12.5	21.9	8.6	0	5.2	8.2	19	3
O. mossambicus	11.7	31.5	0.7	21	15.4	0.1	8.2	0.4	0.9	1.4	5.9	0.9
Stolephorus sp.	0	0	44	7.4	55.5	12	18.1	1.2	6.7	7.2	9.9	1.6
Silago sihama	2	0.1	11	7.3	6.3	6.9	13.6	3.7	9.1	18.5	5.1	0.8
Caranx sp.	8.1	1.4	19.7	20.3	10.4	5.4	3.7	2.9	7.9	45.9	8.2	1.3
Lutjanus sp.	3	2.4	27	17.7	11.2	16.9	29.3	11.2	7.1	18.8	9.4	1.5
Mullets	31.1	26.7	113.4	89.9	95.9	36.5	30.8	1.6	22.9	41.7	32	5.1
Flat fishes	6.4	0.4	3.9	59	2.3	9.2	20.1	0.6	4.8	10.4	7.6	1.2
Half beeks	0.5	0	1.3	0.9	0.5	12.9	4.7	0	2.6	1.4	1.6	0.3
Others	24.4	13.7	271.4	206.6	143.6	41.8	110.3	16.3	56.9	145.8	67.2	10.7
Fish Total	141.6	154.3	906.3	841.7	514.3	251.5	335.7	72.5	240.7	521.2	259.4	41.2
Prawns												
Metapenaeus dobsonii	18.8	30	1255	453.4	1021	172.4	60.7	9	154	214.7	220.9	35.1
M. monoceros	1.9	4.3	238.4	98	146	22.3	19.4	2.8	58	40.7	41.2	6.5
Penaeus indicus	4.7	14.9	210.6	119.7	179	37.6	29.1	8.7	224	60.1	57.9	9.2
P. monodon	0.4	0	24.3	38.7	2	1.2	49.6	0	1.5	0.2	7.7	1.2
Other penaeids	0	0	12.2	15	0.1	2.5	15.9	0	0.6	15.6	4	0.6
Non-penaeids	0.4	0.9	14.9	16.7	0.2	2.5	7.1	0	0.8	0.5	2.9	0.5
Prawn Total	26.2	50.1	1755.4	741.4	1348.2	238.6	181.7	20.5	438.9	331.9	334.6	53.1
Crabs	20.7	10	237	64.8	49.6	20.2	33.3	5	77.4	38	36.2	5.8
Grand Total	188.5	214.4	2898.7	1648	1912.1	510.3	550.6	96.8	757	891	630.1	100
CPUE (kg unit^{-1} day^{-1})	5.9	3.7	7.2	8.1	13.5	5.4	4.9	4.6	7.1	10.8		
Yield (kg ha^{-1})	543.6	410.7	451.2	997.5	2747.3	714.7	606.4	537.8	246.3	1080		
No. of gears km^{-2}	65	120	52	83	158	100	83	85	51	97		
No. of fisherman km^{-2}	104	145	58	137	152	84	89	99	39	68		

generally composed of marine and freshwater organisms, which can adapt to waters of different and varying salinities and truly resident estuaries species (Jhingran and Gopalakrishnan, 1971). Nineteen of them used to be predominantly from the rivers/reservoirs are migrated to backwater during monsoon seasons. Definite zonation can be possible due to the distribution of these species *viz., Puntius filamentosus, P. sarana, Labeo dussumieri, Mystus malabaricus, Anabas testudineus, Channa* sp. *Oreochromis mossambicus* and *Mastacembeles armatus.* On the other hand, several marine forms were recorded from the backwaters during the summer season. *Rhinobatus halavi, Congresox talabonoides, Lobotes surinamensis, Acanthurus strigosus, Eleotris fusca, Lepturocanthus savala, Platax orbicularis* etc., were also recorded from certain systems, though in stray numbers. Thus it is clear that in the backwaters majority of the fish fauna have been recruited from the adjoining Arabian Sea, which shows that the fishes' is mainly, depended on the ingress from the sea.

Of the 94 species recorded 67 are commercially important (Jhingran, 1975) and 12 contribute substantially to the fisheries (Table 3.3). The pearl spot, mullets, Lutjanids among fishes and *Penaeus indicus* are the most predominant forms having considerable market value and contribute appreciably to the landings. None of the commercially important groups belong to the truly estuarine. The fishery is mainly seasonal as it depends on the recruitment from the sea. *Etroplus* spp., *Nandus* sp. is available for all the seasons. *Valamugil seheli* and *V. speigleri* are available round the year. Among the Mullet, *Valamugil seheli* is the most common form. The major groups/species constituting the fishery with their relative contribution were estimated. *Etroplus suratensis* (2.0–13.9 per cent), *Penaeus indicus* (2.5–29.6 per cent) and *Metapenaells monoceros* (1.0–8.2 per cent) represented in the total catch. The mullets (1.5–16.5 per cent), Lutjanids (1.1–11.4 per cent), carangids (1.0–5.2 per cent), the tiger perch (*Therapon jarbua*), the reef cod *Epinephalus* spp., *Sphyraena jello* etc. the *Gerres filamentosus* (1.0–6.1 per cent) and the flathead *Platycephalus indicus* also formed a commercially important species in the fishery. The less known commercial fishes like silver bellies (*Leiognathus* spp.), the half beaks, the marine catfish (*Tachysurus* spp.), the anchovies (*Stolephorus* sp.), the flat fishes and the Ambassids gained importance due to their bulk contribution to the fishery.

Table 3.3: Finfish and Shellfish Recorded from Selected Backwaters During 1996–97

Sl.No.	Species Name	Habitat	Backwaters									
			KDK	ANJ	AST	KYM	AZK	CTV	PNI	MHE	VPM	NEL
1.	*Rhinobats halavi* (Day)	M						*				
2.	*Megalops cyprinoides* (Broussonet)	E	**	**	**	**	**	**	**	**	**	**
3.	*Anguilla bengalensis* (Gray)	R	*	*	*	*	*	*	*	*	*	
4.	*Ophichthys altipinnis* (Kaup)	E										
5.	*Congresoz talabonoides* (Bleeker)	M			**	**		***		**		
6.	*Nematalosa nasus* (Bloch)	M	**	**	**	**	**	**	**	**	**	**
7.	*Pellona ditchela* (Valenciennes)	M	**	**	**	**	**	**	**	**	**	**
8.	*Dussumieria hasseeltii* (Bleeker)	M			*			*	*			
9.	*Stolephorus indicus* (van Hasselt)	M	***	***	***	***	***	***	***	***	***	***
10.	*Thryssa mystaz* (Schneider)	M	***	***	***	***	***	***	***	***	***	***
11.	*Chanos chanos* (Forskal)	E	**	**	**	**	**	**	**	**	**	**
12.	*Puntius filamentosus* (Valenciennes)	R	***	***	***	***	***	***	***	***	***	***
13.	*Puntius sarana* (Hamilton-Buchanan)	R	**	**	**	**	**	**	**	**	**	**

Contd...

Table 3.3–Contd...

Sl.No.	Species Name	Habitat	Backwaters									
			KDK	ANJ	AST	KYM	AZK	CTV	PNI	MHE	VPM	NEL
14.	*Labeo dussumieri* (*Valenciennes*)	R				*	**		**			
15.	*Mystus malabaricus* (*Jerdon*)	R				**	*		**			
16.	*Horabagrus brachysoma* (*Gunther*)	E	*			*		*		*	*	
17.	*Chaca chaca* (*Hamilton-Buchanan*)	R						*		*	*	
18.	*Arius platystomus* (*Day*)	E	****	****	****	****	****	****	****	****	****	****
19.	*Arius rostratus* (*Day*)	E	***	***	***	***	***	***	***	***	***	***
20.	*Hyporhamphus limbatus* (*Valenciennes*)	M, E	**	**	**	**	**	**	**	**	**	**
21.	*Hemirhamphus cantori* (*Day*)	M, E	**	**	**	**	**	**	**	**	**	**
22.	*Hemirhamphus far* (*Day*)	M, E	**	**	**	**	**	**	**	**	**	**
23.	*Strongylura strongylura* (*van Hasselt*)	M, E	**	**	**	**	**	**	**	**	**	**
24.	*Tylosuruscrocodilus* (*Le Sueur*)	M, E	*	*	*	*	*	*	*	*	*	*
25.	*Platycephalus indicus* (*Linnaeus*)	E, M	****	****	****	****	****	****	****	****	****	****
26.	*Lates calcarifer* (*Bloch*)	E, M	*	*	*	**	*	*	*	*	*	**
27.	*Ambassis commersoni* (*Cuvier*)	E, R	****	****	****	****	****	****	****	****	****	****
28.	*Parambassis* spp.	R, E	****	****	****	****	****	****	****	****	****	****
29.	*Epinephelus tauvina* (*Forsskal*)	M, E	**	**	**	**	**	**	**		**	**
30.	*Epinephelus malabaricus* (*Schneider*)	M, E	**	**	**	**	**	**	**		**	**
31.	*Terapon jarbua* (*Forsskal*)	M, E	**	**	**	**	**	**	***	**	**	**
32.	*Sillago sihama* (*Forsskal*)	M, E	****	****	****	****	****	****	****	****	****	****
33.	*Caranx sexfasciatus* (*Quoy and Gaimard*)	M	***	***	***	***	***	***	***	***	***	***
34.	*Caranx nigripinnis* (*Bleeker*)	M, E	**	**	**	**	**	**	**	**	**	**
35.	*Leiognathus equulus* (*Forsskal*)	M, E, R	****	****	****	****	****	****	****	****	****	****
36.	*Leiognathus decorus* (*Le Vis*)	M	****	****	****	****	****	****	****	****	****	****
37.	*Leiognathus splendens* (*Cuvier*)	M	****	****	****	****	****	****	****	****	****	****
38.	*Lutjanus argentimaculatus* (*Forsskal*)	M, E	**	**	**	**	**	**	**	**	**	**
39.	*Lutjanus johni* (*Bloch*)	M	**	**	**	**	**	**	**	**	**	**
40.	*Lutjanus fulviflamma* (*Forsskal*)	M, E	**	**	**	**	**	**	**	**	**	**
41.	*Lobotes surinamensis* (*Bloch*)	M							*			*
42.	*Gerres filamentosus* (*Cuvier*)	E, M	****	****	****	****	****	****	****	****	****	****
43.	*Gerres oyena* (*Forsskal*)	M, E	**	**	**	**	**	**	**	**	**	**
44.	*Monodactylus argenteus* (*Linnaeus*)	E										
45.	*Drepane punctatus* (*Linnaeus*)	M			**	**						

Contd...

Table 3.3–Contd...

Sl.No.	Species Name	Habitat	Backwaters									
			KDK	ANJ	AST	KYM	AZK	CTV	PNI	MHE	VPM	NEL
46.	*Scatophagus argus* (*Linnaeus*)	E, R	***	***	***	***	***	***	***	***	***	***
47.	*Acanthurus bleekeri* (*Gunther*)	M, E	**	**	**	**	**	**	**	**	**	***
48.	*Acanthurus strigosus* (*Bennett*)	M	**	**	**	**	**	**	**	**	**	**
49.	*Nandus nandus* (*Hamilton-Buchanan*)	R	****	****	****	****	****	****	****	****	****	****
50.	*Nandus marmoratus* (*Hamilton-Buchanan*)	R, E			*	*						
51.	*Etroplus suratensis* (*Bloch*)	E, R	****	****	****	****	****	****	****	****	****	****
52.	*Etroplus maculatus* (*Bloch*)	E, R	****	****	****	****	****	****	****	****	****	****
53.	*Oreochromis mossambicus* (*Peters*)	R	****	****	**	****	***	**	**	**	**	**
54.	*Liza tade* (*Forsskal*)	M, E	****	****	****	****	****	****	****	****	****	****
55.	*Liza parsia* (*Hamilton-Buchanan*)	M, E	****	****	****	****	****	****	****	****	****	****
56.	*Migil cephalus* (*Linnaeus*)	M, E, R	**	**	**	**	**	**	**	**	**	**
57.	*Valamugil seheli* (*Forsskal*)	M, E	****	****	****	****	****	****	****	****	****	****
58.	*Valamugil speigleri* (*Bleeker*)	M, E, R	***	***	***	***	***	***	***	***	***	***
59.	*Sphyreena jello* (*Cuvier*)	M	***	***	***	***	***	***	***	***	***	***
60.	*Gossogobius giuris* (*Hamilton-Buchanan*)	M, E, R	***	***	***	***	***	***	***	***	***	***
61.	*Oxyurichthys microlepis* (*Bleeker*)	M, E	***	***	***	***	**	**	**	**	**	**
62.	*Oxyurichthys tentacularis* (*Valenciennes*)	M	***	***	***	***	**	**	**	**	**	**
63.	*Oxyurichthys formosanus* (*Nichols*)	M, E			**	**		*				
64.	*Eleotris fusca* (*Day*)	E, R	*						**		*	
65.	*Lepturacanthus savala* (*Cuvier*)	M, E			*	*			*		*	*
66.	*Anabas testudineus* (*Bloch*)	R	*			**	*		**			
67.	*Channa marulius* (*Hamilton-Buchanan*)	R	*		*	**			**			
68.	*Channa* sp.	R	*			**			**			
69.	*Macrognathus guentheri* (*Day*)	R	*		*	*			*			
70.	*Mastocembelus armatus* (*Lacepede*)	R	*	*	*	*	*	*	**	*	*	*
71.	*Pseudorhombus javanicus* (*Bleeker*)	E, M	**	**	**	**	**	**	**	**	**	**
72.	*Pseudorhombus arsius* (*Hamilton-Buchanan*)	M, E	*	*	*	**	*	**	*	*	*	*
73.	*Cynoglossus cynoglossus* (*Hamilton-Buchanan*)	M, E	**	**	**	**	**	**		**	**	**
74.	*Cynoglossus punticeps* (*Richardson*)	M	**	**	**	**	**	**	**	**	**	**
75.	*Cynoglossus lingua* (*Hamilton-Buchanan*)	M, E	**	**	**	**	**	**	**	**	**	**
76.	*Euryglossa orientalis* (*Bloch and Schneider*)	E	***	***	***	***	***	***	***	***	***	***

Contd...

Table 3.3–Contd...

Sl.No.	Species Name	Habitat	Backwaters									
			KDK	ANJ	AST	KYM	AZK	CTV	PNI	MHE	VPM	NEL
77.	*Triacanthus biaculeatus* (*Bloch*)	M			**	*			*		*	*
78.	*Tetraodon leopardus* (*Day*)	M, E	*	*	*	*	*	*	*	*	*	*
79.	*Teuthis vermiculata* (*Day*)	M, E	**	**	**	**	**	**	***	**	**	***
80.	*Teuthis java* (*Day*)	M	**	**	**	**	**	**	**	**	**	**
81.	*Stromaleus cinereus*	M	**	**	***	**	**	**	**	**	**	***
82.	*Platax orbicularis* (*Forsskal*)	M										*
Prawns												
83.	*Macrobrachium rosenbergii*	R, E		**	**	**	**	*	*		*	*
84.	*Macrobrachium idella*	R, E	**	**	**	**	**	**	**	*	**	**
85.	*Macrobrachium lamerrei*	R, E	**	**	**	**	**	**	**	*	*	**
86.	*Penaeus indicus*	M, E										
87.	*Penaeusmonodon*	M	***	***	***	***	***	***	***	***	***	***
88.	*Penaeus semisulcatus*	M	**	**	**	**	**	**	**	*	*	*
89.	*Metapenaeus monoceros*	M, E	****	****	****	****	****	****	****	****	****	****
90.	*Metapenaeus dobsonii*	M, E	****	****	****	****	****	****	****	****	****	****
Stomatopod												
91.	*Oratosquillanepa*	M, E	****	****	****	****	****	****	****	****	****	****
Crabs												
92.	*Scylla serrata*	M, E	****	****	****	****	****	****	****	****	****	****
93.	*Portunus pelagicus*	M, E	****	****	****	****	****	****	****	****	****	****
94.	*Portunus sanguinolentus*	M, E	****	****	****	****	****	****	****	****	****	****

****: Abundant; ***: Moderate; **: Rare; *: Occasional.

Among the prawns, the most significant contributor to the fishery was *Metapenaeus dobsonii,* to the tune of 10.0–53.4 per cent to the total landings. *Penaeus indicus* also considerably contributed to the fishery (2.5–29.6 per cent) due to its attractive prize followed by *M. monoceros* (1.0–8.2 per cent).

The crab fishery also substantially contributed to fishery (appro. 1100 t; 2.6–11.1 per cent) by the spotted crab (*Portunus sanguinolentus*), the blue crab (*P. pelagicus*) and the mud crab (*Scylla serrata*) serving both the domestic and international markets.

The total landings from different backwaters varied from 96.8 t (Mahe) to 2899 t (Ashtamudi; Table 3.3). The average yield per ha varied from 410 kg at Anchuthengu to 2747.3 t from Azhicode estuary. The low yield per unit area at Ashtamudi was due to its large area (6424 ha). The high density fish/prawns population at Azhikode is due to its wide access to the sea and the narrow stretch of the backwater spread. The system was well filtered by the Stake net (445 units) that brought huge quantity of *M. dobsonii, M. monoceros, P. indicus* and a whole lot of less commercial species. The Anchuthengu backwater brought less catch due to the denial of direct access to the sea and being a highly strained ecosystem resultant to pollution from coconut husk retting. The average value for the whole system was 630.1 kg $ha^{-1}yr^{-1}$.

The nets used in different backwaters are Gill net, Cast net, Seine net, Stake net, Chinese dip net, Scoop net, Ring net, Trawl net, Trap net and Hook and lines. Of these, cast nets and seine nets are commonly used; the latter is mainly used for prawn fishing. Occasionally anchovies, chandids and gobiids are also getting. Dragnets also used in the shallow region, which destroys large number of young ones of *Etroplus suratensis,* the most commercially important species in the state. Stationary dip nets installed on wooden frames popularly known as Chinese dip net are also used. The different gear observed for the fishing operation during the survey was broadly categorised into 10 different types. The gear enumeration indicates that the density (No. km^{-2}) different systems ranged from 52 numbers in Ashtamudi to 174 in Mahe. On an average, one gear unit has different composition and days of operation roughly averaging for the whole year. Operation or Stake nets and Chinese dip nets are influenced directly by the tidal strength. Hence their operation is generally restricted to a few days adjacent to full moon and new moon.

The CPUE exhibited wide variation from gear to gear (Table 3.4), but a general trend for most the backwaters. The high rate CPUE for the Stake net is largely offset by the limited days of operation and the low market value fish forming bulk of the catch. Seine net too brought high CPUE, but again the catch composition is composed largely of small fishes, young fishes and low priced *M. dobsonii.* The catch from this gear was subjected to wide variation.

Table 3.4: Average CPUE and Percentage Contribution by Different Gear to the Total Landings in the Backwaters During 1996–97

Gear	*CPUE (kg unit^{-1} day^{-1})*	*Contribution to Total Landings (%)*
Gill net	3.9–12.1	26.31
Cast net	2.5–6.8	10.94
Seine net	11.3–744.7	17.76
Stake net	7.5–18.1	29.39
Chinese dip net	5.6–10.2	11.6
Hooks and line	2.3–10.3	2.13
Scoop net	2.3–2.9	0.89
Ring net	1.0–5.6	0.91
Trap net	3.8–8.7	0.46

The general observation is that the fishery of the backwater is less dependent on the local species and more dependent on the migratory marine species like the prawns, crabs, catfishes, silver bellies, baracudas, anchovies, carangids, perches and so on. However, considerable contribution has done by the local resident species like *Etroplus suratensis* and mullets etc.

Acknowledgement

The authors are thankful to the Director, Central Inland Capture Fisheries Research Institute, Barrackpore for his constant encouragement for the preparation of the article.

References

Balakrishnan Nair, N., Krishnakumara, K., Rajasekharan Nair, J., Abdul Aziz, P.K., Dharmaraj, K. and Arunachalam, M. 1983. Ecology of Indian estuaries–XI. A preliminary survey of the Ashtamudi estuarine system. *Fish. Tech.*, 20: 75–83.

Day, F. 1967. *The Fishes of India.* Jagmander Book Agency, New Delhi.

George, A.I. and Sebastian, M.J. 1970. Review of the backwater fisheries and brackishwater fish culture in Kerala State. In: *Symposium on Coastal Aquaculture. Indo-pacific Fish Coun.*, IPFC/C70/SYM, 19 (Mimeo).

Jhingran, V.G. 1975. *Fish and Fisheries of India.* Hindustan Publishing Corporation, New Delhi, India.

Jhingran, V.G. and Gopalakrishnan, V. 1971, Estuarine fisheries resources of India in relation to adjacent seas. In: *Symposium on Indian Ocean and Adjacent Seas,* Cochin. Marine Biological Association of India, 12–18, January.

Madhusoodhanakurup, B., Sankaran, T.M., Rabindranath, P. and Sebastian, M.J. 1993. Seasonal and spatial variations in fishing intensity and gear-wise landings of the Vembanad lake. *Fish. Tech.,* 30: 15–20.

Nair, K.K.C., Gopalakrishnan, T.C., Venugopal, P., George, P.M., Jayalakshmi, K.V. and Rao, T.S.S. 1983. Pollution dynamics of estuarine amphipods in Cochin backwaters. *Marine Ecology Progress Series,* 10: 289–295.

Shetty, H.P.C. 1965. Observations on the fish and fisheries of the Vembanad backwaters, Kerala. *Proc. Acad., Sci., India,* 35: 115.

Thalwar and Jhingran. 1991. *Inland Fishes of India and Adjacent Countries.* 1&2: 1–1158.

Thampi, P.R.S. 1973. In: *Proceedings of the Symposium on Living Resources of the Seas Around India.* CMFRI Special Publication, Cochin, pp. 390.

Chapter 4

Application of ITK in Fish Farm Management

Apurba Bhuyan and A.K. Panigrahi

Department of Fishery Extension, Faculty of Fishery Sciences, West Bengal University of Animal and Fishery Science, Mohanpur Campus – 741252, W.B., India

ABSTRACT

The ITK in general is informal in nature and less effort have been made in this regard to document and their subsequent scientific analysis to find out the rationality behind them. An attempt was made to document application of ITK in hatchery management practices among the hatchery owners in the district of North 24 Parganas of West Bengal. The study reveals the use of ITK in fish farm management, which bears scope for further scientific analysis.

Keywords: *ITK, Fish farmer type, Gender concern.*

Introduction

The term Indigenous Knowledge (IK) refers to the unique, traditional, local knowledge existing within and developed around the specific conditions of women and men indigenous to a particular geographic area (Grenier, 1998). Das *et al.* (2002) stated, Indigenous knowledge as such refers not only to knowledge of indigenous peoples, but to that of any other defined community, while Indigenous Technical Knowledge (ITK) is specifically concerned with actual application of the thinking of the local people in various operations of agriculture and allied areas.

Indigenous Knowledge of farmers has two powerful advantages over outside knowledge are–it has little cost and is readily available. Indigenous knowledge systems and technologies are found to be socially desirable, economically affordable, sustainable and involve minimum risk to rural farmers and producers, and above all, they are widely believed to conserve resources (Grenier, 1998).

The fishers and farmers, through their direct interaction with the fishing and farming situations, have developed lots of working knowledge base and practices. These self gathered knowledge, through experiences and repeated practices, have their unexplored rationality. Albeit the emergence of growing concern of indigenous knowledge worldwide, relevant studies regarding fisheries in this line is less evident. Against this background, the present study has been aimed to document and conduct systematic studies on the indigenous knowledge in fisheries and aquaculture with the following objectives:

1. To recognize and document ITK regarding fish farming.
2. To short out gender concern in different practices.
3. To short out practices according to farmer type.

Materials and Methods

The district 24 Paraganas (N) is selected for investing existing Indigenous Technical Knowledge system as the district is found to be rich in terms of fishery resources along with a good number of fisher folk communities belonging to Schedule Caste and Schedule Tribe. Purposive as well as Random sampling techniques with applying lottery method, were adopted for the present study. For selection of state, district and blocks purposive sampling technique was carried out. Two Gram Panchayats were selected randomly out of eight. Out of nineteen villages, only six villages were randomly selected. From each village, a twenty number of respondents selected randomly with a total sample respondents of 120.

In the present investigation, interview schedule and direct observation methods were selected from the set of methods in accordance with the objectives of the study after considering the nature of the fishery resources; type of people and of course the nature of the study.

An interview schedule was prepared with the help of various type of ITK related books, journals etc. Direct observation is systematically observing objects, events, processes, relationship or people and recording these observations. Direct observation is a good way to crosscheck respondents' answers. Information were recorded in notebook. Schematic diagrams, photographs of the materials and indigenous practices were taken.

Categorization of Farmer

The respondents were categorized in to four groups *viz.*, Marginal, Small, Medium and Rich farmers on the basis of their land holding size. This was done as per the convenience of the study and the schedule was developed accordingly. The categorization of farmers are made according to their land holding size which includes total land possessed by farmer for their livelihood *i.e.*, Agricultural land, Land for fisjery activity, home stead land and others.

Table 4.1

Land Holding (Acre)	*Farmer Type*
< 1 Acre	Marginal
1–2.5 Acre	Small
2.5–5 Acre	Medium
> 5 Acre	Rich

Results and Discussion

Table 4.2

Sl.No.	ITK Found	Purpose of Use	Farmer Type	Gender Concern
1.	Provision of shade in culture pond	To provide shelter to cultured fishes especially during hot summer	Medium farmers	Male
2.	Application of raw cow-dung to fishpond	To enhance natural feed production	Small and medium farmers	Male and female
3.	Application of Cattle urine in fishpond	To control algal bloom	Small and medium farmers	Male and female
4.	Placing fish feeds in gunny bags immersed in pond water	To feed fishes and to avoid wastage of feed	Small and medium farmers	Male and female
5.	Application of Banana stem pieces in to fishpond	To purify pond water in occurrence of foul smell, gas bubbles from pond water and to combat illegal fish poisoning	Small, Medium and rich fish farmers	Male
6.	Broadcasting of ash in to fishpond	As prophylactic measure against fish diseases	Small and Medium farmers	Male and female

Interpretation of Findings

Provision of Shade in Culture Pond

Long and broad leaves of Coconut tree (*Cocos nucifera*) or Khajur tree (*Phoenix sylvestris*) are tied with 2–3 bamboo poles. The poles are fixed to bottom of the pond. The leaves are tied in such a position that float or partially submersed into the pond water. The structure is made to provide shelter to cultured fishes specially to overcome hot summer. Locally available materials are used and easy for maintenance.

Application of Raw Cow-dung to Fishpond

Raw cow dung is broadcasted in the fishpond to produce natural fish feed and tempering of water for proper growth of fish. But according to scientific interpretation, cow dung serves as an ideal source of organic manure, which enhances planktonic feed items for fishes in pond.

Application of Cattle Urine in Fishpond

Farmers spray cattle urine in occurrence of algal bloom in the culture fishpond. This is a simple method and cattle urine may control algal bloom.

Placing Fish Feeds in Gunny Bags Immersed in Pond Water

Feedstuff such as Mustard oil cake, Mohua oil cakes are put in to gunny bag and the mouth of the bag is tied up with a jute rope tightly. One to three such bags with feeds are now kept submersed under water near pond bank in 2–3 corners of the pond. Sometimes the bags are tied up with Bamboo pole and fixed to bottom such that the bags remain under water. The feeds are released from the pores of the bags slowly. Thus it checks wastages of feeds.

Other feedstuff includes wheat flour mixed with water, boiled rice, leaves of Pumpkin and Banana. The later two items are specially applied as feed for Grass carp. These items are applied by broadcasting to pond water. Method is simple and may be efficient to feed the fishes.

Application of Banana Stem Pieces in to Fishpond

Banana stems are cut into small pieces on the bank of the fishpond and sprayed into the fishpond. Farmers follow this practice widely when the there is occurrence of fish mortality without any external signs, occurrence of foul smell and gas bubbles from pond water. It is also practiced to purify pond water when farmer's pond has been poisoned for poaching by anonymous. Method is easy, material is readily available, cheap and Banana pseudo-stem may act upon harmful gases and toxicants in pond water.

Broadcasting of Ash in to Fishpond

Paddy straws are burnt and the ash of it is broadcasted into fishpond or sometimes heaped in one corner of the cultured pond. The practice is used as prophylactic measure to check fish disease in cultured pond. The practice involves least cost and materials are readily available. Applying burnt ash of paddy straw may keep pond water quality congenial for fish growth.

The paper has indicated the rich extent of Indigenous Technical Knowledge and practices in fisheries in the study area. Farmers and fishers have long been bearing up these knowledge and practices and most often use them in their farming situations. The study also reveals involvement of women in different fishing and farming activities. Their participation call for women specific approaches in fisheries technology development, which would stand by, the growing concern of women empowerment.

Formation of Indigenous Knowledge Centers nationwide and its effective network to the grass root level is the need of hour. We have the opportunity to conserve and validate Indigenous Technical Knowledge from the indigenous people to make them more productive. It needs back up from the scientific analysis and interpretations. However, the outcome of these interpretations needs to be generalized and disseminated to its end users. Mere documentation and validation of ITK will not be a wise step if the same aren't returned to its owners.

References

Chander, M. and Mukherjee, R. 1994. Traditional agricultural and animal husbandry practices for sustainable agriculture in Kumaun hills of U.P. *Journal of Rural Development*, 13(3): 443–449.

Das, P., Das, S.K., Arya H.P.S., Reddy, G.S. and Mishra, A. 2002. Inventory of Indigenous Technical Knowledge in Agriculture. Mission Mode Project on Collection, Documentation and Validation of Indigenous Technical Knowledge Document–I, ICAR, p. 3–8.

Grenier, L. 1998. Working with indigenous knowledge: A guide for researchers. Ottawa: IDRC. In. *Conservation of Biodiversity: Through Indigenous Knowledge*, (Eds.) Paul, S. and Ramanathan, A. 2002, Kurukshetra. 5(3): 3–7.

Nogard, R.D. 1984. Traditional agricultural knowledge. Pest performance, future prospects and institutional implications. *American Journal of Agricultural Economy*, 66: 874–878.

Pattnaik, B.C. 2000. Marketing of traditional value-added fish products: A profitable venture. *Fishing Chimes*, 20(6): 52–54.

Ray, G.L. 2004. *Research Methods in Social Sciences and Extension Education*. Kalyani Publishers. p. 236–237.

Chapter 5

Requirement of Dietary Vitamin E in Relation to Growth, Feed Conversion and Deficiency Symptoms for the Fingerlings of *Labeo rohita* (Hamilton)

Ashok K. Gupta

*Department of Limnology and Fisheries**

Maharana Pratap University of Agriculture and Technology, Udaipur – 313 001, India

ABSTRACT

In the present investigation vitamin E requirement was evaluated for the fingerlings of *Labeo rohita* using varying levels of vitamin E (as α-tocopherol) supplemented purified diet for a period of 60 days. Significant weight gain and absolute growth, specific growth rate and feed conversion ratio for the fingerlings were observed at 10–20 mg vitamin E/kg of diet ($p < 0.0\ 1$). Whereas relative efficiency of various levels of dietary vitamin E on growth, specific growth rate and feed conversion ratio for the fingerlings was estimated as 20 < 10 < 40 < 80 < 100 < 00 (Control) mg/kg d diet. The dietary vitamin E deficiency symptoms observed in the fingerlings were poor growth, low conversion efficiency, low protein productive value and energy conversion efficiency, anemia, ascites and muscular dystrophy.

Keywords: *Fingerlings, Labeo rohita, Vitamin E.*

---

* Now College of Fisheries.

Introduction

Fat soluble vitamins (ADEK) have been found to play an important role in maintaining normal growth and physiological functions in animals. Vitamin E (α-tocopherol) are biological antioxidants, active in the water and lipid phases respectively of animal tissue. They protect the organism against uncontrolled oxidation by free radicals, which are generated through normal metabolism, and by oxidative drugs and pollution. Vitamin E is an essential nutrient in fish also. The demand for this nutrient has been demonstrated in relatively few fish species (Watanabe *et al.*, 1970a,b; Cowey *et al.*, 1981; Wilson *et al.*, 1984; Roem *et al.*, 1990; Hamre and Lie, 1995, Hamre *et al.*, 1997; Ruff *et al.*, 2001). There has been limited study on the dietary vitamin E requirement of Indian major carps at early stages of development. Therefore, the present study has been designed to evaluate the vitamin E requirement of the fingerlings of *Labeo rohita* in relation to growth, feed conversion and deficiency symptoms.

Materials and Methods

Modified test diet of Yone (1975) supplemented with vitamin E and mineral mixture was used for vitamin E requirement of *Labeo rohita* fingerlings. In isocaloric test diet, graded levels of vitamin E (as α-tocopherol) were added separately. In each prepared diets, 7–10 per cent of water was added to prepare a thick paste. Vermicelli of diameter 1 mm size were prepared from the paste, air dried and refrigerated. Feeding trials were conducted in cemented pools of 185 lit. capacity. For each treatment three sets of experiment were run. In each replicate ten properly acclimatised healthy fingerlings of approximately similar size and length (average weight 9.16 g and length 8.5 cm) were used. Controls with vitamin E deficient diet were also run simultaneously. The fingerlings were fed at the rate of 2 per cent body weight per day in between 9.00–9.30 a.m. The ration was adjusted fortnightly in accordance with the increment in body weight. Faecal deposits were removed daily by siphoning.

Weight gains of fingerlings was recorded regularly at every fortnightly interval. The feeding trials were conducted for a period of 60 days. The data were computed and statistically analysed, *viz.*, analysis of variance and correlation coefficient (r) as per method of Snedecore and Cochran (1980) for different growth parameters such as absolute growth, specific growth rate (SGR), feed on version ratio (FCR) and gross conversion efficiency (GCE). On the basis of present observations major deficiency symptoms were also recorded and concluded.

Results

The absolute growth of rohu fingerlings was significantly high at low vitamin E levels of 20 and 10 mg/kg of diets (Table 5.2). The dietary levels of 40, 80 and 100 mg vitamin E/kg indicated poor absolute growth. At control diet, no weight gain and absolute growth were recorded in rohu fingerlings ($p > 0.01$).

The data indicates that SGR is directly related to absolute growth of fingerlings, *i.e.*, higher absolute growth resulted in higher SGR values and a positive correlation is observed at 1 per cent of significance level. Average highest SGR's of 0.253 and 0.250 g/day were observed when fingerlings of rohu fed with 20 and 10 mg of vitamin E/kg of diet respectively for a total experimental period of 60 days (Table 5.2). However, in vitamin E deficient diet, SGR of the fingerlings for 60 days was found zero ($p > 0.01$). Poor SGRs were recorded in rohu fed with 40, 80 and 100 mg vitamin E/kg of diet.

Rohu fingerlings show lower values of FCR, *i.e.*, 1.73–3.31 at all the dietary vitamin E levels, excluding control (Table 5.2). This suggests that vitamin E levels to some extent enhances the utilization of purified diet during the feeding period. The highest GCEs, *i.e.*, 0.578 and 0.540 for the absolute growth of 15 and 15.2 g were computed at levels of 10 and 20 mg vitamin E/kg of diet respectively. The zero GCE indicate the essentiality of vitamin E in the diet for better growth of fingerlings. Obviously, it was found non-significant ($P > 0.01$) from other levels of vitamin E.

Table 5.1: Ingredient Composition of the Basal Diet Used for Vitamin E Requirement for the Fingerlings of *L. rohita* (Modified after Yone, 1975)*

Ingredients	%	*Ingredients*	%
Casein vitamin free	47.0	Gelatin	14.0
Corn oil	10.5	White dextrin	8.0
α-cellulose	2.3866	Vitamin mixture	1.10
Mineral mixture	10.0	L-phenylalanine	0.7
L-Arginine HCl	1.4	L-Cystine	0.8
L-Tryptophane	0.2	L-Histidine HCl	0.2
DL-Alanine	1.4	L-Aspartate Na	1.0
L-lysine HCl	0.5	L-Valine	0.6
Glycine	0.3		

* Proximate composition (%): Protein 54; Carbohydrate 25, Fat 10, Moisture 8, Ash 3.

Table 5.2: Growth Parameters of the Rohu Fingerlings Fed with Purified Diet Supplemented with Graded Levels of Vitamin E as α-tocopherol for a Period of 60 Days

Vitamin E (mg/kg diet)	*Initial wt. of Fish (g)*	*Final wt. of Fish (g)*	*Absolute Growth (g)*	*SGR (g)*	*FCR*	*GCE*
00 (Control)	19	19	0	0.00	0.00	0.000
10	18	33	15	0.250	1.73	0.578
20	21	36.2	15.2	0.253	1.84	0.540
40	22	32	10.0	0.166	2.95	0.338
80	20	29.3	9.3	0.155	2.98	0.334
100	26	29.0	3	0.133	3.31	0.302

SGR: Specific Growth Rate; FCR: Food Conversion Ratio; GCE: Gross Conversion Efficiency.

The gross deficiency symptoms observed in fingerlings were poor growth, poor SGR, low FCR and low GCE alongwith blood anaemia, ascites and muscular dystrophy (Table 5.3). Other dietary deficiency symptoms of vitamin E evident were low PPV and ECE leading low protein, carbohydrate and high lipid contents in the fingerlings. In vitamin E deficient diets, fingerlings also showed poor accumulation of vitamin E and high lypoxigenase activity in tissues.

Discussion

The exact requirement of α-tocopherol for any fish species may depend upon the form of vitamin E and amount and type of unsaturated fatty acids in the oil components of the ration (Woodall *et al.*,1964; Watanabe *et al.*, 1970ab). Polyunsaturated labile fish oils may invoke an increased requirement for intracellular antioxidants (Hashimoto *et al.*, 1966).

The vitamin E supplemented diets in the present study were also improved growth, SGR, FCR and GCE in the fingerlings of rohu. A significant coefficient of correlation (r) at 1 per cent was also found between weight gain and SGR; GCE and FCR and weight gain and FCR in the fingerlings fed with vitamin E supplemented diets. Whereas, vitamin E deficient diets have shown poor response for growth parameters and proved to be statistically non-significant ($P > 0.01$). Woodall *et al.* (1964) and Murai and Andrews (1974) have also reported that vitamin E deficient diets in chinook salmon and channel catfish produced poor growth and reduced food conversion with several deficiency lessions.

Table 5.3: Gross Deficiency and Hypervitaminosis Symptoms in the Fingerlings of Rohu Fed with Low and High Supplemented and Deficient Vitamin E Diet

Sl.No.	Deficiency Symptoms	Sl.No.	Hypervitominosis Symptoms
1.	Poor growth SGR, FCR and GCE	1.	Poor growth, SGR, FCR and GCE
2.	Low PPVs and ECEs	2.	Low mortality
3.	Low protein, carbohydrate and high lipid contents		
4.	Low vitamin E contents and high lypoxigenase activity in tissues		
5.	Blood anemia ascites muscular dystrophy and fragility in R.B.C.		

Woodall *et al.* (1964) have observed deficiency symptoms, *viz.*, poor growth, exophthalmia, ascites, anemia, clubbed gills, epicarditis and ceroid deposition in the spleen of chinook salmon. However, Dupree (1969) and Murai and Andrews (1974) have noticed vitamin E deficiency symptoms of poor growth, exudative diathesis, muscular dystrophy, depigmentation, fatty livers, anemia and atrophy of pancreatic tissue in channel catfish. Poston *et al.* (1976) fed vitamin E deficient diet to Atlantic salmon displayed deficiency symptoms of anemia, pale gills, anisocytes, poikilocytosis, exudative diathesis and muscular dystrophy. However, in the present investigation besides poor growth response, low vitamin E contents and high lypoxigenase activity in tissues, anemia, muscular dystrophy etc. have been observed in the rohu. Woodall *et al.* (1964) reported that a dietary vitamin E concentration of 5 to 30 mg α-tocopherol combined with 5 per cent herring oil provided satisfactory growth of chinook salmon and also prevented the occurrence of clinical deficiency symptoms. The present results suggest that the range of optimum dietary requirement for vitamin E as α-tocopherol for rohu fingerlings are comparatively higher despite test diet having corn oil with low levels of polyunsaturated fatty acids. Further, the present study reveals that 10 per cent corn oil in the test diet with varying, concentrations of vitamin E displayed a significant role for maintaining maximum growth, feed conversion efficiency, PPV and ECE in rohu fingerlings. At 10 per cent corn oil plus 10–20 mg vitamin E/kg diet, rohu fingerlings showed highest growth and feed conversion efficiency. The investigations of Watanabe *et al.* (1981) emphasized that diets supplemented with 50 mg α-tocopherol/kg dry diet promotes growth rate in rainbow trout fingerlings equal to those fish fed with 100, 300 or 500 mg α-tocopherol/kg of dry diet.

Cowey *et al.* (1981) have established that larger rainbow trout fingerlings (initial mean weight 10g) required 20 or 30 mg DL α-tocopherol acetate of 50 mg/kg of diet should be considered a minimum level since the requirement may vary depending on size, age, water temperature, water ionic strength, fish density, water quality etc. Further, Wilson *et al.* (1984) have also recommended that higher levels of vitamin E may be needed in practical diets that contain higher levels of polyunsaturated fatty acids or oxidized lipids. Cowey *et al.* (1984) have suggested that vitamin E requirement may increase as the water temperature decrease. Therefore, minimum dietary vitamin E requirement measured in the present study at a range of 20.6–22.4°C may be decrease if water temperature increases.

Acknowledgement

Appreciation is extended to the ICAR, New Delhi for financing the project.

References

Cowey, C.B., Adron, J.W., Walton, M.J., Murray, J., Youngsun, A., Knose, D. 1981. Tissue distribution, uptake and requirement for α-tocopherol of rainbow trout (*Salmo gairdneri*) fed diets with minimol contents of unsaturated fatty acids. *J. Nutr.*, 111: 1556–1567.

Cowey, C.B., Elizabeth, D., Tacon, A.G.J., Youngson, A., Bell, J.G., 1984. The effect of vitamin E oxidized fish oil on the nutrition of rainbow trout (*Salmo gairdneri*) grown at natural, varing water temperatures. *British J. Nutr.*, 51: 443–451.

Dupree, H.K. 1969. Vitamin E requirements of channel catfish. *U.S. Bull. Sport Fish.*, Wildlife Resour. Publ. 77: 220–221.

Hamre, K., Lie, ϕ. 1995. α-tocopherol levels of different organs of Atlantic salmon (*Salmo salar* L.): Effect of smoltification, dietary levels of n-3 polyunsaturated fatty acids and vitamin E. *Comp. Biochem. Physiol.*, 111A: 547–554.

Hamre, K., Waagbo, R., Berge, R.K., Lie, ϕ. 1997. Vitamin C and E interact in juvenile Atlantic salmon. *Free Radical Biol. Med.*, 22: 137–149.

Hashimoto, Y., Ikaichi, Y., Watanabe, T., Furukawa, A., Umezu, T. 1966. Muscle dystrophy of carp due to oxidized oil and the preventive effect of vitamin E. *Bull. Jpn. Soc. Sci. Fish*, 32: 64–69.

Murai, T. and Andrews, J.W. 1974. Interactions of dietary α-tocopherol, oxidized menhaden oil and ethoxyquin on channel cat fish (*Ictalurus punctatus*). *J. Nutr.*, 104: 1416–1431.

Poston, H.A., G.F. Combs Jr., Leibovitz, L. 1976. Vitamin E and Selenium interrelation in the diet of Atlantic salmon (*Salmo salar*): Gross, histological and biochemical deficiency signs. *J. Nutr.*, 106: 892–904.

Roem, A.J., Kohler, C.C. and Stickney, R.R. 1990. Vitamin E requirements of the blue tilapia *Oreochromis aureus* (Steindachner), in relation to dietary lipid level. *Aquacult.*, 87: 155–164.

Snedecor, S.W. and Cochran, W.G. 1980. *Statistical Methods*, 7th edn., IOWA State University Press.

Watanabe, T., Takashima, F., Ogino, C. and Hibiya, T. 1970a. Requirement of young carp for α-tocopherol. *Bull. Jpn. Soc. Sci. Fish*, 36: 935–946.

Watanabe, T., Takashima, F., Ogino, C. and Hibiya, T. 1970b. Effect of α-tocopherol deficiency on carp. *Bull. Jpn. Sci. Fish*, 36: 972–976.

Watanabe, T., Takeuchi, T., Wada, M. and Uehara, R. 1981. The relationship between dietary lipid levels and α-tocopherol requirement of rainbow trout. *Bull. Jpn. Sci. Fish*, 47: 1463–1471.

Wilson, R.P., Bowser, P.R. and William, P. 1984. Dietary vitamin E requirement of fingerlings channel catfish. *J. Nutr.*, 114: 2053–2058.

Woodall, A.N., Ashley, L.M., Halver, J.E., Olcott, H.S. and VanDer Veen, J. 1964. Nutrition of salmonid fishes. XIII. The α-tocopherol requirement of chinook salmon. *J. Nutr.*, 184: 125–135.

Yone, Y. 1975. Nutrition studies on red sea bream. In: *Proceedings of the International Conference on Aquaculture Nutrition*, (Eds.) Price, K.S., Shaw, W.N., Danbert, K.S. Lewes/Rehoboth, University of Delaware, p. 34–64.

Chapter 6

Role of Sediments in Relation to Productivity and Fishery Potentiality of Mangroves Swamp of Akshi Creek

S.S. Yeragi and S.G. Yeragi
Mangrove Research Laboratory, K.J. Somaiya College of Science, Vidyavihar, Mumbai – 400 077

ABSTRACT

The sediment is the treasure house of nutrients in Mangrove habitat. The benthos population density is always correlated with the nature of the sediments. The characteristic feature of the wetland ecosystem is in trapping the varieties of sediments and made available to the primary producers. The present investigation was conducted in the mangrove mud flats of Akshi creek of Alibaug taluka (Lat. 17°, 72′ N., Long. 73°, 64′ E.). The benthos always prefers the soft floor for their quick burrowing to avoid predation. The eggs as well as larvae have more protection of soft sediments by which new recruitments of resource organisms increased quickly. Due to the availability of high quality of nutrients, safety to the young ones and safe nursery, spawning ground, by which the yield production increase within short span of time compared to other ecosystems.

Keyword: *Sediment, Fishery, Mangroves.*

Introduction

An aquatic productivity is totally dependent on the type of benthic floor, on which the organisms live. The three most important basic elements which support an aquatic productivity are carbon, nitrogen and phosphorus. These elements are trapped in sediments and made available to the primary producers. The wetland ecosystems have great ability to store significant amount of carbon in vegetation and sediment. This ecosystem plays a vital role in cycling of these nutrients. According to Jordan (1989) decomposition of plant litter is an important component of nutrient uptake and retention in wetland as the mangrove plant litter alternately takes up and releases nutrients as it decomposes. These nutrients cycle is one in mangrove swamps and exchange with its surrounding areas.

The release of nitrogen from the sediments is related to a great event by biological process like the death and decay of living organisms. Substratum is a surface on which the organisms rest, move, feed as well as burrow. The population density of any organisms is totally depending on the nature of substratum. The bulk of larvae get natural protection through the substratum The surface film of substratum is smooth, easier for burrowing and for feeding hence recruitment of organisms also more in that area. The substratum is the main controlling factor in an aquatic ecosystem for varieties of population. The shelter and nourishment are the basic needs of organisms which are provided by the substratum. The distribution of flora and fauna are related to the nature of substratum The selectivity of substratum is differing from organism to organism. The selection of the substratwn is done by the animal in its larval stage itself (Patwardhan, 1990).

Materials and Methods

All the sediment samples were collected from the surface of 10–20 cms depth from low level water mark to high level water mark from the mangrove stations. The texture of sediment was analyzed by Beaker method as suggested by Piper (1942).

Results

Table 6.1 shows the percentage composition of mangrove soil samples. The values noted in the table are the average of three samples.

Table 6.1: Mean Percentage Composition of Soil

Months	*Moisture (%)*	*Clay (%)*	*Silt (%)*	*Fine Sand (%)*	*Coarse Sand (%)*
December 2003	49.50	49.21	9.31	41.49	–
January 2004	46.20	51.36	20.23	29.51	–
February 2004	46.10	42.39	4.35	54.36	–
March 2004	45.25	38.50	10.11	51.64	–
April 2004	40.12	44.20	11.61	45.19	–
May 2004	38.24	37.07	8.25	5.25	–
June 2004	50.20	77.34	5.26	16.51	–
July 2004	52.11	70.51	10.52	15.78	3.09

The moisture content of sediment of Akshi mangrove stations was varied from 38.24 per cent in May and 52.11 per cent in the month of July. This change in the range of moisture can be attributed to the direct exposure of mud flats of mangrove region. The highest rate of desiccation in May resulted in lowering the moisture. According to Robinson (1936), clay fraction of sediment has colloidal in nature, which determines the holding capacity for water. The clay is ranging between 37.07 per cent in May and increased to 70.51 per cent in July. This is mainly because of the floods and more flow of riverine and terrestrial water with greater load of clay. The coarse sand is also recorded in monsoon period only.

During pre-monsoon and post-monsoon seasons the fine sand component increased reaching maximum in February 54.36 per cent. The percentage composition of silt was maximum in January 20.23 per cent and lower in February 4.35 per cent.

Different species of mangrove plants are tolerant of different degrees of acidity and anaerobic conditions of the soil, a factor that naturally influences distribution and zonation of the species of trees at each size in Akshi creek.

The ubiquitous burrowing fauna is represented by decapod crustaceans like, *Uca* spp., the fiddler crab, *Scylla serrata,* the edible crab and other crabs living in or near the surface of the mud. The mud lobster *Thallassina squilla* and other crustaceans occur in large number in mangroves and are absent from others, due to variations in soil contents. Mud skippers of genus *Periophthalmus, Boleophthalmus* and others may be so abundant.

Conclusion

Development of substratum in mangrove ecosystem is mainly due to clay, silt and fine sand. The substratum in the mangrove environment having fine grained sediments with calcareous debris is characterized by high salt content, low oxygen and high organic content with the dominance of cellulose and ionic fractions.

Mangrove soils are generally slightly acidic due to oxidation of ferrous sulphates in the soil. Carbon dioxide arising from decomposition of organic matter and from animal respiration may also lower the pH values of the soil. Different mangrove species, according to their salt tolerant ability and substratum preference occupy different zones along the creek. The accretion of sediment of the mouth of Akshi creek is quite common.

The availability of the fauna in abundance is fully dependent on how much nutrients are present on the feeding ground (Shet, 1994). The shallow region is always washed out with all the nutrients. In the monsoon due to inflow of fresh water from the adjacent areas and downward currents are formed. The mangrove region is the most protective region, so that no nutrient could be washing off from the central zone.

References

Jordan, T.C. 1989. The role of litter in nutrient cycling in a brackish tidal marsh. *Ecology,* 70(6): 1906–1915.

Patwardhan, D.D. 1990. A study of estuarine ecology of South Konkan. *Ph.D. Thesis,* University of Mumbai.

Chapter 7

Observation on Opercular Movement and Heartbeat Rate in Freshwater *Glossogobius giuris* (Ham.) After Exposed to Malathion

V. Sreenivasa, V. Aravindhan, M.B. Nadoni and P.S. Murthy
Physiology of Reproduction Unit, Department of Zoology, Bangalore University, Bangalore – 560 056

ABSTRACT

The Opercular movement indicates the respiratory efficiency of the animal, indicating the flow of water over the gills and rate of heartbeat which could read the blood flow in the heart. It served as good indicator of various physiological processes of the fishes under malathion stress. The rate of opercular movement and rate of heartbeat increased in *G. giuris* after exposure to sub-lethal concentrations of malathion (0.05, 0.25 and 0.5 ppm) for 24 hrs. Further the activity was found to be decreased during 72 hrs of exposure. The activity reaches nearer to the normal after prolonged duration (96 hrs) of exposure to pesticide. The alteration in the activity might be the neurotoxic effect of malathion.

***Keywords**: Malathion, Opercular movement, Heartbeat, G. giuris.*

Introduction

The pesticide sprayed over the plants is drained in to the water bodies and causes many disorders in aquatic fauna. The rate of opercular movement and rate of heartbeat are highly conspicuous in fishes with reference to toxic substances (Thiede, 1965). The mechanism of respiratory system with reference to gills and bronchial circulation has been reported in teleost (Christopher, 1979). The Opercular movements in teleost after treatment with pesticide have been studied with the correlation between opercular activity and uptake of oxygen (Bashamohideen and Onnuruppa, 1986). The rate of heartbeat is the source of energy for cellular metabolism (Unminger, 1971a) and also acts as an index of oxygen intake (Sate *et al.*, 1978). The rate of heartbeat with reference to environmental stress like salinity, temperature and atmospheric pressure has been reported in teleost (Rutharane *et al.*, 1980; Bashmohideen and Faslealikhan, 1983). The attempt has been made in the present investigation to study the changes in the opercular movement and rate of heartbeat of fish, *G. giuris* after exposure to sub-lethal concentrations of malathion.

Materials and Methods

The freshwater gobiid fish, *Glossogobius giuris* (Ham) were randomly collected from in and around Bangalore, using cast and gill net. They brought alive to the laboratory and were acclimatized for 15 days in 50 liters glass aquaria containing aerated tap water. They were fed daily with earthworms. Mature fish weighing about 50–60 gms and 110–150 cm in length were selected for experiment. The malathion was dissolved in Acetone and added to the test water to obtain the desired concentrations by adopting the dilution technique as outlined in APHA, AWWA, WPCF (1985).

The acclimatized fishes were divided into four experimental groups of ten each. The first three groups were placed in sub-lethal concentrations (0.05, 0.25 and 0.5 ppm) of malathion. The fourth group in freshwater served as control. The rate of opercular movements and heartbeat were calculated in control and treated fishes by using the formula 10/t after 24, 48, 72 and 96 hours of exposure. Where 't' is time taken for 10 opercular movement or heartbeat, 10 is number of opercular movement or heartbeat in each fish. An average values were calculated for standard deviation and percentage change.

Results and Discussion

The rate of opercular movement and heartbeat of treated and control fishes were presented in the Tables 7.1–7.2 and Figures 7.1–7.2. The result indicates, the opercular activity increased at 24 hrs (1.14 ± 0.01, 0.89 ± 0.03 and 1.42 ± 0.026). In contrast to this, opercular activity gradually decreased to 0.3 ± 0.02, 0.71 ± 0.02 and 1.02 ± 0.01 at 48 hrs and maintained a steady activity during 72 hrs (0.7 ± 0.03, 0.75 ± 0.01 and 0.76 ± 0.01), Further the rate of opercular activity was increased after 96 hrs (0.94 ± 0.02, 0.78 ± 0.01 and 0.94 ± 0.01) of exposure in all the concentrations of malathion when compared to control (0.75 ± 0.01). In teleosts, the pesticide toxicity causes variation in the opercular movements. The variations in the opercular activity suggest the fish looses its physiological balance due to the toxicity of pesticide on central nervous system, which is the regulating center of various physiological activities. Similar results are observed when the fish exposed to pesticide (Rafi Ahamed and Basha Mohideen, 1985; Malla Reddy 1988; Basha Mohideen and Sailabala 1992 and Jagadeesan and Vijayalaxmi, 1999).

Table 7.1: Rate of Opercular Movement in *G. giuris* (Ham.) Exposure to Sub-lethal Concentrations of Malathion (Mean ± SE per cent Change)

Control	*Concentration in ppm*	*Exposure Period in Hours*			
		24	*48*	*72*	*96*
	0.05	1.14±0.1	0.63±0.02	0.7±0.03	0.94±0.02
0.75±0.01	0.25	0.89±0.03	0.71±0.02	0.75±0.01	0.78±0.01
	0.5	1.42±0.02	1.02±0.01	0.76±0.01	0.94±0.01

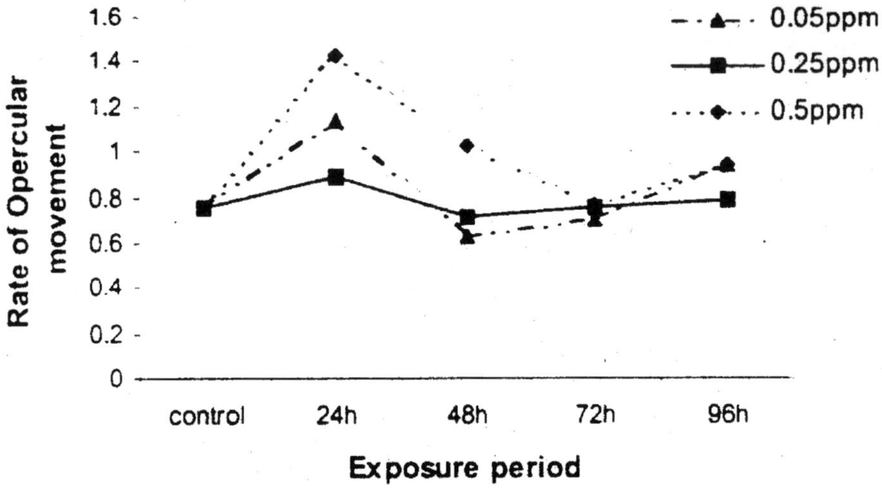

Figure 7.1: Rate of Opercular Movement in *G. giuris* (Ham.) Exposure to Sub-lethal Concentrations of Malathion

Table 7.2: Rate of Heartbeat in *G. giuris* (Ham.) Exposure to Sub-lethal Concentration of Malathion (Mean ± SE per cent Change)

Control	*Concentration in ppm*	*Exposure Period in Hours*			
		24	*48*	*72*	*96*
	0.05	1.67±0.02	1.5±0.03	0.87±0.02	1.29±0.02
1.45±0.07	0.25	1.20±0.02	1.09±0.02	0.8±0.02	0.91±0.02
	0.5	1.58±0.02	1.57±0.05	1.06±0.05	1.14±0.02

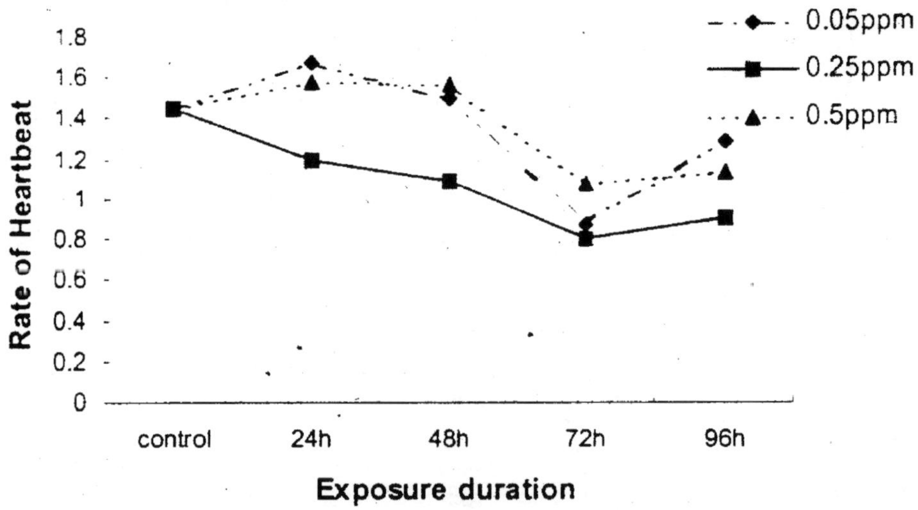

Figure 7.2: Rate of Heartbeat in *G. giuris* (Ham.) Exposure to Sub-lethal Concentration of Malathion

The rate of heartbeat during exposure to malathion shows greater variability. Initially it was increased in 0.05 and 0.5 ppm of malathion during 24 and 48 hrs (1.67 ± 0.02, 1.5 ± 0.03 and 1.58 ± 0.02, 1.57 ± 0.0) of exposure when compared to control (1.45 ± 0.07). In contrast, the rapid decrease in the rate of heartbeat was observed in 0.25 ppm during 24 to 72 hrs. Further, it is also observed that there was decline in the rate of heartbeat at 96 hrs of exposure when compare to control. During 24 hrs of exposure, an increase in the heartbeat rate was observed in sub lethal concentrations when compare to control. But the percentage of heartbeat rate gradually decreased.

The rate of heartbeat was decreased in the fish exposed to sublethal concentration of malathion. This is due to the fluctuation in the rate of opercular activity and oxygen consumption during the exposure to the pesticide. There are few reports over the decrease in the heartbeat rate in fishes exposed to pesticides (Jayantha Rao, 1982). Malathion and its toxic metabolic breakdown product, malaoxon enters the fishes circulation causes more fluctuation in cardiac muscle activity and also affects on the synthesis of energy releasing metabolites (Wayne Sindair, 1979).

References

APHA, AWWA, WPCF. 1985. *Standard Methods for the Examination of Water and Wastewater*. Washington, DC.

Bashamohideen, Md and Faslealikhan. 1983. The rate of heartbeat as a function of salinity and size in the freshwater euryhaline teleost *Saratherodon mossambicus. Comp. Physiol. Ecol.*, 8: 365–369.

Bashamohideen, Md and Onnurappa. 1986. Oxygen consumption and rbc number as good indicators of pollutional (phosphomidon) stress in *sarotherodon mossambicus. First National Symposium Environ. Biol.*, S.K. University Ananthpur.

Bashamohideen, Md and Sailabala 1992. Oxygen consumption and opecular activity as good indicators of pollutional (malathion and methyl parathion) stress in Indian major carp, *Catla catla. I. J. Comp. Animal Physiol.*, 10(1): 20–24.

Christopher, L.A. 1979. Water exchange across gills: The significance of treated water flux measurements. *J. Exp. Biol.*, 79: 147–162.

Jagadeesan, G and Vijayalaxmi, S. 1999. Alteration in the behavior patterns in *Labeo rohila* (Ham.) fingerlings induced by mercury. *I. J. Environ. Toxic.*, 9(1): 45–52.

Jyantha Rao, K. 1982. Effect of synthetic pesticide, phosphomidon on some aspects of metabolism in the freshwater fish *Tilapia mossambica. Ph.D. Thesis*, S.V. University, Tirupati, India.

Malla Reddy, P. 1988. Effect of malathion on the oxygen consumption and rate of opercular movement of *Cyprinus carpio. Geobios*, 15(1): 7–10.

Rafii, Ahmed and Bashamohideen, Md. 1985. *Proc. Symp. Environ. Pollution and Biochemical Systems*, September 7–9, Tirupathi, pp. 29.

Rutharane, Batcheller and Pithin. 1980. Heart rate of adult red spottend neuts. *Nolopthalmus birideseens* iari and submerged in normaxic and hypoxic; water at different temperature. *Comp. Biochem. Physiol.*, 65: 493–496.

Sate, M., Yoshinake, R. and Keda, S. 1978. Biochemical studies on L-ascorbic acid in aquatic animals, II-dietary ascorbic acid requirements of rainbow trout for growth and collages formation. *Bull. Japan Soc. Sci. Fish.*, 44(9): 1029.

Thiede, W. 1965. Uber die hormonale basis der adaptation on extreme temperature bein schweltlraer XI Phosphormishelleni. Heckel (peecilidae) Pisces.

Umminger, B.L. 1971a. Osmoregulatory role of serum glucose in freshwater adapted Killifish, (*fundulus heterocliters*). At temperatures near freezing. *Comp. Biochem. Physiol.*, 38a: 141–143.

Wayne Sindair, M.D. 1979. Heart defects in fish from low levels of malathion. *Teratology*, 19: 51–62.

Chapter 8

Food and Feeding Habits of the Gobiid Fish *Pseudapocryptes lanceolatus* (Bloch and Schneider, 1801) of the Vasista Godavari Estuary, East Coast of India

K.V.C.S. Appa Rao and K. Sreeramulu*

Department of Zoology, Andhra University, Visakhapatnam

ABSTRACT

Investigations on the food and feeding habits of *Pseudapocryptes lanceolatus* (Bloch and Schneider, 1801) show that the first mainly feed on planktonic organism. The diet of males and females did not differ. Qualitative analysis of the diet composition of *P. lanceolatus* revealed that the fish is phytoplanktonic bottom feeder.

Keywords: *Pseudapocryptes lanceolatus, Vasista Godavari Estuary, Food and Feeding habits.*

Introduction

Pseudapocryptes lanceolatus (Bloch and Schneider) is an important food fish of the east coast estuaries of India. Food and feeding habits of gobiid fishes were studied by Blaber (1977), Sarker and Al-daham (1980). In Indian waters, similar works were carried out among others by Venkateswarlu (1966),

* Lecturer in Zoology, Sri Y.N. College, Narsapur – 534275, West Godavari District.

Geevarghese (1983 and 1984), Mehta *et al.*, (1990) and Sankar Rao (2001). In gobioids, herbivores have a long and coiled gut thrice the length of their body, whereas carnivores possess short gut (Venkateswarlu, 1966). No attempt has been made on the food and feeding habits of the *P. lanceolatus*. The present study deals with the food and feeding habits of the lanceolate goby *P. lanceolatus* of the Vasista Godavari estuarine region.

Materials and Methods

Generally the fish is captured by employing stake nets, laid across the Vasista Godavari estuary and also by using traps in canals and creeks around. Samples (30–50 numbers) for the present investigations were procured from such fish stocks once in a month from Narsapur during December 2000 to November 2001. On bringing the fish to the laboratory belly of each fish was cut open, alimentary canal removed and stomach dissected and its contents were emptied into a petri dish. The food items were examined under a binocular microscope.

Various methods are employed in the food analysis of fishes. Pearse (1915) first used volumetric method. Point method was adopted by Swynnerton and Worthington (1940). Frequency of occurrence method (Hynes, 1950) was adopted to decipher the relative importance of various items in the food number of stomachs (containing food) during each month. A number of modifications and combinations of different methods are used to express feeding intensity, employing different indices (Natarajan and Jhingran, 1962). One of such important method is index of preponderance (IP). This method combines both volumetric and frequency of occurrence methods giving rise to accuracy in grading various food elements. Details of methods are: (1) Per cent volume and percent occurrence of each food item were calculated and respective indices (V_1–Volumetric index and O_1–Occurrence index) for all the food items are assessed at (2) Subsequently, a combined index for each food item is calculated using the relation.

$$I_1 = \frac{V_1O_1}{\Sigma V_1O_1} \times 100$$

where,

I_1 is a composite index, *i.e.*, Index of preponderance, V_1 is volumetric index of food item and 1 and O_1 is the occurrence index of food item 1.

(3) Another method of indices is the Gastrosomatic index (GSI) which is expressed as:

$$GSI = \frac{\text{Weight of Stomach Contents}}{\text{Weight of Fish}} \times 100$$

Results and Discussion

A total of 493 stomachs of *P. lanceolatus* (Bloch and Schneider) were examined during the entire course of study. Out of them, 368 stomachs contained food though in varying quantities and stages of digestion where as the remaining 125 guts were empty (Tables 8.2 and 8.3).

Feeding Intensity

Seasonal variation in feeding intensities was ascertained based on monthly analysis of the data. Percent occurrence of empty stomachs was found to be quite high during almost all the months (Tables 8.2 and 8.3).

Table 8.1: Total Points and Gastro Somatic Index (GSI) in *P. lanceolatus*

Month	*Specimens Examined*	*Specimens with food*	*Total Points*	*Average Points*	*Average Feeding*	*GSI*
December 2000	45	42	42	93.33	0.466	3.747
January 2001	50	33	33	64.00	0.250	1.984
February	45	42	42	93.33	0.375	3.760
March	50	27	27	54.00	0.287	0.888
April	50	28	28	56.00	0.268	1.118
May	30	16	16	53.33	0.281	1.502
June	40	34	34	85.00	0.449	5.910
July	45	28	28	62.22	0.348	1.551
August	40	31	31	77.50	0.379	2.567
September	40	35	35	87.50	0.329	2.358
October	40	27	27	67.50	0.315	2.919
November	30	25	25	83.33	0.330	2.934

Table 8.2: Occurrence of Fullness in the Stomachs of *P. lanceolatus*

Month	*Specimens Examined*	*Condition of Stomachs*						
		Gorged	*Full*	*¾ Full*	*½ Full*	*¼ Full*	*Trace*	*Empty*
December 2000	45	6 (13.33)	–	–	9 (20)	27 (60)	–	3 (6.67)
January 2001	50	–	2 (4.44)	–	–	33 (66)	–	17 (34)
February	45	1 (2.22)	–	–	11 (24.44)	28 (62.2)	–	3 (6.67)
March	50	–	–	–	4 (8)	23 (46)	–	23 (46)
April	50	–	–	–	6 (12)	22 (44)	–	22 (44)
May	30	–	–	–	2 (6.66)	14 (46.67)	–	14 (46.67)
June	40	2 (5.00)	–	–	19 (47.5)	13 (32.5)	–	6 (15)
July	45	–	–	1 (2.22)	8 (17.78)	19 (42.22)	–	17 (37.78)
August	40	1 (2.50)	–	3 (7.5)	6 (15)	21 (52.5)	–	9 (22.5)
September	40	–	–	2 (5)	7 (7.75)	56 (65)	–	5 (12.5)
October	40	–	–	–	7 (7.75)	20 (50)	–	13 (32.5)
November	30	–	–	–	8 (26.67)	17 (56.67)	–	5 (16.66)

Table 8.3: Feeding Condition in *P. lanceolatus*

Month	*Well Fed (Active)*	*Moderately Fed*	*Poorly Fed*	*Empty*
December 2000	13.33	20.00	60.00	6.67
January 2001	–	–	66.00	34.00
February	6.67	24.44	62.22	6.67
March	–	8.00	46.00	46.00
April	–	12.00	44.00	44.00
May	–	6.66	46.67	46.67
June	5.00	47.50	32.50	15.00
July	2.22	17.78	42.22	37.78
August	10.00	15.00	52.50	22.50
September	5.00	17.50	65.00	12.50
October	–	17.50	50.00	32.50
November	–	26.67	56.67	16.66
Total	**42.22**	**213.05**	**623.78**	**320.95**
Average	**3.50**	**17.75**	**51.98**	**26.75**

Food Composition

Analysis of the gut contents revealed that *P. lanceolatus* consumes only planktonic organisms and is thus a planktovorous fish. Diet of the fish generally comprised of phytoplankton belonging to algae and diatoms besides sand grains. No difference was observed between the food of male and female fishes.

Qualitative Variation

Qualitatively no variation was noticed between the male and female fishes. The gut contents could be identified into five categories, *i.e.*, *Pleurosigma* sps., *Navicula* sps., *Nitzchia* sps., *Oscillatoria* sps., sand grains and digested material *Pleuro sigma* sps. and *Nitzchia* sps. constituted main food (Table 8.4).

Analysis of Points

Month wise pooled data of the gut contents indicated that a maximum of 93.33 points per stomach occurred in the month of December, 2000 and February, 2001, followed by 87.50 points in September while minimum points of 53.00 were obtained during May, followed by 54.00 points in March (Table 8.1). The study also showed that on average only 17.75 per cent of the fishes existed in moderately fed condition. Based on the fullness of stomach, it was found that fishes with well fed stomachs were relatively few than remaining categories. A maximum of 13.33 per cent well fed stomachs were encountered during December 2000 followed by 10 per cent in August, 2001 (Table 8.3).

Analysis of Volume

Volume of each item in the food material was calculated and they were 43.32 per cent for *Pleurosigma* sps., 30.30 per cent *Nitzchia* sps., 7.67 per cent for *Navicula* sps., 0.53 per cent for *Oscillatoria,* 4.15 per cent for sand grains and 14.03 per cent for digested matter (Table 8.4).

Table 8.4: Food Composition of *P. lanceolatus* in Terms of Volume (V) Frequency of Occurrence (O) and Index of Preponderance (IP)

Food Items	Volume		Frequency of Occurrence		Index of Preponderance	
	Per cent	Rank	Per cent	Rank	Per cent	Rank
Pleurosigma sp.	43.32	1	27.67	2	49.66	1
Navicula sp.	7.67	4	18.62	3	5.90	4
Nitzchia sp.	30.30	2	28.45	1	35.58	2
Oscillatoria sp.	0.53	6	2.41	6	0.05	6
Sand grains	4.15	5	10.09	5	1.68	5
Digested matter	14.03	3	12.76	4	7.13	3
Total	**100**		**100**		**100**	

Table 8.5: Index of Preponderance of Food Items of *P. lanceolatus*

Food Item	V_1	O_1	V_1O_1	IP	Rank
Pleurosigma sp.	43.32	27.67	1208.35	49.66	1
Navicula sp.	7.67	18.62	143.56	5.90	4
Nitzchia sp.	30.30	28.45	865.73	35.58	2
Oscillatoria sp.	0.53	2.41	1.28	0.05	6
Sand grains	4.15	10.09	40.97	1.68	5
Digested matter	14.03	12.76	173.54	7.13	3
	100	**100**	**2433.43**	**1000**	

Analysis by Frequency of Occurrence

Different food items occurred in different frequencies in the food material and they are *Nitzchia* sps. 28.45 per cent, *Pleurosigma* sps. 27.67 per cent, *Navicula* sps. 18.62 per cent, digested matter 12.75 per cent, sand grains 10.09 per cent and *Oscillatoria* sps. 2.41 per cent (Table 8.4).

Analysis of Index of Preponderance

By Index of Preponderance, *Pleurosigma* sps. ranked first with 49.66 per cent, *Nitzichia* sps. second with 35.58 per cent, digested matter with 7.13, *Navicula* sps. fourth with 5.90 per cent, sand grains, fifth with 1.68 per cent and *Oscillatoria* sps., least with 0.05 per cent (Table 8.4).

Gastrosomatic Index

In gastrosomatic index is higher in the months of December, 2000, February, June, October and November months of 2001. It is lower in the month of January, March, April, May, July, August, September months of 2001 (Table 8.1.)

Generally, fishes feed intensively during post-monsoon and winter periods. This habit is attributed to the abundance of food supply (Hardy, 1936) and to the combined effect of food supply and temperature (Ricker, 1937).

Gastro somatic indices (GSI) of the fish for different months revealed that the feeding intensity in both the sexes attained peaks twice during a year. The first peak was observed (Figure 8.1) in the month of February while the second peak in the month of June. Stomachs of a majority of fishes were with ¾ and ½ full during these two months. The first peak corresponded to the post spawning period and the second peak to the abundant production of phytoplankton.

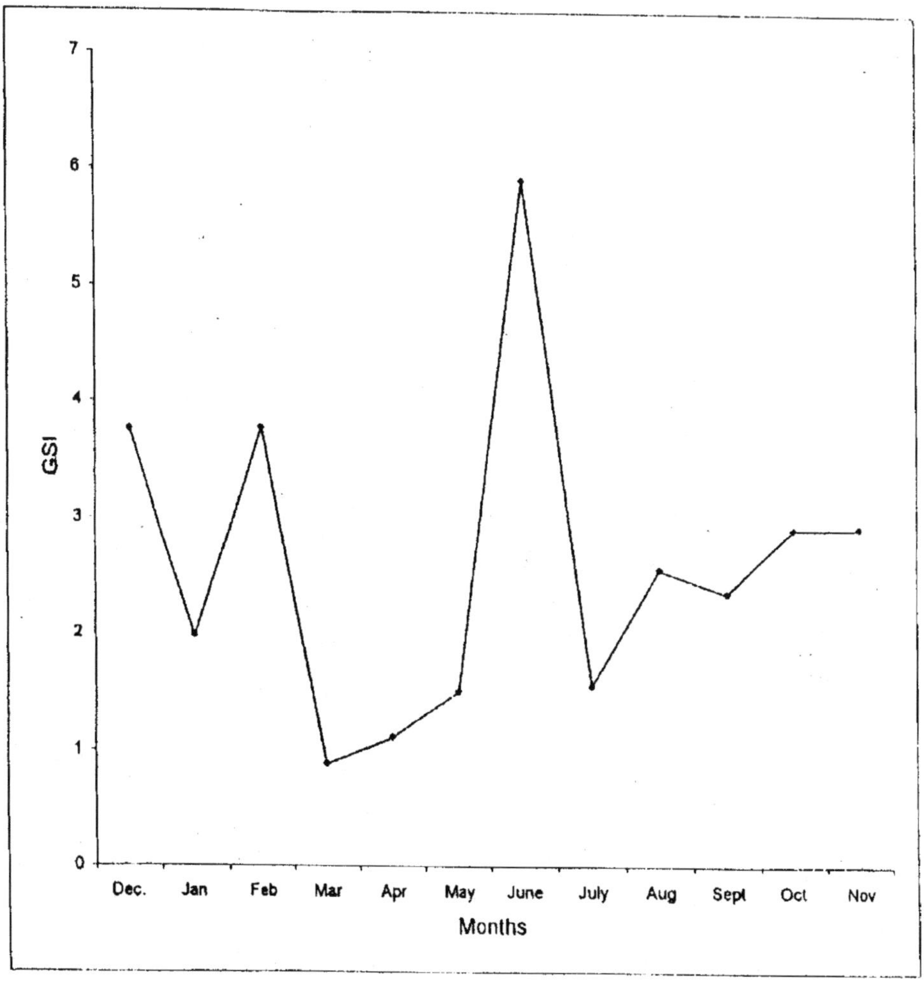

Figure 8.1: Gastro Somatic Index (GSI) in *P. lanceolatus*

An analysis of the food and feeding habits of *P. lanceolatus* based on various methods clearly indicated that this fish is a pure herbivore chiefly feeding on phytoplankton both during juvenile and adult stages of its life history. *Pleurosigma* sps. and *Nitzchia* sps., were the two diatoms that constituted its main food. All the three methods of food analysis, namely volumetric, frequency of occurrence and index of preponderance methods confirmed this finding. However, according to Hora (1935) *P. lanceolatus* inhabiting the Gangetic estuaries is a planktonic feeder during juvenile stages but a mud feeder during adult stages. Such variation in the food and food feeding habits of a single species from two distinct geographical regions is rather uncommon and may be attributed to competition for food coupled with isolation.

Acknowledgement

The authors are grateful to Andhra University for providing facilities. One of the authors is grateful to UGC for the award of a Teacher Fellowship during the tenure.

References

Blabber, S.J.M. 1977. The biology of the burrowing goby *Croilia mossambica* (Smith) (Teleostei, Gobiidae). *Environ. Boil. Fishes,* 1(2): 197–204.

Geevarghese, C. 1983. Morphology of the alimentary tract in relation to diet among gobioid fishes. *J. Nat. Hist.,* 17(5): 731–741.

Geevarghese, C. 1984. Feeding intensity in relation to breeding cycle in gobiid fish, *Oligolepsis acutipennis* (Cuv. and Val.). *Indian J. Mar. Sci.,* 13(1): 37–38.

Hardy, A.E. 1936. The ecological relation between the herring and the plankton investigated with the plankton indicator. *J. Mar. Biol. Ass., U.K.,* 21: 147–291.

Hora, S.L. 1935. Ecology and bionomics of the gobioid fishes of Gangetic Delta. *Des. Comptes. Rendus Du XII. Congress International De Zoologie-Lisbonne,* p. 265–290.

Hynes, H.B.N. 1950. The food of the freshwater stickle backs (*Gasterosteus acculeaus* and *Pygosteus pungitis*) with a review of methods used in the studies of the food fishes. *J. Anim. Ecol.,* 19: 36–58.

Mehta, R., H.S. Mehta and K. Devi. 1990. Structure and shape of mouth in some gobioid fishes. *Environ. Ecol.,* 8(2): 668–671.

Natarajan, A.V. and V.G. Jhingran. 1962. Index of preponderance, a method of grading the food elements in the stomach analysis of fishes. *Indian J. Fish,* 8: 54–59 (for 1961).

Pearse, A.S. 1915. On the food of the small shore fishes in the waters near Madison Wisconsin. *Bull. Wis. Nat. Hit. Soc. Milkwaukee,* 13: 7–22.

Ricker, W.E. 1937. The food and food supply of Sockeye Solmon (*Onchorhynkus Nerkar* Walbum) in Cultus Lake, British Colombia. *J. Biol. Bd., Canada,* 5: 450–468.

Sankara Rao, P. 2001. Ecobiology and *Glossogobius giuris* of Gosthani estuary, Visakhapatnam, East Coast of India. *Ph.D. Thesis,* Andhra University, Waltair.

Sarker, A.L. and N.K. Al-Daham. 1980. Food habits of mud skipper, *Pseudapocryptes dentatus* (Val.). *J. Fish. Biol.,* 17(6): 635– 639.

Swynnerton, G.H and E.B. Worthington. 1940. Notes on the food of fish in Hawes waters. *J. Anim. Ecol.,* 9: 283–287.

Venkateswarlu, T. 1966. A note on the oxygen carrying capacity of the blood in three gobioid fishes. *Naturwiss,* 12: 313.

Chapter 9

Biochemical Studies of Insulin Effect on Freshwater Fish, *Oreochromis mossambicus*: 1. On Liver Tissue

M.R. Delphine Rose*, W. Premasugantha**, A. Jeyaseeli*, S. Anitakannagi* and M. Narayanan***

**Department of Zoology, Jayaraj Annapackiam College (Autonomous), Periyakulam – 625 601, India*
***Department of Zoology, Holy Cross College (Autonomous), Trichy – 620 002, India*
****Department of Advanced Zoology and Animal Biotechnology, St. Xavier's College (Autonomous), Palayamkottai – 627 002, India*

ABSTRACT

The effect of intraperitonial administration of different levels of insulin (0, 1 µg, 5 µg and 10 µg) on the glycogen, protein and RNA content in the liver tissue of *Oreochromis mossambicus* was evaluated at different time intervals (6, 9, 12, 15, 18, 24 and 48 hrs). Insulin has a stimulatory effect on both glycogen, and protein content, but inhibitory effect on RNA content at all the concentration except at 1 µg. It is axiomatic that there is cyclic pattern of glycogen synthesis and utilization in the liver tissue with a duration of 6 hr. A significant ($P < 0.05$) dose dependent increase in glycogen and protein was noticed. An overall time related decrease in RNA content in the treated liver was observed.

***Keywords**: Glycogen, Protein, RNA, Insulin.*

Introduction

Insulin, a peptide hormone, facilitates the utilization of glucose by the cells, promotes glycogenesis and prevents the excessive breakdown of glycogen stored in the liver tissue. It is a powerful hypoglycemic agent and promotes positive nitrogen balance and net protein synthesis. Besides, it stimulates aminoacid transport into cells, accelerates the synthesis of ribonucleic acid and biosynthetic activity of ribosomes (Turner and Bagnara, 1976). In fishes, plasma glucose level variations may be due to the mode, gonadal stage, availability of food, season and many other factors (Chavin and Young, 1970). Insulin contributes minimally, but significantly, to regulation of plasma glucose levels during stressful conditions or the fasting state in raptors (Minick, 1986). The effect of insulin on blood glucose, glycogen and protein content were studied in some cyclostomes and in cartilaginous and bony fishes (Leibson and Plisetskaya, 1968 and Jackim and La Roche, 1973). In the present study, the duration dependent effects of different levels of insulin on the liver glycogen, protein and RNA content of *O. mossambicus* was investigated.

Materials and Methods

The experimental fishes *O. mossambicus* weighing 15.00 ± 1.00 g were collected from a local pond in Trichy. These fishes were acclimated for ten days to the laboratory conditions at room temperature (30±1°C) before carrying out the experiments. During this acclimation period they were maintained in plastic containers of 20 litres capacity. Water was changed every day to get rid of nitrogenous and non-nitrogenous wastes. The fishes were fed daily with algae *ad libitum*.

The experiments were carried out in four sets, one set served as control, whereas the other three sets were treated with insulin at three levels *viz.*, 1 µg, 5 µg and 10 µg. Insulin was injected intraperitenially using a syringe. In the control samples were analysed at 0, 6, 9,12, 15, 18, 24 and 48 hrs, whereas in the treated, it was at 6, 9, 12, 15, 18, 24 and 48 hrs after treatment. In all, five replicates were used. The fishes in each set were weighed separately and sacrificed for the study. The liver tissues were dissected out, weighed (wet wt.) and stored in the freezer for estimation of glycogen, protein and RNA content.

Glycogen, protein and RNA were estimated respectively by anthrone method (Hassid and Abraham, 1957), Biuret method (Gornall *et al.*, 1949) and Orcinol method (Ceriotti, 1955). Extraction of proteins and RNA was carried out by using the procedure of Mittermayer *et al.* (1964). The obtained results were analysed statistically (Zar, 1984).

Results and Discussion

Fishes exposed to different concentration of insulin induced remarkable changes in glycogen, protein and RNA content at different time intervals of liver tissue. Fishes administered with 1 µg, 5 µg and 10 µg doses showed a decrease in glycogen content at 6 and 9 hr when compared to the control. For instance, at 6 hr it was found to be 1.06±0.104, 2.01±0.348 and 0.93±0.185 mg/g, whereas, at 9 hr it was 1.21±0.237, 1.95±0.085 and 0.72±0.104 mg/g for 1, 5 and 10 µg insulin respectively. Thereafter, an increase in glycogen content was noticed at 12, 15,18 and 24 hr. Furthermore, such an increase was found to be dose dependent (Table 9.1). However, there was a slight decrease (3.20±0.289 mg/g) with 10 µg at 24 hr and 48 hr the decrease was a little less (3.99±0.327 mg/g). Always, the increase in the glycogen was followed by a decrease, thereby showing that the liver tissue is the organ for glycogenesis, storage and utilization (William, 1974). It is axiomatic that there is a peak at every 6 hr, revealing the cyclic pattern of glycogen synthesis and utilization. Two way ANOVA testing of data within the treatment was insignificant ($F = 0.31$; $P > 0.05$) and between the treatment was significant ($F = 83.5$;

P < 0.05). In the treated liver tissue an increase in the glycogen content might be due to the hypoglycemic action and due to decrease in glycogen breakdown thereby enhancing glycogenesis (Greenspan and Forsham, 1983). This increased trend has already been reported by Carniero and Amaral (1983), Plisetskaya *et al.* (1984), and Ottelenghi *et al.* (1985).

Table 9.1: Glycogen Content (mg/g) in the Liver Tissue of *O. mossambicus* Exposed to Insulin at Different Time Intervals. Each value represents mean ± SD of five replications.

Duration (hrs)	*Control*	*Insulin Concentration*		
		1 μg	*5 μg*	*10 μg*
0	3.27±0.317	–	–	–
6	2.20±0.121	1.06±0.104	2.01±0.348	0.93±0.185
9	3.85±0.634	1.21±0.237	1.95±0.085	0.72±0.104
12	1.07±0.092	2.58±0.381	3.43±0.487	3.97±0.297
15	2.71±0.292	3.44±0.508	3.78±0.464	6.88±0.272
18	1.23±0.212	1.76±0.309	3.99±0.665	5.61±0.858
24	1.11±0.124	2.99±0.175	4.51±0.372	3.20±0.289
48	5.89±0.438	1.92±0.543	3.37±0.475	3.99±0.327

In all the tested insulin concentration there was a decrease in the amount of protein at 6 hr and 48 hr. when compared to the control. However, an overall increase in this protein content was observed in an experimental animals. A recalculation of datas presented in Table 9.2 into percentage clearly supports the increased protein level in the treated animals (1 μg = 139 per cent; 5 μg = 123 per cent and 10 μg = 120 per cent). This supports the earlier findings the insulin promotes protein anabolism (Brobeck, 1975). This might be due to increase in membrane transport of certain free aminoacids or decrease in proteolysis (Frye, 1967) or increase in the production of enzymatic glycogen phosphorylase in glycogenolysis (Brimacombe and Stancey, 1962) or intracellular enzymes (Granner, 1988) synthesized by the liver.

Table 9.2: Protein Content (mg/g) in the Liver Tissue of *O. mossambicus* Exposed to Insulin at Different Time Intervals. Each value represents mean ± SD of five replications.

Duration (hrs)	*Control*	*Insulin Concentration*		
		1 μg	*5 μg*	*10 μg*
0	103.13±0.881	–	–	–
6	117.58±5.238	79.29±1.666	73.31±3.193	86.65±2.571
9	176.32±5.468	176.74±6.602	186.00±5.026	193.42±6.085
12	72.06±1.924	134.31±5.301	158.19±1.945	199.19±2.914
15	156.69±2.705	277.36±32.337	161.31±5.275	157.72±1.573
18	101.83±3.128	127.74±1.905	127.80±2.252	131.44±1.476
24	77.86±4.456	186.70±3.698	156.57±1.280	76.36±1.667
48	185.07±10.648	116.26±5.361	158.05±3.499	126.25±5.369

The liver protein separated by electrophoresis after 12 hr treatment is shown in Table 9.4. It was observed that the separation of bands were not remarkably altered by different concentrations of insulin treatment.

It is evident from the Table 9.3 that the RNA content in all the insulin treated experimental samples invariably showed lower values than the control at zero hour. The sudden increase in the RNA content after 12 insulin treatment might explain a cyclical pattern of RNA synthesis. The F-value within the treated was insignificant ($F = 3.06$; $P > 0.05$) and between the treatment was higher ($F = 95.51$) indicating the significant dose dependent effect of insulin ($P < 0.05$). The RNA synthetic pattern for liver has its peak at 12 hr. It is evident that the RNA synthesized in the liver tissues could be of mRNA (Short lived) which might have been utilized by the protein synthesizing machinery (Granner, 1988 and Kleinsmith and Kish, 1988). There was an overall time related decrease in RNA content in the treated liver tissue could be attributed to the utilization of the available transcripts in protein synthesis, as shown by the increase in the protein content.

Table 9.3: RNA Content (mg/g) in the Liver Tissue of *O. mossambicus* Exposed to Insulin at Different Time Intervals. Each value represents mean ± SD of five replications.

Duration (hrs)	*Control*	*Insulin Concentration*		
		1 μg	*5 μg*	*10 μg*
0	5.45±0.116	–	–	–
6	2.50±0.289	2.56±0.089	3.30±0.257	2.32±0.425
9	2.40±0.100	3.69±0.124	1.92±0.065	2.48±0.371
12	3.50±0.132	3.42±0.205	2.89±0.180	1.97±0.098
15	2.25±0.099	2.26±0.212	2.22±0.161	2.20±0.093
18	2.92±0.166	1.40±0.240	1.97±0.099	2.66±0.163
24	2.68±0.045	3.96±0.290	2.52±0.108	3.00±0.149
48	2.01±0.069	2.29±0.210	2.18±0.100	2.16±0.274

Table 9.4: Separation of Proteins from the Liver Tissue of *O. mossambicus* by SDS–PAGE, After 12 hr of Insulin Treatment

No. of Bands Obtained	*Control*	*1 μg*	*5 μg*	*10 μg*	*BSA*
1	95,200	95,200	95,200	95,200	
2	79,333	79,333	79,333	79,333	
3	59,500	59,500	59,500	59,500	
4	52,889	52,889	52,889	52,889	
5	47,600	47,600	47,600	47,600	68,000
6	43,273	43,273	43,273	43,273	
7	39,666	39,666	39,666	39,666	
8	36,615	36,615	36,615	36,615	
9	34,000	34,000	34,000	34,000	

Insulin has been known to increase polyamine synthesis (O'Riordan *et al.*, 1982). In *Physarum* the polyamine mediated phosphorylation of protein (Molecular wt. 70,000 daltons) regulates ribosomal RNA in the process of transcription (Kuehn *et al.*, 1979). Thus the decrease in RNA in the liver tissue of *O. mossambicus* could be attributed to the regulatory role of insulin on rRNA transcription.

References

Brimacombe, J.S. and Stacey, M. 1962. In: *Comparative Biochemistry*, (Eds.) Florkin, M. and Mason, S.H. Academic Press, New York and London, 4: 27–58.

Brobeck, J.R. 1975. *Best and Taylor's Physiological Basis of Medical Practice*. The Williams and Wilkins Company, Baltimore.

Carneiro, N.M. and Amaral, A.D. 1983. Effects of insulin and glucagon on plasma glucose levels and glycogen content in organs of fresh water teleost, *Pimelodes maculates. Gen. Comp. Endocrinol*, 49: 115–121.

Ceriotti, G. 1955. Determination of nucleic acids in animal tissues. *J. Biol. Chem.*, 214: 59–70.

Chavin, W. and Young, J.E. 1970. Factors in the determination of normal serum glucose levels of gold fish, *Carossius auratus. Comp. Biochem. Physiol.*, 33: 629–653.

Frye, B.E. 1967. *Hormonal Control in Vertebrates*. The Macmillian Company, New York, Collier Macmillian Ltd., London.

Gornall, A.G., Bardawill. C.J. and David, M.M. 1949. Determination of total serum protein by means of biuret reaction. *J. Biol. Chem.*, 177: 751–766.

Granner, D.K. 1988. Regulation of gene expression. In: *Harper's Biochemistry*, 21st edn., (Eds.) Murray, R.K., Granner, D.K., Mayes, P.W. and Rodwell. V.W. Appleton and Lange, California, p. 428–444.

Greenspan, F.S. and Forsham, P.H. 1983. *Basic and Clinical Endocrinology*. Lange Medical Publications, USA.

Hassid and Abraham. 1957. Chemical procedures for analysis of polysaccharides. In: *Methods in Enzymology*, (Eds.) Colowick, P. and Kaplan, O. Academic Press Inc., New York, 3: 34.

Jackim, E., and LaRoche, G. 1973. Protein synthesis in *Fundulus heterolitus* muscle. *Com. Biochem. Physiol.*, 44: 851–856.

Kleinsmith, L.J. and Kish, V.M. 1988. *Principles of Cell Biology*. Harper and Row Publishers, New York.

Kuehn, G.D., Affolter, H.V., Atmar, V.J., Seeback, T., Gubler, U. and Brawn, R. 1979. Polyamine mediated phosphorylation of a nucleolar protein from *Physarum polycephalum* that stimulates rRNA synthesis. *Proc. Natl. Acad. Sci.*, 76: 2541–2545.

Leibson, L.G. and Plisetskaya, E.M. 1968. Effect of Insulin on blood sugar level and glycogen content in organs of some cyclostomes and fish. *Gen. Comp. Endocrinol.*, 11: 381–392.

Minick, C.M. 1986. Correlation of simultaneously determined plasma insulin and glucose levels in fasting and mildly or severely stressed raptors. *Comp. Biochem. Physiol.*, 83: 301–325.

Mittermayer, C., Braun, R. and Rusch, H.P. 1964. RNA synthesis in the mitotic cycle of *Physarum polycephaum. Biochem. Biophys. Acta*, 91: 399–405.

O'Riordan, F.L.H., Malan, P.G. and Gould, R.P. 1982. *Essentials of Endocrinology*. Blackwell Scientific Publications, London.

Ottolanghi, C., Puviani, A.C., Gavioli, M.E. and Brighenti, L. 1985. Effects of insulin on glycogen metabolism in isolated and perfused catfish liver. *Comp. Biochem. Physiol.*, 80: 135–138.

Plisetskaya, E.M., Bhattucharya, S., Dickoff, W.W. and Gorbman, A. 1984. The effect of insulin on amino acid metabolism and glycogen content in isolated liver cells of juvenile coho salmon, *Oncorhynchus kisutch*. *Comp. Biochem. Physiol.*, 78: 773–778.

Turner, D.C. and Bagnara, J.T. 1976. *General Endocrinology*, 6th edn. W.B. Saunders Company, Philadelphia, London.

Zar, J.H. 1984. *Biostatistical Analysis*. Prentice Hall, London, U.K.

Chapter 10

Toxicological Studies of Nuvan in the Intestine of Freshwater Catfish *Clarias batrachus*

Smita Shukla*, Sujata K. Dass** and D.N. Saksena***

**Department of Zoology, Bundelkhand University, Jhansi – 294 128*
***B.R. Ambedkar Chair Professor, Department of Medicine, Vallabh Bhai Patel Chest Institute, University of Delhi, Delhi – 110 007*
****School of Studies in Zoology, Jiwaji University, Gwalior – 474 0011*

ABSTRACT

Toxicological studies of Nuvan (organophorus pesticide) on freshwater catfish *Clarias batrachus* using bioassay experiments to determine acute toxicity and prolong toxicity by evaluating histopathological aspects has been done 96 hrs LC_{50} value was 1.58±0.0287 mgl^{-1}. For the histopathological studies the concentration was selected 1/15 and 1/10 (0.10 and 0.16) of 96 hrs LC_{50} value. After on exposure to 0.10 mgl^{-1} for 10 days enlargement and increase in number of mucous cells in the villi and for 20 days exposure. The epithelium cell also lost their regular appearance. An exposure to 0.16 mgl^{-1} for 10 days lesions in intestinal epithelium was observed at certain places, for 20 days the epithelium cells of intestine villi lost their cellular integrity and were marked by their nuclei only.

Keywords: *Nuvan, Clarias batrachus.*

Introduction

Toxicity of pesticides takers into account problems relating to both, their injurious effects directly on man and animals and their long-term effect in the environment. The pesticides include a large variety chemical nature and biological activity grouped together only on the basis that they are used to destroy or eliminate pests pesticides are used for pest control but they seep into the ecosystem and food chain to affect the non target species.

Some pesticides may produce tumours in experimental animals but may not be associated with risk of cancer in human populations. DDT is known to cause hepatic cell tumours in rat but there is no information concerning hepatic cancer in humans population. Likewise, mirex induces cancer and ethylene, thiourea produces thyroid tumours in experimental animals (Omkar, 1984).

Toxicity of pesticides in an intestine were studied by many workers Mathur (1962), Saxena and Mathur (1985), Mishra and Jain (1988), Rana *et al.* (1999).

Materials and Methods

The fish were obtained from the Department of Fisheries, Gwalior Madhya Pradesh. The fish having 10±3 cm length were selected for performing experiments. The fish were brought to the laboratory and placed in the fish tanks of the School of Studies in Zoology normal condition of day length and temperature. They were kept in the water tanks for acclimatization as suggested by Bhida (1998) for approximately 20 days before the experiment was taken up. After placing the fish in the fish tanks, they were treated with 0.1 per cent $KMNO_4$ to avoid any infection. The fish were fed with chopped goat liver (100 g) and boiled eggs (one egg/tank) on alternate days except during the acute toxicity bioassay experiments. Fish of approximately equal size were selected for experimentation and transferred to glass aquaria (40 × 30 × 35 cm). The experimental aquaria were set up with a parallel control. Food (as above) was given to both the experimental and control fish and water was renewed immediately after feeding. Six healthy specimens each were taken for the control as well as experimental groups.

The Nuvan solution was prepared by dissolving Nuvan 76 per cent w/w obtained from M/s Ciba-Giegy India Ltd., Bombay. Before the beginning of the experiment the dilution water drawn from a tube well was used to prepare desired concentration of Nuvan by adding known quantity of stock solution and making a total volume of 10 litres. The toxicant solution was changed every day.

Results and Discussion

Histological Effect on Intestine

Intestine of Control Fish

Histologically, the intestine consists of serosa which is outer most peritoneal layer beneath with layers of longitudinal muscles, circular muscle, and submucosa. The intestinal epithelium is disposed into villi. The epithelial layer consists of epithelial cells and mucous goblet cells (Figure 10.1).

Intestine After an Exposure to 0.10 mgl^{-1} of Nuvan (1/15th of 96 h LC_{50} Value) for 10 Days

Except for the enlargement and increase in number of mucous cells in the villi, no other alterations in the histology were observed in 0.10 mgl^{-1} Nuvan exposed fish (Figure 10.2).

Intestine After an Exposure to 0.10 mgl^{-1} of Nuvan (1/10th of 96 h LC_{50} Value) for 20 Days

The epithelial cells had also lost their regular appearance. In other layers no remarkable change was observed although there was further enlargement and increase in the number of mucous cell in the intestinal villi, when fish were exposed to 0.10 mgl^{-1} Nuvan for 20 days (Figure 10.3).

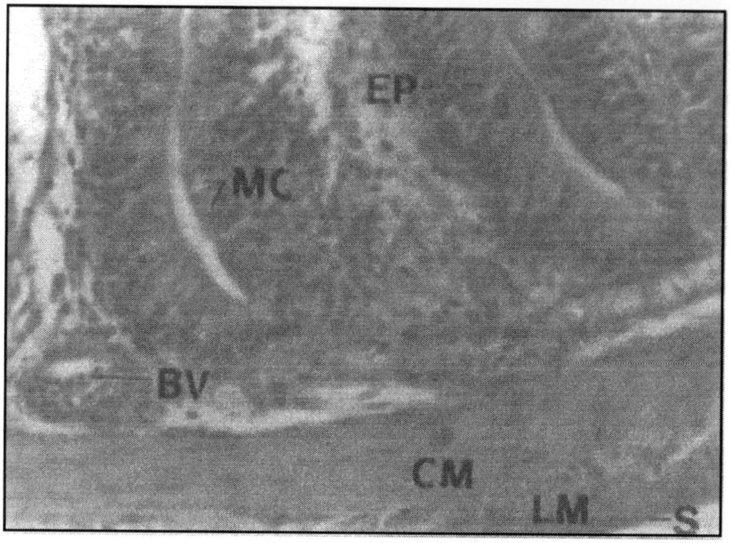

Figure 10.1: Photomicrograph of Transverse Section of Intestine of Freshwater Catfish, *Clarias batrachus* Showing Normal Histological Structure × 400

BV: Blood vessel; CM: Circular muscular layer; EP: Epithelial cell; LM: Longitudinal muscular layer; MC: Mucous cells; S: Serosa; SM: Submucosal Layer.

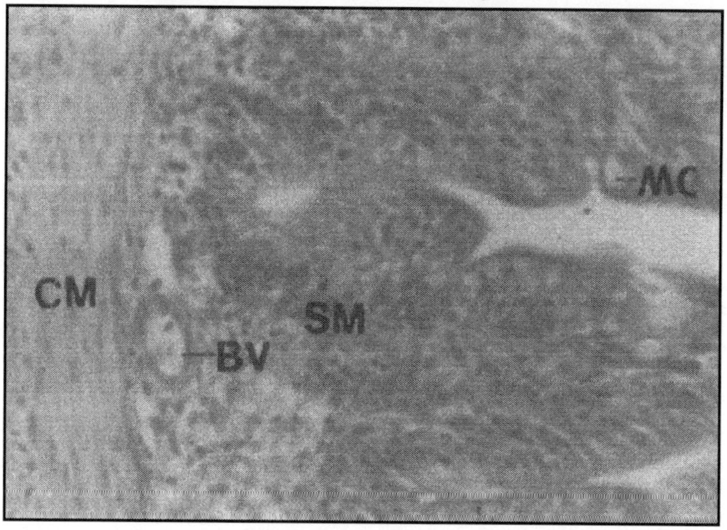

Figure 10.2: Photomicrograph of Transverse Section of Intestine of Freshwater Catfish, *Clarias batrachus* Showing Increase in Number of Mucous Cell After an Exposure to 0.10 mgl^{-1} of Nuvan for 10 Days × 400

BV: Blood vessel; CM: Circular muscle layer; MC: Mucous cells; SM: Submucosal layer.

Intestine After an Exposure to 0.16 mgl^{-1} of Nuvan (1/10th of 96 h LC_{50} value) for 10 Days

There was no spectacular change in he muscular layer as well as in the submucosal layer of intestine in treated fish. However, a multifold increases in the mucous cells of villi was observed. The lesions in the intestinal epithelium were observed at certain places. Such of histological changes were seen when fish were exposed to 0.16 mgl^{-1} Nuvan for 10 days (Figure 10.4).

Intestine After an Exposure 0.16 mgl^{-1} of Nuvan (1/10th of 96 h LC_{50} value) for 20 Days

The exposure of fish to 0.16 mgl^{-1} of Nuvan for 20 days induced intestinal changes. The epithelial cells of intestinal villi lost their cellular integrity and were marked by their nuclei only. The epithelial

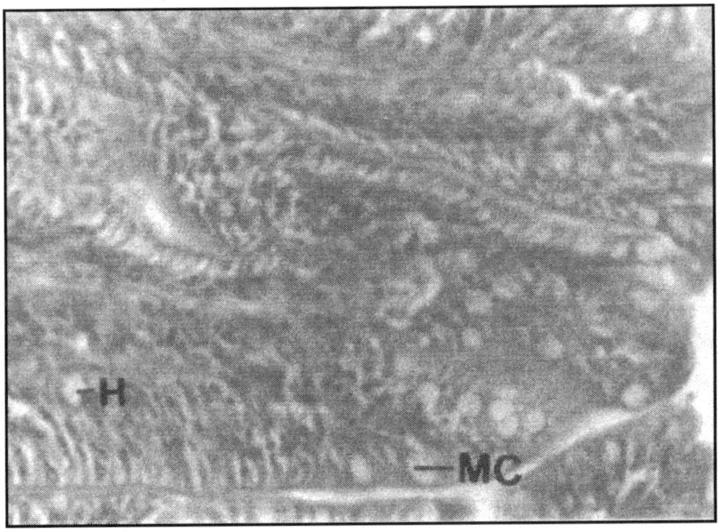

Figure 10.3: Photomicrograph of Transverse Section of Intestine of Freshwater Catfish, *Clarias batrachus* Showing Enlargement and Increase Number of Mucous Cell and Epithelial Cells Loosing their Regular Appearance After an Exposure to 0.10 mgl^{-1} of Nuvan for 20 Days × 400

MC: Mucous cells; SM: Submucosal layer.

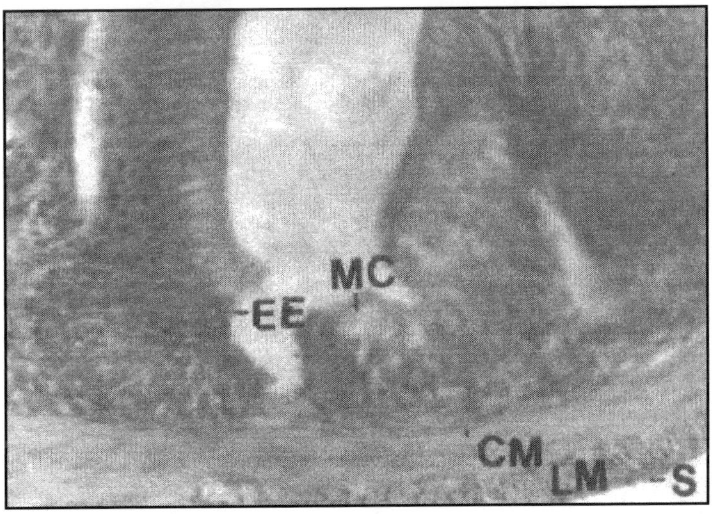

Figure 10.4: Photomicrograph of Transverse Section of Intestine of Freshwater Catfish, *Clarias batrachus* Showing Erosion of Epithelial Cell an After an Exposure to 0.16 mgl^{-1} of Nuvan for 10 days × 400

EE: Erosion of epithelial cell; MC: Mucous cell.

cells along with the mucous cells of villi got detached from the submucosal tissues. The submucosal haemorrhage was also wide spread in this group of fish (Figures 10.5 and 10.6).

From the above observation, it is clear that the histological changes in liver, kidney, gills and intestine were more intense in the fish, which were treated with 0.16 mgl^{-1} of Nuvan for 20 days, than

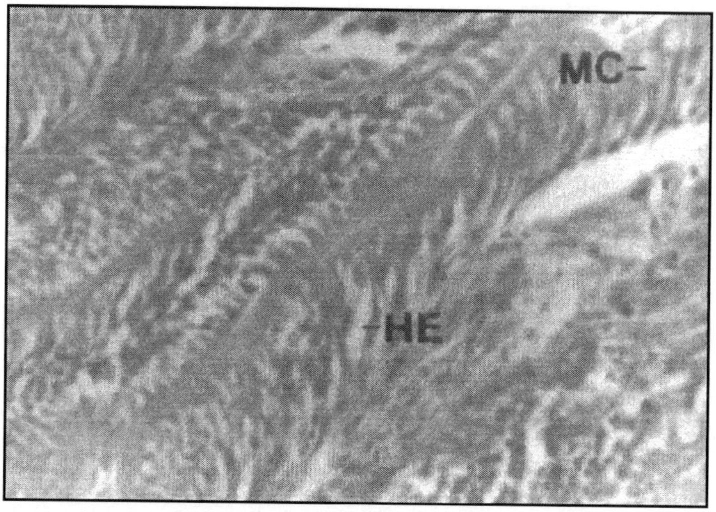

Figure 10.5: Photomicrograph of Intestine of Freshwater Catfish, *Clarias batrachus* Showing Haemorrhage in Submucosa After an Exposure to 0.16 mgl^{-1} of Nuvan for 20 Days × 400

HE: Haemorrhage in submucosa; MC: Mucous cell.

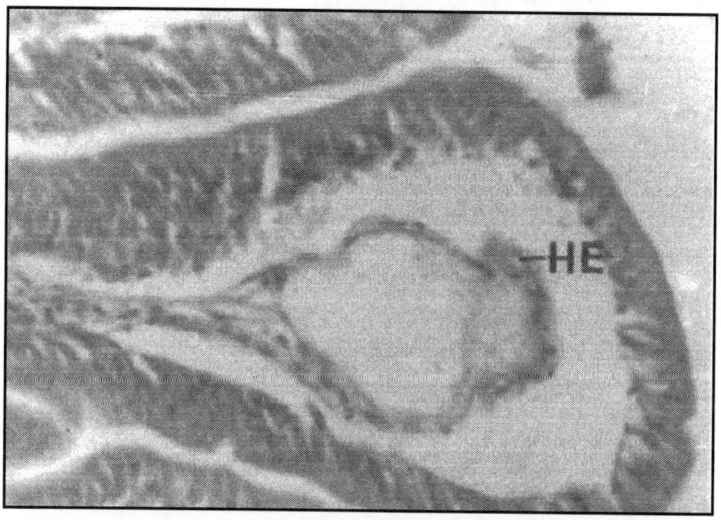

Figure 10.6: Photomicrograph of Intestine of Freshwater Catfish, *Clarias batrachus* Showing Haemorrhage in Submucosal Haemorrhage After an Exposure to 0.16 mgl^{-1} of Nuvan for 20 Days × 400

HE: Haemorrhage in submucosa.

other groups. Therefore, it may be concluded that the changes in vital organs of *Clarias batrachus* due to Nuvan are concentration and duration of exposure dependent.

Intestine is an organ of digestion and absorption of food materials and is greatly affected by toxic substances reaching to gut through food directly or via liver which transfer waste materials to the intestine for ultimate excretion. Mathur observed complete disappearance of intestinal villi. According to Jauhar and Kulshrestha (1985) sublethal closes of sevin and thiodon on *Channa striatus* produced vacuolation of columnar cells, increase in the number of mucous cells and degeneration of the basement membrane in mucosa and vacuolation in submucosa followed by detachment of serosa. Saxena and Mathur (1985) exposed *Channa striatus* to sublethal concentration of rogor and observed degenerative changes in serosa, villi muscles and submucosa. Mishra and Jain (1988) found damage to muscles, shrinkage of mucosal folds, increase in goblet cells, rupture of villi and release of mucus in lumen and *Glossogobius giuris* on exposure to sublethal doses of marathon. In *Clarias batrachus* exposed to Nuvan in the present study on increase in number and size of mucous cells was evident due to exposure to increased concentration of Nuvan. The lesion in the intestinal epithelium detachment of intestinal epithelium with that of submucosa and haemorrhage in the villi were observed in the intestine due to increase concentration and duration of exposure of Nuvan.

Besides food, the intestine gets detoxified waste material from liver for ultimate elimination. Mathur (1962) observed complete disappearance of intestinal villi on DDT exposure in some fish. Konar (1977) observed rupture and disintegration of villi. According to Jauhar and Kulshrestha (1985) sublethal doses of sevin and thiodon on *Channa striatus* produced vacuolation of columnar cells, increase in the number of mucous cells and degeneration of the basement membrane in mucosa and vacuolation in submucosa followed by detachment of serosa. On prolonged exposure, carbaryl produced more damage than endosylfan which is contrary to situation in the stomach where endosulfan is more damaging. Saxena and Mathur (1985) exposed *Channa striatus* to sublethal concentration of rogor and observed degenerative changes in serosa, villi, muscles and submucosa. Mishra and Jain (1988) found damage to muscles, shrinkage of mucosal folds, increase in goblet cells, rupture of villi and release of mucus in lumen in *Glossogobius giuris* on exposure to sublethal doses of malathion. Liver tissue of fish was exposed to effluents and various histopathological alterations were observed (Rane *et al.*, 1999).

In *Clarias batrachus* exposed to Nuvan in the present study an increase in number and size of mucous cells was evident due to exposure to increased concentration of Nuvan. The lesions in the intestinal epithelium, detachment of intestinal epithelium with that of submucosa and haemorrhage in the villi were observed in the intestine due to increased concentration and duration of exposure to Nuvan.

The present observations are in accordance with findings of various other authors cited above. The result of the present investigation revealed histopathological changes as a consequence of Nuvan exposure. Fish exposed to sublethal concentration of Nuvan for a period of 10 and 20 days showed degenerative changes in kidney, gill and intestine that indicated a deviation from normal functions. The extent of damage was highest in the gills followed by liver, kidney and least in the intestine.

Acknowledgement

I register my deepest regards and thanks to Prof. Ramesh Chandra, Vice-Chancellor, Bundelkhand University, Jhansi for the encouragement.

I am thankful to Head, School of Studies in Zoology, Jiwaji University, Gwalior for providing laboratory facilities.

References

Mathur, D.S. 1962. Studies on the histopathological changes induced by DDT in the liver, kidney and intestine of certain fishes. *Experientia,* 178: 506–509.

Jauhar, L. and S.K. Kulshrestha. 1985. Histopathological effects induced by sublethal dose of Sevin and Thiodon on the gills of *C. stiatus. Acta Hydrochim Hydrobiol.,* 13: 395–400.

Omkar. 1984. *Concept of Toxicology*. Shoban Lal Nagin Chand and Co., Jalandhar, India.

Saxena, S. and R. Mathur. 1985. Histopathological changes in intestine of fish *Channa striatus* exposed to rogor. 5(1): 42–45.

Mishra, A. and B. Jain. 1988. Toxicological studies of malathion on fish *Glossogobius giuris. Water Res.,* 13: 123–129.

Rana, K.S. and R. Sudhir. 1999. Acute toxicity of tannery and textile dye effluents on a common teleost *Labeo rohita:* histological alteration in intestine. *J. Environ. Biol.,* 20(1): 33–36.

Chapter 11

Toxicity of Cadmium to the Rostral Pars Distalis of *Rasbora daniconius*

Monica K. Kale, P.P. Joshi and G.K. Kulkarni
Department of Zoology, Dr. Babasaheb Ambedkar Marathwada University, Aurangabad – 431 004, Maharashtra, India

ABSTRACT

Among the 3 zones of adenohypophysis the rostral pars distalis (RPD) is the smallest component and occupies the anterior most part of the pituitary gland. It is divided into small lobules separated by connective tissue septa and is mainly composed of acidophils. It contains two types of secretory cells. The fish, *Rasbora daniconius* exposed to 110 ppm of cadmium chloride ($CdCl_2.2½H_2O$) for 24 h showed changes in the secretory cell structure. Size of cell type 1 was decreased and nuclei showed shrinking and decreased in size. Vacuoles appeared in the cytoplasm which became degranulated and the nuclei appeared hypertrophied. There was a 25 per cent reduction in the nuclear size of cell type 2. The granular cytoplasm became clumpy with reduced staining intensity. These changes are discussed in the light of functional significance of rostral pars distalis.

Keywords: *Cadmium, RPD, Rasbora daniconius.*

Introduction

Heavy metal contaminants arising out of industrial effluents enter into ecosystem through aquatic reservoirs. They cause adverse effects on aquatic productivity and living resources. Heavy metal salts constitute a serious type of pollution to aquatic biota because they are stable compounds and are not readily removed by oxidation, precipitation or other means and affect the activity of animals (Costa, 1965). Joshi and Dubey (1984) also reported on histopathological lesions in the brain of a another teleost, *Oxygaster bucaila* exposed to industrial effluents. Bano and Hasan (1992) studied inorganic mercurial encephalopathy in discrete brain region of catfish, *Heteropneustes fossilis*. Whereas Banerjee and Bhattacharya (1994) studied histopathology of kidney of *Channa punctatus* exposed to chronic nonlethal level of elsan, mercury and ammonia. Jagtap and Wagh (1999) observed histopathological alternations in *Cyprinus carpio* under pesticidal stress.

Materials and Methods

The freshwater fishes, *Rasbora daniconious* were collected from Godavari river at Paithan, 55 km south of Aurangabad. They were acclimatized to the laboratory conditions for 10 days. Fishes having approximately equal size (7 to 9 cm body length) and weight (5 to 7 g) were selected for experimentation. The experimental fishes were exposed to 110 ppm cadmium chloride ($CdCl_2.2½H_2O$) (LC_{50} for 24 h) in 10 litre capacity glass aquaria, 10 fishes in each. Aeration was not provided. Control fishes were maintained in freshwater aquarium with aeriation. After 24 h of exposure fishes were carefully and quickly dissected. The pituitary gland was taken out, washed and fixed in Bouin's fluid for 24 h. After fixation tissue was passed through alcohol grades for dehydration and cleared in xylene. The blocks were prepared in paraffin wax (M.P. 58 to 60°C), 7 micron sections were cut and stained with aldehyde fuchsin (Meola, 1940).

Results and Discussion

The pituitary gland is a dull whitish coloured labile structure in *Rasbora daniconius*. It lies midventral to the infundibulum. Rostral pars distalis is composed of faintly stained acidophils and large number of prolactin cells which are arranged in definite pattern with lot of intracellular space. The rostral pars distalis is the smallest adenophypophysial component and occupies the anterior most part of the gland in this fish (Figure 11.3). It is divided into small lobules separated by connective tissue septa and is chiefly composed of acidophils. However, after the application of aldehyde fuchsin, 2 cell types are distinguishable in the RPD. Cell type 1 are situated as a palisade along the border between the neurohypophysis and RPD and occur in layer of 3–8 cells around the neurohypophysis (24–27 μm in diameter). They are polygonal or elongated in shape having no clear boundary. The cell type 2 constitute a major part of the RPD. They are negative to aldehyde fuchsin and stained light red. The cells are mostly spherical (12–13 μm in diameter) having no clear cell boundries. The nuclei are round more or less central and are comparatively small (Figure 11.1). The cell type 1 showed vacuolation in the cytoplasm, the nucleus measured about 27 μm in diameter and possessed one or more prominent nuclei. The type 2 cells show profound changes in their nuclei and cytoplasm after treatment. In treated fish the nuclei size reduced by 25 per cent as compared to those of controls. There is a reduction in staining intensity of type 2 cells. The granular cytoplasm became clumpy, and nuclear shrinkage is noticed in their cells and the nuclei became atrophied (Figure 11.2). It was observed that the swimming speed of fish decreased with the increased period of exposure to cadmium chloride. It suggests that the disfunction of RPD portion of pituitary gland brings about disturbances in balance, circummovements, abnormal eye movements and pupillary reflexes.

Bhattacharya and Mukherjee (1978) reported on optic tectum of *Claris batrachus* and *Channa punctatus*. They observed histopathological changes including vacuolation, dialation of blood capillary, fibrosis, agglutination of neurons and loss of definite demarcation between layers of the brain (optic

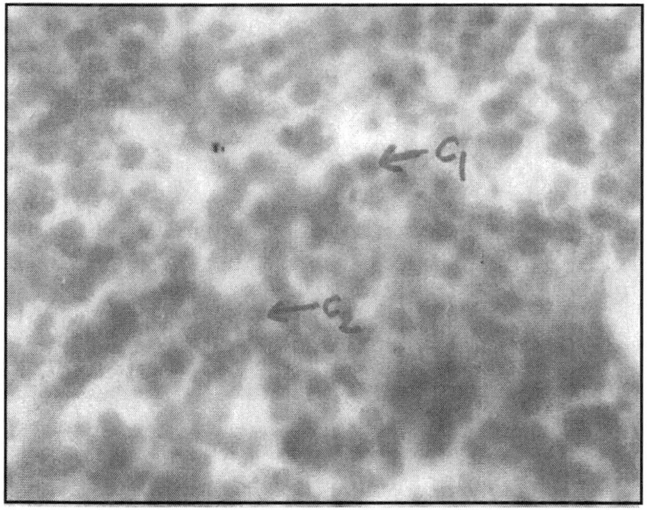

Figure 11.1: T.S. of Pituitary Gland of Normal (Control) Fish, *Rasbora daniconius,* Stained with Aldehyde Fuchsin and Counter Stained with Halmi's Mixture × 540

Note: The normal structure of pituitary gland of *R. daniconius.* The rostral pars distalis with distinct cell type 1 and 2. The cell type 1 are large polygonal or elongated, and are situated palisade along the border. The type 2 cells are mostly spherical having round nuclei and are located centrally (scale : 1 cm = 9 μm)

C_1: Cell type 1; C_2: Cell type 2.

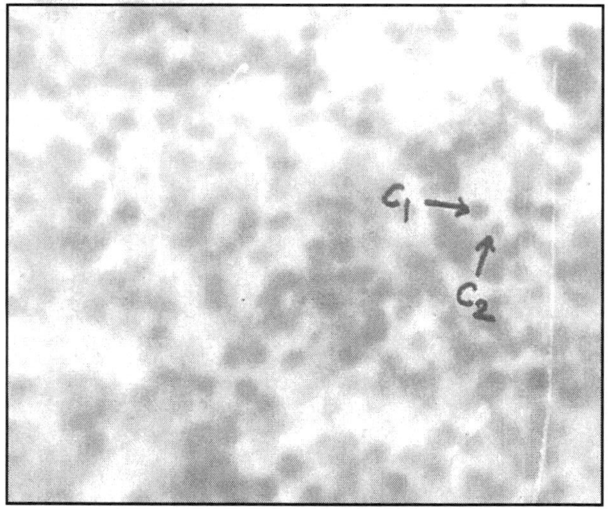

Figure 11.2: T.S. of Pituitary Gland of Fish, *Rasbora daniconius,* Showing the Effect of Cadmium Chloride Exposed to 110 ppm for 24 h. The cell type 1 showed vacuolation in the cytoplasm. The nuclei appear hypertrophied. The cell type 2 showed that its granular cytoplasm became clumpy. Nuclear shrinkage, and reduction in size, nuclei were noted. Stained with Aldehyde fuchsin and counter stained with Halmi's mixture × 540.

(Scale : 1 cm = 9 μm)

C_1: Cell type 1; C_2: Cell type 2.

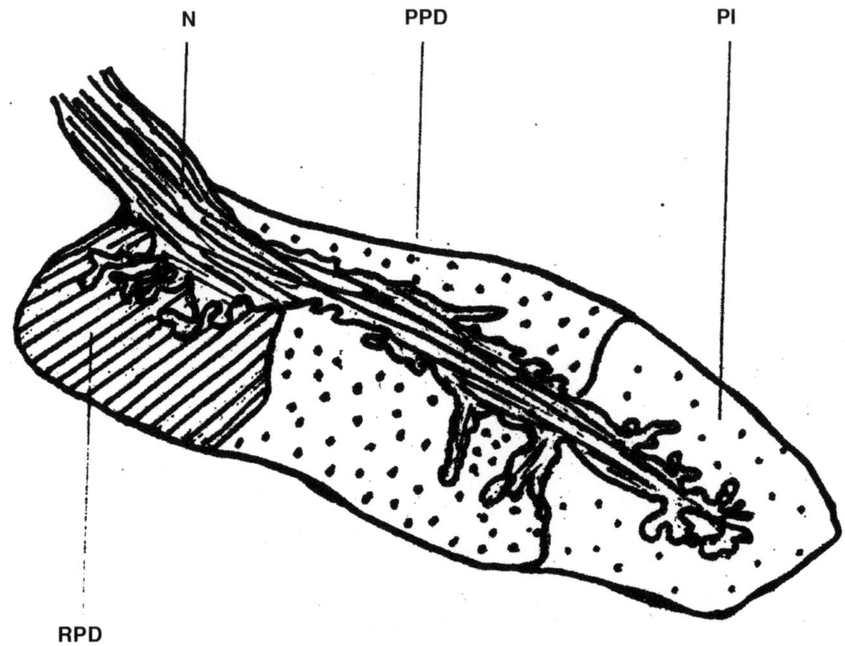

Figure 11.3: The Rostral Proximal Pars Distalis and Pars Intermedia are Arranged in a Linear Fashion with the Rostral Pars Distalis at the Anterior Extremity in *Rasbora daniconius*

Diagram of Medium Segittal Section of Pituitary Gland Showing

- **Neurohypophysis (N)**
- **Rostral Pars Distalis (RPD)**
- **Proximal Pars Distalis (PPD)**
- **Pars Intermedia (PI)**

tectum) of teleost exposed to industrial pollutants. Srivastava *et al.* (1977) also showed histomorphology of the tinctorial cells in pituitary gland of *Channa striatus* (Bloch). Karuppasamy (2000) studied histopathological changes in *Channa punctatus* treated with phenyl mercuric acetate.

References

Banerjee, S. and Bhattacharya, S. 1994. Histopathology of kidney of *Channa punctatus* exposed to chronic nonlethal level of elsan, mercury and ammonia. *Ecotoxico. Environ. Safety*, 29(3): 265–275.

Bano, Y. and Hasan, M. 1992. Inorganic mercurial encephalopathy in discrete brain regions of catfish, *Heteropneustes fossilis*. *J. Biomed. Res.*, Aligarh, 3(1): 25–34.

Bhattacharya, S. and Mukherjee, S. 1978. Histopathology of the brain (optic tectum) of teleost exposed to industrial pollutants. *Indian Biologists,* 10(1): 31–36.

Costa, H.H. 1965. Oxygen consumption of Gammarus in relation to toxic substances. *Ceylon J. Sci. (Biol. Sci.)*, 5(2): 87–96.

Jagtap, U.L. and Wagh, S.B. 1999. Histopathological alterations in *Cyprinus carpio* under pesticidal stress. *Nat. Seminar on Pollution: Its Control and Disaster Planning,* Abst., pp. 35.

Joshi, R.K. and Dubey, G.K. 1984. Histopathological lesions in the brain of a teleost. *Oxygaster bucaila* exposed to industrial effluents. In: *Proceedings of Seminar on Effect of Pesticides on Aquatic Fauna.* The Academy of Environmental Biology, Muzaffarnagar, India, p. 45–49.

Karuppasamy, R. 2000. Tissue histopathology of *Channa punctatus* (Bloch) under phenyl mercuric acetate toxicity. *Bull. Pure and Appl. Sci.,* 19A(2): 109–116.

Meola, S.M. 1940. Sensitive paraldehyde fuchsin technique for neurosecretory system of mosquitoes. *Trans. Amer. Microsc. Soc.,* 89(1): 66–71.

Srivastava, S., Prakash, R.B. and Swarup, K. 1977. Histomorphology of the tinctorial cells in the pituitary gland of *Channa striatus* (Bloch). *Arch. Biol. (Bruxelles),* 88(1): 101–116.

Chapter 12

Study of the Length-Weight Relationship and Population Structure of Chocolate Mahseer *Neolissocheilus hexagonolepis* (McClelland) in Shally Lake in Arunachal Pradesh

Boni Amin Laskar*, Azen Pujen* and Pronob Nath*

**Senior Research Fellow, NRCCWF (ICAR) Project, Fisheries Experimental Farm, IDULI, Lower Dibang Valley District – 792 110, Arunachal Pradesh*

***Deputy Director, Department of Fisheries, Government of Arunachal Pradesh*

ABSTRACT

"Mahseer" the premier sport fish of the country which was once the major fishery of the state of Arunachal Pradesh is on the decline. Among the three varieties available in this region, *Neolissocheilus hexagonolepis* (Chocolate mahseer synonyms of *Accrosocheilus hexagonolepis*) is in the verge of extinction and has been regarded as a threatened species of this region of the Himalayas.

As a step towards conservation measure, the state department has stocked the fishes from time to time in 'Shally Lake' near Roing of Lower Dibang Valley District of Arunachal Pradesh. In order to assess the pattern of growth of this fish in the confined water, systematic studies including the studies on population dynamics has been carried out in the monsoon seasons. The studies showed that the fishes has adapted to the confined condition of the lake, and the growth pattern is negatively allometric while the average condition factor being 1.005 showed the well-being of the fishes.

The article deals with the result of the population studies of the different age groups to focus the importance of mahseer and its future prospect to form a component in the culture fishery.

Keywords: *Length-weight relationship, Lake, Chocolate mahseer, Allometric growth and natural population.*

Introduction

Studies on the relationship between length and weight of fishes constitute an important tool in fish biology. Nevertheless, such study are usually undertaken to know whether variations from expected weight for the known length groups are the indicators for fatness, well-being, gonadal development with relation to the environment (Beckman, 1948; Lecren, 1951; Bagenal, 1978) and fluviatile condition of their habitat.

Length-weight (L–W) relationship of fishes has been developed through the use of general equation of $W = CL^n$. The general expectation is that the weight increases as the cube of the length (Rousenfell and Everhart, 1953; Lagler, 1956; Brown, 1957). As the fishes passes through several stages, the simple cube law does not hold good through out the life span and equilibrium constant shows certain variations (Martin, 1949) in the growth pattern of fish.

Interestingly, several works had been done on the L–W relationship of commercial freshwater fishes from different water bodies in India (Srivastava and Singh, 1964; Mitra and Das, 1967; Jhingran,1968; Venkateswarlu and Banerjee, 1971; Talukdar and Das, 1975; Pathak, 1976; Sen, 1982; Kar and Barbhuiya, 2000; Patgiri *et al.*, 2001); but hardly any record is available on Chocolate mahseer in this regard.

Place of Experiment and Location

The study was carried out at the Shally lake in Lower Dibang Valley District of Arunachal Pradesh lying between 28°15′N and 95°45′E, located at an altitude of 395 m (msl).The lake is of tectonic origin and perhaps created during the year 1950 with water area of 2 (Two) hector. The fishery activities were taken up by the department of fisheries, Govt. of Arunachal Pradesh from the year 1970–1971 onwards with provision of sluice gate to check the outflow of water at the outlet. The IMCs were initially stocked and during 1994–1995 fires of mahseer (Chocolate mahseer) were thereafter stocked. The lake presently attaches great importance not only for composite fish practices but also for tourist attraction.

Materials and Methods

Chocolate mahseer were the principal materials for the study. The fishes has been collected by using dragnet (mesh size 5 mm). Altogether, 131 number of Chocolate mahseer (Length group 70–285 mm) were collected from the lake to undertake the studies. Total length (mm) of all the specimens were measured from tip of snout to longest axis of caudal fin, through measuring board. Body weight (g) of each individual fish was also recorded along with the corresponding length.

The L–W relationship was established by fitting in the logarithmic form of general equation:

$$W = CL^n$$

where,

L = Measured length,

C = Constant,

n = Equilibrium and

W = Expected weight.

The logarithmic form of the equation is Log W = Log C + n Log L.Log C was calculated by the following formula:

$$\text{Log C} = \frac{\Sigma \text{LogW}\,(\Sigma \text{LogL}^2) - \Sigma \text{LogL} * \Sigma(\text{LogL} * \text{LogW})}{N * (\Sigma \text{LogL}^2) - (\Sigma \text{LogL})^2}.$$

The value of 'n' was calculated by the following formula:

$$n = \frac{\Sigma \text{LogW} - N * \text{LogC}}{\Sigma \text{LogL}}$$

Condition factor (K) was calculated by the following formula:

$$K = \frac{\text{Observed Weight}}{\text{Expected Weight}}$$

Different statistical methods has been used to evaluate the population structure.

Results and Discussion

The present study on L–W relationship indicated a wide variation from the cube law. Such variations has also been reported by several workers (Jhingran, 1952; Rao and Rao,1972; Choudhuri *et al.*, 1982).The calculated L–W relationship has been LogW = –4.402 + 2.736 Log L. This shows that the growth although was allometric; yet it could have been isometric (*i.e.*, n = 3) and the weight of each fish would have been more from that as observed. The average observe weight was 83.4 g while the average expected weight was 110.27 g. Also, the calculated weight per mm of length considering the n value as 3 (three) was 1.738 g, but the observed weight per mm of length was 0.42 g. Therefore, following the cube law it could be inferred that the body weight of the fishes (*N. hexagonolepis*) would have been increased by 1.32 gm/mm than as observed; if proper nutrition and fertilization been carried out ensuring proper tropic level in the confined ecosystem.

The variation in the condition factor may be due to the various factors like environmental conditions, food availability and gonadal maturity as reported by Lacren (1951), Jhingran (1972), Bassirullah (1975) and Dasgupta (1982). From the present studies it has been recorded that the fish population was with average condition factor = 1.005. The weight of 43 per cent of the population belonging to the length group 130–285 (mm) was more than the expected weight. This is perhaps due to the extra-weight of the gonads of the matured/brood fishes. The observation also correlates with the findings of Dasgupta (1982) that even among the members of the population sampled on the same day may show considerable variation in condition factor in relation to length.

The proportion of different age groups and sexes showed that the population has a definite structure with a broad base, when plotted the number, of fishes against different size groups (Figure 12.1) which further reveals the expanding nature of population of chocolate mahseer in the lake.

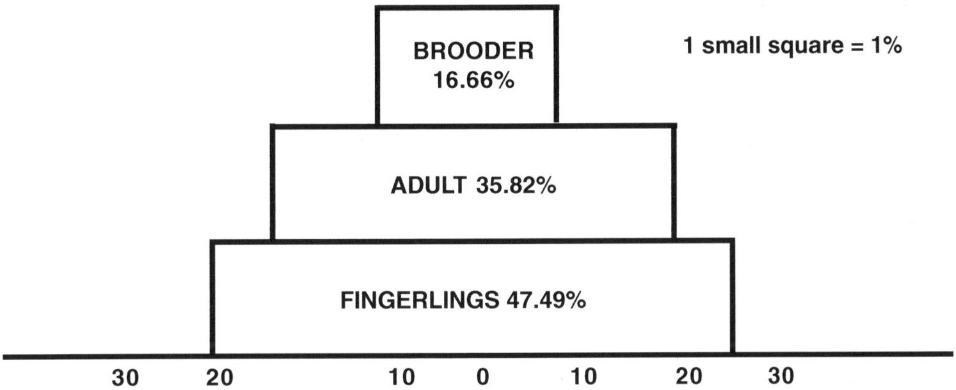

Figure 12.1: Population Structure of Chocolate Mahseer

The total sample of fishes has been segregated depending on the maturity status in to Fingerlings (47.49 per cent), Adult (35.82 per cent) and Brood fish or Brooder (16.66 per cent) and reveals that the population of chocolate mahseer in the lake has suitably adapted with prominent growth since its introduction in 1994–1995.

Table 12.1: Details of Length-Weight Relationship of *Neolissocheilus hexagonolepis*

Source	*Number*	*Length Group (mm)*	*Average Observed Weight (gm)*	*Observed Weight per mm Length*	*Expected per mm Length when n = 3*	*N Value*	*Average Condition Factor*
Shally lake	131	70–285	83.40	0.420	1.738	2.736	1.005

The above studies lead us to conclude that although the Chocolate mahseer or *N. hexagonolepis* is a semi-rheophilic fish, yet can be suitably culture in the closed environment either as monoculture or as composite culture along with the IMCs.

References

Beckman, W.C. 1948. The Length-weight relationship, factors for conservations between standard and total length and co-efficient of condition for seven Michigan fishes. *Trans. Am. Fish. Soc.*, 75: 237–256.

Bagenal, T. 1978. *Methods for Assessment of Fish Production in Freshwaters*, 3rd edn. Blackwell Scientific Publications Ltd., Australia, pp. 365.

Bashirullah, A.K.N. 1975. Biology of *Lutjanus griseus* (L.) of the Cubagua Island Venezuela. Length-weight, body length-gut length relationship and condition factor. *Bio. Inst. Oceanogr. Uni. Oriente.*, 14(1): 101–107.

Choudhuri, M., Kolekar, V. and Chandra, R. 1982. Length-weight relationship and relative condition factor of four Indian major carps of river Brahmaputra. *Assam Journal of Inland Fishery Society of India,* (2): 42–48.

Dasgupta, M. 1982. An investigation on some aspects of the biology of mahseer from North-East India. *Ph.D. Thesis,* North-East Hill University, Shillong.

Jhingran, A.G. 1968. The length-weight relationship and K factor of *Gudusia chapra* (Ham.) from the Ganga river. In: *Proc. Nat. Acad. Sci.,* India, 18: 149–163.

Jhingran, V.G. 1952. General length-weight relationship of three major carps of India. *Proc. Nisc., India,* 18: 449–455.

Lagler, K.F. 1956. *Freshwater Fish Biology,* Dubugue, IOWA, WMC Brown.

Lecren, E.D. 1951. The length-weight relationship and seasonal cycle in gonad weight and condition in perch *Perca fluviatilis. J. Animal Ecology,* 20: 201–209.

Martin, W.R. 1949. The mechanism of environmental control of body form in fishes. *Univ. Toronto. Stu. Biol.,* 58: *Ont. Fish. Res. Lab.,* 70: 1–91.

Mitra, B. and Das, C. 1967. On the interrelationship between length and weight of *Gudusia chapra* (Ham.). *Proc. 54th Indian Sci. Cong.,* Kolkata, (3): 476 (Abstract).

Patgiri, A., Goswami, M.M., Kar, D. and Barbhuiya, M.H. 2001. Comparative study on the length-weight relationship and relative condition factors in major and exotic carps in ponds of Guwahati. *Indian J. Environ. and Ecoplan.,* 5(1): 179–180.

Pathak, S.C. 1976. Length-weight relationship and condition factor and food study of *Labeo calbasu* (Ham.) from Loni reservoir (MP). *J. Inland Fish. Soc.,* India, 1: 58–64.

Rao, G.R. and Rao, L.H. 1972. On the biology of *Labeo calbasu* (Ham–Buch) from the river Godavari. *Journal of Inland Fishery Society of India,* p. 74–86.

Rousenfell, G.A. and Everhart, W.H. 1953. *Fishery Science: The Method and Application.* John Wiley and Sons, New York.

Sen, N. 1982. Studies on the systematics, distribution and ecology of the ichthyofauna of Meghalaya and their bearing on the fish and fisheries of the state. *Ph.D. Thesis,* University of Guwahati, Assam, pp. 536.

Srivastava, C.B. and Singh, V.R.P. 1964. Inter relationship between standard length and body weight of *Cirrhinus mrigala* (Ham.). *Proc. Nat. Acad. Sci.* Allahabad, India, 34(2): 37–41.

Thakur, N.K. and Das, U.K. 1975. Length-weight relationship of *H. fossilis* (Bl). *J. Inland Fish Soc.,* India, 6(1974): 95–96.

Venkateswarlu, T. and Banerjee, V. 1971. Length-weight studies in *Gudusia chapra* (Ham.). *Proc. Nat. Acad. Sci.,* India, 41(1): 43–46.

Chapter 13

A Study of Arsenic Effect on Blood and Tissue Glucose Concentrations in a Fish Model

Nimai Chandra Saha, Trilochan Midya and Nirmal Kumar Sarkar
Postgraduate Department of Zoology, Presidency College,
Kolkata – 700 073, West Bengal, India

ABSTRACT

Glucose concentrations of blood, liver, kidney and muscle were determined in the live-fish, *Clarias batrachus* (L.) exposed to sodium arsenite in water for one month. The values were found to be significantly lower in the arsenic-treated fishes, as compared to the controls maintained in arsenic-free water. The changes were moderate in blood and liver, most pronounced in kidney and least in muscle. The overall findings indicate an adverse influence of chronic arsenic toxicity on glucose metabolism.

***Keywords**: Glucose, Blood, Liver, Kidney, Muscle, Clarias batrachus.*

Introduction

Chronic arsenic toxicity has variously been reported to cause different physiological abnormalities like anaemia, keratosis and pigmentation of skin, hepatic malfunction, renal tissue damage, degenerative changes of gonad and cancers of different organs in human subjects (Bates, 1992; Morton and Dunnette, 1994; Guha Mazumder *et al.*, 1997; Buchancova *et al.*, 1998), laboratory mammals (Flora *et al.*, 1997; Guha and Sarkar, 2000), farm animals (Galloway *et al.*, 1992) and fish models (Shukla and Pandey, 1984; Kotsanis and Iliopoulou-Georgudaki, 1999; Sarkar and Sarkar, 2001). However, the relationship between arsenic toxicity and blood or tissue glucose levels remains a controversial issue as yet. Acute arsenic treatment has been reported to adversely affect gluconeogenesis and blood glucose level in laboratory mammals (Reichl *et al.*, 1988; Szinicz and Forth, 1988). Another report indicates a marked decrease of glucose concentrations in liver, muscle, gill and brain of the freshwater fish, *Tilapia mossambica,* following its exposure to sodium arsenite in water for 24 to 96 hours (Shobha Rani *et al.*, 2000). In another study, normal blood sugar level was found in 156 persons who were drinking arsenic-contaminated tube-well water (0.05–3.2 mg of arsenic/litre) for years in different rural areas of West Bengal, India (Guha Mazumder *et al.*, 1997). On the other hand, a high incidence of diabetes mellitus has been reported amongst people who were ingesting arsenic with artesian well water in a locality of Taiwan (Lai *et al.*, 1994).

In the present study, we have estimated glucose concentrations of blood, liver, kidney and muscle of the freshwater live-fish, *Clarias batrachus* (Linnaeus), following its exposure to sodium arsenite in water for one month. The objective of the study is to add to the existing knowledge of arsenic effect on blood and tissue glucose concentrations. We have chosen a fish model for the study, because a possibility of exposure of fishes to arsenic in nature cannot be nullified in ponds which are filled up in the summer with arsenic-contaminated groundwater by using pumps. It may be recalled here that the arsenic concentration of groundwater samples from rural areas of West Bengal has already been reported to reach a highest value of 3.7 mg/litre (Roy Chowdhury *et al.*, 1997).

Materials and Methods

Adult, male specimens of *Clarias batrachus* (20–22 cm in length and 95–110 g in weight) were maintained in some aquaria for one month, with 10 mg of sodium arsenite ($NaAsO_2$) per litre of water (total volume of water was 10 litre per aquarium). $NaAsO_2$ was chosen for the study, because a major portion of arsenic detected in groundwater as well as surface water-bodies in nature occurs in a soluble arsenite form (Thornton and Farago, 1997). The water of the aquaria were periodically changed and with each change, $NaAsO_2$ was added afresh. The concentration of $NaAsO_2$ was kept same as that used by an earlier group of workers in case of *Tilapia mossambica* (Shobha Rani *et al.*, 2000). Control fishes were maintained in absence of $NaAsO_2$ in separate aquaria containing chlorine-free tap water. The water of all aquaria used in either the experimental or the control study had a pH of 7.1 to 7.2, a dissolved oxygen concentration of 6.5 to 6.6 ppm and a free carbon dioxide concentration of 1.4 to 1.5 ppm. All fishes were provided with commercial fish pellet and *Tubifex* as food *ad libitum.* Moreover, all fishes were allowed to acclimatise to the laboratory conditions for one week prior to the beginning of the study.

Both experimental and control fishes (five specimens per group) were sacrificed by striking their heads on a laboratory table and subjected to a puncture at the caudal peduncle. Blood samples were allowed to drip into clean watch-glasses from the caudal vein of the fishes. The samples were prevented from clotting by allowing a thin film of sequestrene (0.1 ml of a 1.0 per cent solution) to dry on each watch-glass one hour before collecting the samples. Tissue samples (liver, kidney and muscle) were homogenised in presence of chilled phosphate buffer (pH 7.0) (2.0 ml of the buffer was added to 1.0 g of each tissue). Glucose concentration of blood and tissue samples were then determined colorimetrically following a modified version (Nath and Nath, 1990) of the well-known arsenomolybdate method (Somogyi, 1952). In the adopted method, 5.0 per cent zinc sulphate and 0.5N sodium hydroxide solutions have been used for deproteinization of blood and tissue samples while 5.0 per cent zinc sulphate and 0.3 N barium hydroxide solutions were used in the original method.

All studies were carried out twice for confirmation of results, initially in October–November, 2004 and then, in November–December, 2004. The temperature of water at the mid-day varied between 26.5 to 28.5°C during the first study and between 24.5 to 27.2°C during the second study. The differences between the results obtained from the experimental and the control groups of fishes were analysed statistically by means of Fisher's two-tailed t-test (Goon *et al.*, 1981).

Results

In either the first study (October–November, 2004) or the second study (November–December, 2004), glucose concentrations of blood, liver, kidney as well as muscle were found to be significantly decreased in the experimental group of fishes exposed to sodium arsenite in water for one month, as compared to the corresponding values observed in the control group of fishes. The most remarkable reduction of glucose concentration was noted in case of kidney (35.3–39.3 per cent reduction) while the least reduction was noted in case of muscle (14.6–16.3 per cent reduction). Moderate reductions of glucose concentrations were noted in blood and liver (25.6–27.9 per cent and 22.7–24.1 per cent reductions, respectively) of the treated fishes, as compared to the controls. Glucose concentrations recorded in blood and different tissues of the experimental as well as the control groups of fishes in two of our successive studies are presented in Tables 13.1 and 13.2, respectively.

Table 13.1: Blood and Tissue Glucose Concentrations (mg/100 ml of blood and mg/g of tissue) in the Fish Model During the First Study (October–November, 2004)

Tissue	*Glucose Concentration*	
	Control Group	*Arsenic-treated Group*
Blood	127.4±8.98	94.8±7.17*
Liver	17.49±1.01	13.51±0.95*
Kidney	6.12±0.41	3.96±0.40*
Muscle	16.54±0.87	14.12±0.75*

* Significantly different from control value at 1 per cent level.

Values represent means ± S.D. of five fishes in each group.

Table 13.2: Blood and Tissue Glucose Concentrations (mg/100 ml of blood and mg/g of tissue) in the Fish Model During the Second Study (November–December, 2004)

Tissue	*Glucose Concentration*	
	Control Group	*Arsenic-treated Group*
Blood	123.8±6.33	89.2±5.72*
Liver	17.11±0.92	12.98±0.94*
Kidney	6.16±0.42	3.74±0.38*
Muscle	16.24±0.82	13.59±0.81*

* Significantly different from control value at 1 per cent level.

Values represent means ± S.D. of five fishes in each group.

Discussion

Our findings on tissue glucose concentrations in arsenic-treated and control groups of the fish model, *Clarias batrachus*, are in conformity with that reported in another fish model (*Tilapia mossambica*) by an earlier group of workers (Shobha Rani *et al.*, 2000) who, however, did not determine blood glucose concentration. In so far as blood glucose concentration is concerned, our finding in the fish model is comparable to that in laboratory mammals by some other earlier workers (Reichl *et al.*, 1988; Szinicz and Forth, 1988). Considering our findings in conjunction with that of all the aforesaid workers, we conclude that chronic arsenic toxicity may considerably lower blood glucose concentration, probably by adversely affecting the rate of gluconeogenesis (Reichl *et al.*, 1988; Szinicz and Forth, 1988) in liver and kidney, which are the main sites of gluconeogenesis in the vertebrate body (Martin *et al.*, 1981). Muscle is not an active site of gluconeogenesis (Martin *et al.*, 1981). Therefore, a decreased blood glucose concentration might have been responsible for a decreased supply of glucose to the tissue, resulting in a fall of muscle glucose concentration. It may be added here that the arsenic-treated fishes are likely to be weaker than the control fishes owing to decreased blood and tissue glucose concentrations.

Finally, it seems necessary to account for the apparent discrepancy between our finding on blood glucose concentration in an arsenic-treated fish model and that of others in arsenic-affected human subjects. On account of their gill-breathing habit, fishes are continually exposed to arsenic in water. Moreover, arsenic concentration of water was 10 mg/litre in our study in *Clarias batrachus* as well as another study in *Tilapia mossambica* by an earlier group of workers (Shobha Rani *et al.*, 2000). On the other hand, human subjects studied by some earlier workers (Guha Mazumder *et al.*, 1997) were drinking water with an arsenic concentration ranging between 0.5 to 3.2 mg/litre. Therefore, it is likely that a relatively larger quantity of arsenic entered everyday into the body of a fish than into the body of a human subject. This is why the blood glucose concentration remained unaffected in the aforesaid human subjects. However, it is difficult to understand the relationship between ingestion of arsenic with artesian well water and a high incidence of diabetes mellitus (hyperglycaemia), as reported in case of some people in a locality of Taiwan by another group of workers (Lai *et al.*, 1994). In view of the facts that (*i*) a decrease of blood glucose concentration has repeatedly been reported in arsenic-treated mammalian models (Reichl *et al.*, 1988; Szinicz and Forth, 1998) and (*ii*) hyperglycaemia has not been found in any of the many studied human subjects drinking arsenic-contaminated water in our country (Guha Mazumder *et al.*, 1997), we express that whether arsenic or any other substance present in groundwater was responsible for diabetes mellitus in some human subjects of a particular locality requires to be re-examined by means of as many sensitive methods as possible.

References

Bates, M.N. 1992. Arsenic ingestion and internal cancers: A review. *Amer. J. Epidemiol.*, 135: 462–472.

Buchancova, J., Klimentova, G., Knizcova, M., Mesko, D., Gallikova, E., Kubik, J., Fabianova, E. and Jakubis, M. 1998. Health status of workers of a thermal power station, exposed for prolonged periods to arsenic and other elements from fuel. *Eur. J. Pub. Hlth.*, 6: 29–36.

Flora, S.J.S., Pant, S.C., Malhotra, P.R. and Kannan, G.M. 1997. Biochemical and histopathological changes in arsenic-intoxicated rats co-exposed to alcohol. *Alcohol.* 14(6): 563–568.

Galloway, D.D., Wright, P.J., de Krester, D. and Clarke, I.J. 1992. An outbreak of gonadal hypoplasia in a sheep flock: Clinical, pathological and endocrinological features, aetiological studies. *Veter. Rec.*, 131(22): 506–512.

Goon, A.M., Gupta, M.K. and Dagupta, B. 1981. *Basic Statistics*, 3rd Edn. The World Press, Kolkata, p. 306–310.

Guha, A. and Sarkar, N.K. 2000. Chronic arsenic toxicity as a causative agent of anaemia and echinocytosis: An experimental study in Swiss mice. *Ind. J. Env. Ecoplan.*, 3(1): 19–26.

Guha Mazumder, D.N., Dasgupta, J., Santra, A., Pal, A., Ghose, A., Sarkar, S., Chattopadhyaya, N. and Chakraborti, D. 1997. Non-cancer effects of chronic arsenicosis with special reference to liver damage. In: *Arsenic-Exposure and Health Effects*, (Eds.) Abernathy, C.O., Calderon, R.L. and Chappell, W.R. Chapman and Hall, London, p. 112–123.

Kotsanis, N. and Iliopoulou-Georgudaki, A. 1999. Arsenic induced liver hyperplasia and kidney fibrosis in rainbow trout (*Oncorhynchus mykiss*) by microinjection technique: A sensitive animal bioassay for environmental metal toxicity. *Bull. Environ. Contamin. Toxicol.*, 62(3): 169–178.

Lai, M.S., Hsueh, Y.M. and Chen, C.J. 1994. Ingested inorganic arsenic and prevalence of diabetes mellitus. *Amer. J. Epidemiol.*, 139: 484–492.

Martin, D.W., Mayes, P.A. and Rodwell, V.W. 1981. *Harper's Review of Biochemistry*, 18th Edn. Lange Medical Publications, California, p. 160–185.

Morton, W.E. and Dunnette, D.A. 1994. Health effects of environmental arsenic. In: *Advances of Environmental Science and Technology*, (Ed.) Nriagu, J.O. John Wiley, New York, 27: 17–34.

Nath, R.L. and Nath, R.K. 1990. *Practical Biochemistry in Clinical Medicine*, 2nd Edn. Academic Publishers, Kolkata, p. 38–41.

Reichl, F.X., Szinicz, L., Kreppel, H., Fichtl, B. and Forth, W. 1988. Effect of As_2O_3 on carbohydrate metabolism after single or repeated injection in guinea pigs. *Toxicologist*, 8(1): 20–23.

Roy Chowdhury, T., Mondal, B.K., Samanta, G., Basu, G.K., Chowdhury, P.P., Chanda, C.R., Karan, N.K., Lodh, D., Dhar, R.K., Das, D., Saha, K.C. and Chakraborti, D. 1997. Arsenic in groundwater in six districts of West Bengal, India: The biggest aresenic calamity in the world: The status report upto August, 1995. In: *Arsenic-Exposure and Health Effects*, (Eds.) Abernathy, C.O., Calderon, R.L. and Chappell, W.R. Chapman and Hall, London, p. 93–111.

Sarkar, J. and Sarkar, N.K. 2001. Haematological profile in the freshwater fish, *Channa punctatus* (Bloch) following chronic experimental exposure to sodium arsenite in water. *Trans. Zool. Soc. East. India.* 5(1): 10–21.

Shobha Rani, A., Sudarshan, R., Reddy, T.N., Reddy, P.U.M. and Raju, T.N. 2000. Effect of sodium arsenite on glucose and glycogen levels in freshwater teleost fish, *Tilapia mossambica. Poll. Res.*, 19(1): 129–131.

Shukla, J.P. and Pandey, K.O. 1994. Impaired spermatogenesis in arsenic-treated freshwater fish, *Colisa fasciatus* (Bl. and Sch.). *Toxicol. Lett.*, 21(2): 191–195.

Somogyi, M. 1952. In: *Practical Biochemistry in Clinical Medicine*, 2nd Edn. (Eds.) Nath, R.L. and Nath, R.K. Academic Publishers, Kolkata, p. 38–41.

Szinicz, L. and Forth, W. 1988. Effect of As_2O_3 on gluconeogenesis. *Arch. Toxicol.* 61: 444–449.

Thornton, I. and Farago, M. 1997. The geochemistry of arsenic. In: *Arsenic-Exposure and Health Effects*, (Eds.) Abernathy, C.O., Calderon, R.L. and Chappell, W.R. Chapman and Hall, London, p. 1–16.

Chapter 14

Effect of Chlorine on Common Carps

C. Bala Murali Krishna

Assistant Professor, Department of Civil Engineering, NMAM Institute of Technology, Nitte – 574 110

ABSTRACT

The fish Common Carps were collected from Ramasamudra lake, Karkala under oxygen packing. The fishes were exposed to different concentration of chlorine at laboratory conditions to find out the acute toxicity of chlorine on Common Carps. The static bioassay method followed in the present study. Concentration of 0.3 mg/l of chlorine was found to effect the fishes, with 0.8 mg/l resulting in 100 per cent mortality after 96 hours. LC_{50} values were found as 0.578 mg/l and 0.563 mg/l for 24 hrs and 48 hrs respectively.

Introduction

Chlorine is one of the eight fundamental blocks for modern society. So important is chlorine to life that virtually every product manufactured today benefits in some way from chlorine. From plumbing pipes to X-ray film, medicines to seat belts, household bleach to rocket fuel, blood bags to computers, tap water to bullet–resistant vests. The products of chlorine chemistry have made life healthier, safer, more enjoyable and more productive.

Chlorine in the form of bleaching powder, liquid chlorine, chlorine gas, etc, is widely using disinfectant in water treatment and wastewater treatment. Chlorine is a carcinogenic agent. Excess use of chlorine gives adverse effect on human health. Cancer risk among people drinking chlorinated water is 93 per cent higher than among those whose water does not contain chlorine. Public concerns about chlorine and chlorinated chemicals were first aroused by Rachel Carson in her 1962 book Silent Spring. Attacks on chlorine and chlorinated chemicals have continued since then and intensified in. the late 1980s with the launch of Green Peace's "Chlorine Free" campaign. In the present investigation an attempt has been made to understand the effect of chlorine on Common Carp, as it is widely cultured fish in the world and contributes significantly to the inland fish production.

Materials and Methods

The test fishes, Common Carps were selected for investigation because of its importance in fresh water aquaculture and its availability throughout the year. The test fishes were collected from Ramasamudra lake, Karkala, Udipi district under oxygen packing and were stocked in a well cleaned tank of dimension 10 × 7 × 5 ft for acclimatization. During the acclimatization period, the fishes were fed with nutritionally balanced food at the rate of 5–10 per cent of their body weight. After 7 days of stocking period when they were found adapted to the environment, 10 active healthy and same size fishes were selected and brought to the laboratory for the bioassay test. The experiments were carried out at NMAM Institute of Technology, Nitte. The test fishes were transferred to clean, transparent glass troughs, starved for 24 hrs before the commencement of experiment and during the experiment. The troughs were of 10 lts capacity and 5 lts of water was taken during tests according to specifications. The fresh water used in the investigation was ground water. Different concentration of chlorine solution was prepared by using Bleaching Powder.

Results and Discussion

Bioassays are indispensable tools in aquatic pollution monitoring since physical and chemical tests individually or in combination would not help in assessment of potential toxicity to organisms. In the present investigation, an attempt has been made to find out the acute toxicity of chlorine to Common Carp. The experiments were designed to find out the lethal effects of chlorine. Experiment conducted was of static bioassay type, based on the standard procedure for the measurement of pollutant toxicity on fresh water organism (APHA, 1989).

Initially a pilot test was conducted to find an approximate range of chlorine concentration to which the fishes were lethal. The results of pilot test was presented in Table 14.1. From pilot test it was found that 0 to 100 per cent mortality lies between 0.1 mg/l to 1 mg/l of chlorine present in the water. Bioassay test was then conducted on Common Carp by exposing them to different range of chlorine solutions *i.e.*, from 0.1 mg/l to 1 mg/l. Two trials of bioassay tests were conducted and average of these two trials was used to find LC_{50} value. Mortality rates of Common Carp exposed to different concentration of chlorine from 0.1 mg/l to 0.8 mg/l with time is presented in Table 14.2.

Table 14.1: Observed Mortality During Pilot Test

	Concentration of Chlorine in mg/l		
	0.1	*1*	*10*
Mortality observed	No mortality observed	100 per cent mortality observed within 6 hrs	100 per cent mortality observed within 30 minutes

No mortality of Common Carps were observed even after 96 hrs exposure to the chlorine concentration of 0.1 mg/l and 0.2 mg/l. The chlorine concentration of 0.3 mg/l was found to effect the Common Carps and with 0.8 mg/l resulting in 100 per cent mortality after 96 hrs exposure. During the first 48 hrs of experiment, the mortality rates were high. After 48 hrs there was no change in mortality rates of the Common Carps. In the control test none of the fishes died till the termination of the experiments. LC_{50} value is calculated using "Probit Method." In this approach to find LC_{50} value, only those readings are considered which mortality increase linearly *i.e.*, only mortality values of 0.5 mg/l, 0.55 mg/l, 0.6 mg/l and 0.7 mg/l of 24 hrs and 48 hrs are considered.

Table 14.2: Mortality Rates of Common Carp Exposed to Different Concentration of Chlorine with Time

Time in hrs	*Concentration of Chlorine in mg/l*																			
	0.10		*0.20*		*0.30*		*0.35*		*0.40*		*0.50*		*0.55*		*0.60*		*0.70*		*0.80*	
	N	*%*	*N*	*%*	*N*	*%*	*N*	*%*	*N*	*%*	*N*	*%*	*N*	*%*	*N*	*%*	*N*	*%*	*N*	*%*
6	0	0	0	0	0	0	1	10	1	10	1	10	1	10	1	10	2	20	10	l00
12	0	0	0	0	1	10	1	10	1	10	1	10	2	20	2	20	5	50	10	l00
24	0	0	0	0	1	10	1	10	1	10	1	10	3	30	5	50	10	100	10	100
48	0	0	0	0	1	10	1	10	1	10	1	10	4	40	5	50	10	100	10	100
72	0	0	0	0	1	10	1	10	1	10	1	10	4	40	5	50	10	100	10	100
96	0	0	0	0	1	10	1	10	1	10	1	10	4	40	5	50	10	100	10	100

Number of fishes tested: 10; N: Number of test fishes died; %: Mortality expressed in percentage.

Tables 14.3 and 14.4 represents the probit value calculation for different percentage of mortality at 24 hrs and 48 hrs respectively. Figures 14.1 and 14.2 represents the probit graph for LC_{50} values for 24 hrs and 48 hrs respectively. From Figure 14.1 the value corresponding to 50 per cent mortality for 24 hrs is 1.7622. This value is equal to log (concentration × 100).

Log (concentration × 100) = 1.7622

Concentration = 0.578 mg/l

LC_{50} 24 hrs = 0.578 mg/l

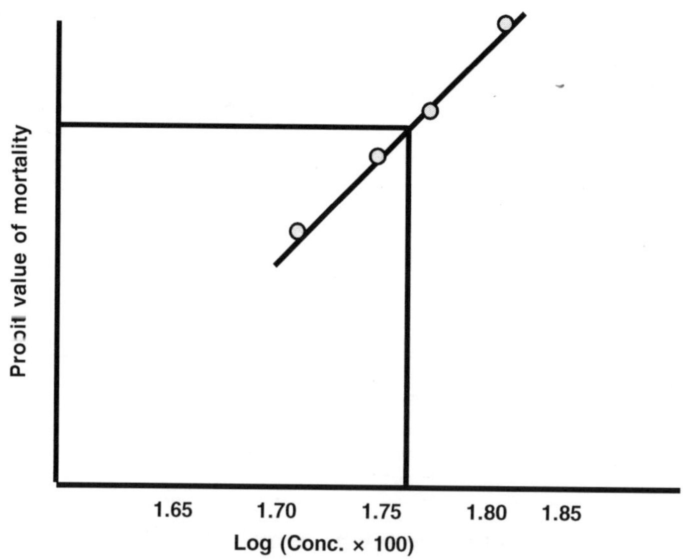

Figure 14.1: Probit Method for LC_{50} 24 hrs

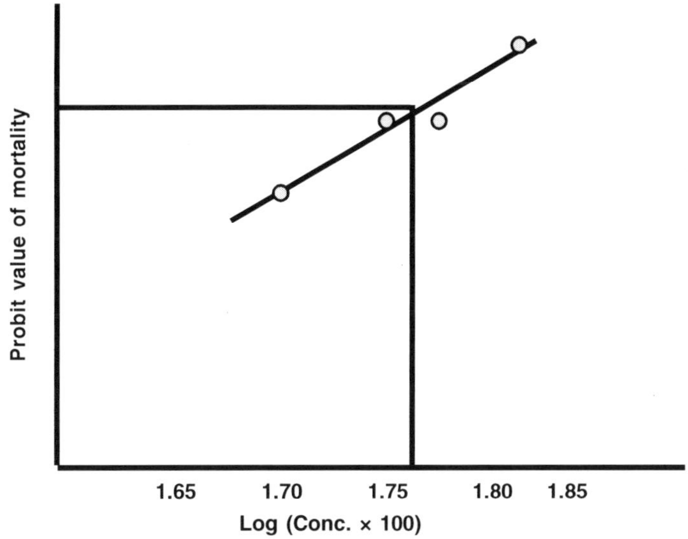

Figure 14.2: Probit Method for LC_{50} 48 hrs

Similarly from Figure 14.2 the value corresponding to 50 per cent mortality for 48 hrs is 1.75044. LC_{50} 48 hrs = 0.563 mg/l.

Table 14.3: Probit Value Calculation for Different Percentage Mortality for 24 hrs

Concentration in mg/l	*Concentration × 100*	*Log (Concentration × 100)*	*Mortality*	*Percentage Mortality*	*Probit Value for Percentage Mortality*
0.50	50	1.698	1	10	3.7184
0.55	55	1.74	3	30	4.4756
0.60	60	1.778	5	50	5.0000
0.70	70	1.845	10	100	8.7190

Table 14.4: Probit Value Calculation for Different Percentage Mortality for 48 hrs

Concentration in mg/l	*Concentration × 100*	*Log (Concentration × 100)*	*Mortality*	*Percentage Mortality*	*Probit Value for Percentage Mortality*
0.50	50	1.698	1	10	3.7184
0.55	55	1.74	4	40	4.7467
0.60	60	1.778	5	50	5.0000
0.70	70	1.845	10	100	8.7190

Conclusion

1. No mortality of Common Carps were observed even after 96 hrs exposure to the chlorine concentration of 0.1 mg/l and 0.2 mg/l.
2. The chlorine concentration of 0.3 mg/l was found to effect the Common Carps and with 0.8 mg/l resulting in 100 per cent mortality after 96 hrs exposure.
3. LC_{50} value were found 0.578 mg/l and 0.563 mg/l for 24 hrs and 48 hrs respectively.

Chapter 15

Pesticide Induced Impairment on the Carbohydrate Metabolism in the Fish *Mystus vittatus*

R. Sonaraj, A.J.A. Ranjit Singh** and A. Pushparaj**

**Reader, Research Department of Zoology, Kamaraj College, Thoothukudi – 628 003, Tamil Nadu*

***Reader, Department of Biology, S.P.K. College, Alwarkurichi*

ABSTRACT

Pesticides stress on the carbohydrate metabolism in the fish *Mystus vittatus* was studied by treating the fish with different sublethal concentrations of parrysulfan (0.1672 and 0.3345 ppm), a organo chlorine pesticide and Sicocil (0.3358 and 0.6717 ppm), a organophosphate pesticide. The total glycogen level in the liver and muscle tissues and also the SDH level and lactic acid level showed significant changes when the fish was exposed to the pesticides for a long duration. This fish exposed to the concentration 0.3345 ppm of parrysulfan for 60 days showed 60 per cent and 50 per cent reduction in glycogen level in the muscle and over respectively. Similarly the fish exposed to 0.6717 ppm concentration of sicocil showed 59 per cent and 48 per cent reduction in the level of glycogen in muscle and liver respectively. The same trend is noted in the presence of SDH also. There is 55 per cent and 59 per cent reduction observed in the muscle and liver of the fish exposed to parrysulfan for 60 days and 48 per cent and 54 per cent reduction in the muscle and liver of the fish exposed for 60 days in sicocil. The presence of lactic acid in the tissues of fish exposed to pesticide showed an increasing trend. The fish exposed to the parrysulfan for 60 days in the 0.3345 ppm concentration showed 69 per cent and 60 per cent increase in lactic acid accumulation in the muscle and liver respectively where as the fish exposed to sicocil at the concentration 0.6717 ppm for 60 days showed 68 per cent and 46 per cent increase in the lactic acid accumulation in the muscle and liver respectively. With regard to carbohydrate metabolism, both the pesticides create a massive stress and inflict physiological disturbances to the fish *Mystus vittatus.*

Keywords: *Pesticide, Sicocil, Parrysulfan, Glycogen, Organophosphate.*

Introduction

In recent years, pesticides are widely used to control the various pests of our crops. Pesticides used ultimately reach the nearby water bodies through rain, irrigation surface runoff, etc. and affect the entire ecosystem. The dominating group of the aquatic system, fish, is the most affected animal. Much emphasis is being made to evaluate the effect of pesticides on the gmeral organization of fishes. The pesticide that is reaching the aquatic system gets entrenched into the aquatic food chain through bio-accumulation, bio-concentration and bio-magnification process. Ramasami (1990) stated that the pesticides accumulated in the tissues of fishes alters the biochemical constitution of the fish. The effect of thiodon, dithane and nuvan on biochemical consituents in muslce, liver and kidney of cyprinid fish *Barilus benedelisis* was studied by Deoray (1987).

Rao and Rao (1981) and Babu *et al.* (1988) had reported a decrease in glycogen, total carbohydrates and lipid contents in tissues of *Sarotherodon mossambicus* exposed to pesticides. Ghosh (1987) reported the toxic effects of three organophosphate pesticides on the carbohydrate metabolism of the freshwater fish, *Channa punctatus.* Ramasamy and Maheswari (1993) reported lactic acidosis in fishes exposed to pesticides.

Shrivastava and Singh (1997) reported enhancement of glucose level in, the intestinal tissues and liver during early period of exposure of the fish, *H. fossilis,* to the pesticide endosulfan. But the trend was completely altered when the exposure duration was extended.

From the available literature it was found that the pesticide exposure impairs the biochemical constituents in the experimental animal, fish. Though much work has been done to trace the quantity of the biochemical lesions in several fishes, less attention has been focussed on the cat fish, *Mystus vittatus,* although they inhabit all freshwater system and are highly edible. Hence the present biochemical investigation was carried out on the toxicity of this chemicals on the fish, *Mystus vittatus.*

Materials and Methods

The fish, *Mystus vittatus,* (weight 5–6 gms) were collected from the local pond, brought to the laboratory and acclimatized in the laboratory condition for two weeks under natural photoperiod and temperature (29±2°C) in 50 L glass aquaria.

From the 96h LC_{50} value, two sublethal concentrations *viz.,* 1/4th and 1/5th of parrysulfan (0.1672 ppm and 0.3345 ppm) and 1/4th and 1/8th of sicocil (0.3358 ppm and 0.6717 ppm) were chosen to expose the fish for biochemical studies. The experiment was run for 60 days keeping control separately. All fish were regularly fed *ad libitum* with nodules prepared from groundnut oil cakes and rice bran during the experimental period. After 10, 30 and 60 days of exposure the fish were taken out and the following tissues *viz.,* muscle and liver, were taken out under aseptic condition. The biochemical parameters *viz.,* total glycogen, SDH and lactic acid were estimated using standard procedures.

Results

Carbohydrate Metabolism

Glycogen

Due to the continuous exposure to the sublethal concentrations of parrysulfan and sicocil, the fish, *Mystus vittatus,* has developed a stress and reduction in food intake resulted in the depletion of muscle and liver glycogen from the beginning to termination of the experiment. The percentage reduction in the glycogen level in both the liver and muscle tissues of parrysulfan and sicocil exposed fish was high with the increase of exposure duration (Table 15.1).

Table 15.1: Changes in the Level of Glycogen (mg/g wet tissue) in the Tissues of *Mystus vittatus* Exposed to Two Sublethal Concentrations of *Parrysulfan* and *Sicocil*. Values in parentheses indicate percentage change over control.

Pesticides	*Tissue*	*Concentration ppm*	*Sublethal Exposure Duration (days)*			*'t' Value*
			10	*30*	*60*	
Parrysulfan	Muscle	Control	0.94±0.07	0.92±0.05	0.90±0.04	
		0.1672	0.76±0.03	0.58±0.06	0.45±0.03	3.5682
			(–19.15)	(–36.96)	(–50.00)	P < 0.05 S
		0.3345	0.69±0.08	0.52±0.08	0.36±0.04	4.1330
			(–26.60)	(–43.48)	(–60.00)	P < 0.05 S
	Liver	Control	9.94±0.53	9.81±0.35	9.85±0.64	
		0.1672	7.80±1.08	6.04±0.34	5.11±0.34	4.4954
			(–21.52)	(–38.43)	(–48.12)	P < 0.05 S
		0.3345	7.01±0.74	5.71±0.27	4.96±0.41	6.6214
			(–29.48)	(–41.79)	(–49.64)	P < 0.05 S
Sicocil	Muscle	Control	0.94±0.07	0.92±0.05	0.90±0.04	
		0.3358	0.87±0.01	0.73±0.01	0.48±0.06	1.9770
			(–7.45)	(–20.65)	(–46.67)	P > 0.05 NS
		0.6717	0.81±0.05	0.61±0.04	0.37±0.14	2.5317
			(–13.83)	(–33.69)	(–58.87)	P > 0.05 NS
	Liver	Control	9.94±0.53	9.81±0.35	9.85±0.64	
		0.3358	8.41±0.94	7.01±0.53	5.91±0.23	3.8052
			(–15.09)	(–28.54)	(–40.00)	P < 0.05 S
		0.6717	7.93±0.68	6.83±0.11	5.11±0.43	3.9481
			(–20.22)	(–30.38)	(–48.12)	P < 0.05 S

Student 't' analysis revealed that the muscle glycogen content of the fish exposed to sicocil do not show any significant change.

Succinic Dehydrogenase (SDH)

The SDH level in the tissues of both parrysulfan and sicocil exposed fish was considerably reduced with the increase in the duration of exposure. The degree of SDH decrease was comparatively less in sicocil exposed fish than parrysulfan exposed fish. A significant decrease in SDH level in the liver tissues of both parrysulfan and sicocil exposed fish was also observed (Table 15.2).

Table 15.2: Changes in the Level of SDH (μ mole formazon/mg protein/hr) in the Tissues of *Mystus vittatus* Exposed to Two Sublethal Concentrations of *Parrysulfan* and *Sicocil*. Values in parentheses indicate percentage change over control.

Pesticides	*Tissue*	*Concentration ppm*	*Sublethal Exposure Duration (days)*			*'t' Value*
			10	*30*	*60*	
Parrysulfan	Muscle	Control	0.63±0.04	0.65±0.03	0.64±0.07	
		0.1672	0.51±0.02	0.46±0.08	0.36±0.04	4.422
			(–19.05)	(–29.23)	(–43.75)	P < 0.05 S
		0.3345	0.49±0.01	0.41±0.09	0.29±0.06	4.1663
			(–22.22)	(–36.92)	(–54.69)	P < 0.05 S
	Liver	Control	0.89±0.03	0.88±0.05	0.84±0.03	
		0.1672	0.70±0.04	0.55±0.05	0.42±0.03	3.8059
			(–21.35)	(–37.50)	(–50.00)	P < 0.05 S
		0.3345	0.61±0.03	0.50±0.05	0.35±0.08	4.9859
			(–31.46)	(–43.18)	(–58.33)	P < 0.05 S
Sicocil	Muscle	Control	0.63±0.04	0.65±0.03	0.64±0.07	
		0.3358	0.55±0.06	0.50±0.03	0.39±0.04	3.3607
			(–12.69)	(–23.08)	(–39.06)	P < 0.05 S
		0.6717	0.53±0.08	0.45±0.02	0.33±0.01	3.4815
			(–15.87)	(–30.77)	(–48.44)	P < 0.05 S
	Liver	Control	0.89±0.03	0.88±0.05	0.84±0.03	
		0.3358	0.75±0.03	0.57±0.08	0.45±0.04	3.1636
			(–15.73)	(–35.23)	(–46.42)	P < 0.05 S
		0.6717	0.64±0.01	0.51±0.07	0.39±0.04	4.8338
			(–28.09)	(–42.04)	(–53.57)	P < 0.05 S

Lactic Acid

Lactic acid level in the muscle and liver of the pesticide exposed fish registered an increasing trend. The increasing trend was more pronounced in parrysulfan exposed than sicocil exposed fish. The increment in lactic acid level was higher when the fish was exposed for 60 days. The level of lactic acid was almost double in both muscle and liver tissues after 60 days of exposure. The increase in lactic acid level in the muscle tissues of both parrysulfan and sicocil exposed fish was not statistically significant whereas the increase in lactic acid level of liver was statistically significant in both the pesticide exposed fish (Table 15.3).

Discussion

After exposure to the sublethal concentrations of parrysulfan and sicocil the fish, *Mystus vittatus*, developed alterations in the quantity of various biochemical constituents. The modulations in biochemical set up impose stress to the fish. It has been stated that (Ghosh, 1987) the normal physiological functioning of different tissues is greatly disturbed by the pesticide pollution. The relative potency of toxicity of pesticides differs in relation to tissues, biomolecules and duration of exposure (Sheela *et al.*, 1992; Ganathy *et al.*, 1994).

Table 15.3: Changes in the Level of Lactic Acid (mg/g wet tissue) in the Tissues of *Mystus vittatus* Exposed to Two Sublethal Concentrations of *Parrysulfan* and *Sicocil*. Values in parentheses indicate percentage change over control.

Pesticides	*Tissue*	*Concentration ppm*	*Sublethal Exposure Duration (days)*			*'t' Value*
			10	*30*	*60*	
Parrysulfan	Muscle	Control	11.7±0.38	11.6±0.34	11.7±0.44	
		0.1672	12.4±0.86	13.8±0.48	18.7±0.34	1.7276
			(5.98)	(18.96)	(59.82)	P > 0.05 NS
		0.3345	13.1±0.98	17.5±0.41	19.8±0.34	2.6113
			(11.96)	(50.86)	(69.23)	P > 0.05 NS
	Liver	Control	6.8±0.35	6.8±0.38	6.8±0.49	
		0.1672	7.9±0.34	8.9±0.44	10.4±0.44	3.1201
			(16.18)	(30.88)	(52.94)	P < 0.05 S
		0.3345	8.2±0.74	9.9±0.38	10.9±0.98	3.6374
			(20.59)	(45.59)	(60.29)	P < 0.05 S
Sicocil	Muscle	Control	11.7±0.38	11.6±0.34	11.7±0.44	
		0.3358	12.2±0.64	13.2±0.47	14.8±0.86	2.2870
			(4.27)	(13.79)	(26.49)	P > 0.05 NS
		0.6717	12.9±0.38	16.3±0.78	19.6±0.81	2.3779
			(10.26)	(40.52)	(67.52)	P > 0.05 NS
	Liver	Control	6.8±0.35	6.8±0.38	6.8±0.49	
		0.3358	7.4±0.41	8.5±0.84	9.7±0.44	2.6098
			(8.82)	(25.00)	(42.65)	P > 0.05 NS
		0.6717	7.9±0.48	9.5±0.86	9.9±0.64	3.7643
			(16.18)	(39.71)	(45.59)	P < 0.05 S

In the present investigation the organochlorine pesticide, parrysulfan, and the organophosphorous pesticide sicocil, were found to have induced alterations in the cellular machinery and have altered the level of biochemical constituents.

Liver and muscle are two active sites where storage and metabolism of glycogen take place. The observed glycogen loss in the liver and muscle of the fish, *Mystus vittatus,* is attributed to the toxicogenic effects of parrysulfan and sicocil on tissue energy reserve. There are two possibilities about the mode of action of the pollutant: either, the parrysulfan and sicocil are responsible for cellular disorganization of the internal tissue system affecting the storage metabolism as reported by Ghosh, 1986; Sambasiva Rao *et al.*, 1984; or it forced the exposed fish to heavy physical exercise in the form of erratic and rapid movement and high respiration rate, thus increasing utilization of reserve glycogen of the liver to meet the glycogen hunger of the body followed by less glycogenesis and increased glocogenolysis. Corbett (1974), Matsumera *et al.* (1972) have added that the glycogen loss in muscle may be due to accumulation of the pesticide in the pancreatic islets of the fish, which selectively damages the insulin producing β cells. Mazeaud *et al.* (1977), Anil *et al.* (1980) had reported that the pesticide, methyl parathion stimulated rapid secretion of both glucocorticoides and catecholamines from the adrenal tissues of the fish, *Channa punctatus,* resulted in hyperglycemia.

The succinic dehydrogenase (SDH) level in the muscle and liver of *Mystus vittatus* showed a decreasing trend when exposed to parrysulfan and sicocil. The degree of decrease was correlated to the concentrations of the pesticides and exposure duration. As SDH is one of the enzymes in the TCA cycle, the decrease in SDH activity indicates depression of oxidative metabolism at the mitochondrial level, leading to an overall depression of TCA cycle (Radhaiah *et al.,* 1987). It has been reported that energy crisis induced by the pesticides alters the metabolism in the tissue. In such cases, a shift of aerobic metabolism towards anaerobic side could be presumed with the increase of glycolytic pathway. Joyce *et al.* (1989, 1990) reported SDH reduction results in the switch over of aerobic to anaerobic pathway in *Clarius batrachus* exposed to organophosphorous pesticide. Increase in the glycolytic path way was observed in the present study. The fall in SDH indicates the oxygen stress and energy crisis and to some extent mitochondrial disturbances (Azhar *et al.,* 1991).

In both muscle and liver, significantly higher levels of lactic acid had persisted in all the three intervals in both concentrations of parrysulfan and sicocil when compared to that of control*Parrysulfan* and *Sicocil.* Values in parenthesis indicate percentage change over control. Increase of lactic acid in the muscle and liver tissues of pesticide treated fishes was due to the inadequacy of oxygen supply in cells to cope up with the complete break down of carbohydrate into carbondioxide and water (Seskin and Levine, 1947). Hence the persistence of tissue lactate in fishes exposed to parrysulfan and sicocil in due present study implies hypoxia. Significant decrease in oxygen consumption of the whole animal and corresponding decrease in succinic dehydrogenase (SDH) enzyme activity in liver and muscle of parrysulfan and sicocil exposed fish further support the above implication of tissue hypoxia.

Moreover the persistence of higher levels of lactic acid in both muscle and liver of parrysulfan and sicocil treated fishes may be due to the impairment in the diffusion of lactic acid from the two tissues studied. Such impairment in the diffusion of tissue lactate in fishes was reported by Heath and Pritchard (1965).

References

Azharbaig, M.D., K. Vijay Joseph and K. Jayantha Rao 1991. Heptachlor induced changes in the enzymes associated with carbohydrate metabolism in functionally different muscles of *Channa punctatus. J. Ecotoxicol. Environment,* 1(2): 156–161.

Anil, K. Srivastava and N.N. Singh. 1980. Observations on hyperglycemia in the murrel *Channa punctatus* after acute exposure to methyl parathion. *Comp. Physiol. Ecol.,* 5(2): 100–101.

Babu, P.R.A., G.R. Reddy, G.R.V. Babu and C.S. Chetty. 1988. Glycolytic oxidation in freshwater fish, *Sarotherodon mossambicus* during benthicarp exposure. *Curr. Sci.,* 57(11): 591–594.

Corbett, J.R. 1974. *Biochemical Mode of Action of Pesticides.* Academic Press, New York, USA.

Deoray, B.M. 1987. Effects of pesticides on freshwater cyprinid fish, *Barilus benedelisis* (Ham.) *Ph.D. Thesis,* Marathwada University.

Ganathy, V., D. Srinivasa Reddy, S.L.N. Reddy and K. Shankaraiah 1994. Effect of hexachlorocyclohexane on biochemical composition of the fish. *Channa punctatus. J. Ecotoxicol. Envi. Monit.,* 4(1): 15–20.

Ghosh, T.K 1986. Effect of dimethoate on tissue glycogen content of some freshwater fishes. *J. Environ. and Ecol.*, 4(4): 554–557.

Ghosh, T.K. 1987. Toxic impact of three organophosphate pesticides on carbohydrate metabolism in a freshwater fish, *Channa punctatus. Adv. Biol. Sci.*, 6: 20.

Heath, A.G. and A.W. Pritchard 1965. Effect of severe hypoxia on carbohydrate energy stores and metabolism in two species of freshwater fish. *Physiol. Zool.*, 38: 325–334.

Joyce Shobha Rani, V., P. Venkateshwarlu and C. Janaiah 1989. Changes in carbohydrate metabolism of *Clarias batrachus* when exposed to two organophosphorus insecticides. *J. Environ. Biol.*, 10(2-supp.): 197–204.

Joyce Shobha Rani, V., P. Venkateshwarlu and C. Janaiah. 1990. Impact of sublethal of concentration of malathion on certain aspects of metabolism in freshwater fish in *Clarias batrachus. Comp. Physiol. Ecol.*, 15(1): 13–16.

Matsumera, F., G.M. Boush, and J. Misato. 1972. *Environmental Toxicology of Pesticides*. Academic Press, New York, USA.

Mazeaud, M.M., F. Mazeaud and E.M. Donaldson. 1977. Primary and secondary effects of stress in fish; some new data with general review. *Trans. Am. Fish. Soc.*, 106: 210–212.

Ramasami, M. (1990). Adaptive trends in lipid levels of liver and muscles of *Sarotherodon mossambicus* exposed to sevin. *J. Ecobiol.*, 2(1): 56–61.

Ramasami, M. and R.V. Maheswari. 1993. Comparative lactic acidosis in fishes following pesticides stress. *Proc. 2nd European Conf. on Ecotoxicol.*, 1(2): 877–885.

Rao, P.S.K. and P.KV. Rao. 1981. Lipid derivatives in tissues of the freshwater teleost fish, *Sarotheroden mossambicus.* Effect of methyl parathion. *Proc. Ulnd. Nat. Sci. Acad.*, 47: 53– 57.

Radhaiah, V., M. Girija and K. Jayantha Rao. 1987. Changes in selected biochemical parameters in the kidney and blood of fish, *Tilapia mossambica. Bull. Env. Cont. Toxicol.*, 30: 1006–1011.

Seskin, S. and Levine, R. 1947. *Carbohydrate Metabolism*. University of Chicago Press, Illinois.

Sheela, M., R. Mathivannan and S. Muniandi. 1992. Impact of fenvalarate and phosphamidon on food intake, growth and conversion efficiency in the fish *Channa striactus. J. Env. and Ecol.*, 10(3): 593–596.

Sambasiva Rao, K.R.S., K. Sivaprasada Rao and K.V. Ramana Rao. 1984. Impact of technical and commercial grade phenthoate on some selected parameters of oxidative metabolism in the fish *Channa punctatus* (Bloch). *Ind. J. Ecol.*, 11: 6–11.

Srivastava, A.K. and Singh, N.N. 1997. Effect of endosulfan on carbohydrate metabolism in Indian cat fish. *Ecotoxicol. Env. Safety*, 5: 412–417.

Chapter 16

A Study on Biochemical Changes in Liver Due to Sugar Mill Effluent in Freshwater Fish *Cirrhinus mrigala*

K. Shanthi, N. Saradhamani and J. Smitha

P.G. and Research Department of Zoology, Kongunadu Arts and Science College, Coimbatore – 641 029, Tamil Nadu, India

ABSTRACT

The physico-chemical characteristics of sugar mill effluent, discharged from Ponni Sugars Ltd., Pallipalayam (Tamil Nadu) and their effect on the biochemical changes in blood of the fresh water fish *cirrhinus mrigala* was analysed in the present investigation.

Keywords: *Sugar mill effluent, Physico-chemical characteristics, Cirrhinus mrigala, Biochemical changes.*

Introduction

Water is the basic need of man, without water there would not be life of any kind on the earth. The quantity of effluent discharged rise day-by-day, adding to pollution load of the waters. Aquatic pollution is caused by four kinds of substances *i.e.*, traditional organic waste, generated from industrial process, chemical agents such as fertilizers and pesticides used for crop protection and slit from degraded catchments.

It is estimated that three fourth of the wastewater generated is from municipal sources, industrial waste contribute to over one half of the total pollution and the major portion of this is coming from large and medium industries (Neena Vyas, 1993).

The advances of science and technology, increasing urbanization have brought about changes in the environment. In the urban areas effluents of different industries are the main sources of environmental pollution. Among the industries that contribute the aquatic pollution in India are pulp, sugar, paper and textile industries. Rural folk have to depend largely on the use of river water and therefore are exposed to a variety of health hazards. Almost every river system of India is polluted to a considerable extent (Chawdhury, 1977).

The important raw material of industries is water. It is used as a solvent, cleansing agent, mineral extractant, coolant and waste removal agent. Discharge of industrial effluent such as distillery, tannery, pulp, paper, textile and sugar mill effluents have been of major concern in water pollution abatement (Haniffa and Arul Selvan, 1991).

The body components like protein, carbohydrate and lipid which play a significant role in body construction and energy production are also affected by water pollution (Palanichamy *et al.*, 1986). Protein, carbohydrate and lipid which constitute the major component of the body play an important role in body construction and energy metabolism.

Enzymatic parameter are sensitive indices of changes in ecological conditions and hence constitute important diagnostic tool in pollution studies (Wiester and Hinterleitner, 1980).

Materials and Methods

The physico-chemical parameters of the sugar mill effluent collected at the discharge point (3 times) were analysed. Healthy specimens of *Cirrhinus mrigala* were collected from Aliyar, Aliyar Nagar, Pollachi (Tamil Nadu).

The procured bulk samples of *Cirrhinus mrigala* were acclimatized in laboratory condition 28±2°C for a month by stocking them in large tank previously washed with potassium permanganate to prevent any fungal infection and enough dechlorinated water flow was maintained and fed with ad libitum, rice bran and groundnut oil cake.

Feeding was stopped two days prior to the experiment to maintain same state of metabolic requirements. Fishes belonging to both the sexes were selected for the present investigation. The median lethal concentration was found out by exposing batches of 10 healthy fishes to different concentration of the sugar mill effluent. Simultaneously control was maintained.

$1/10^{th}$ of the 24 hour LC_{50} was used for sublethal studies. The fishes were exposed to diluted effluent in 2.6 per cent, 3.6 per cent, 4.6 per cent for 24, 48 and 72 hours. Control and experimental fishes were sacrificed at the end of the exposure periods. Fishes were dissected and the liver tissue was removed. The liver tissue weighing about 10 gms were homogenizined in 80 per cent ethanol, centrifuged at 3500 rpm for 15 minutes and clear supernatant was used for the analysis of different biochemical parameters such as glucogen, total protein, cholesterol, GOT, GPT, ACP and ALP according to the methods of Kemp and Kits (1945), Lowry *et al.* (1951), Richmond (1973), Reitman and Frankel (1957), King's method (1959) and King and Kind (1954) respectively. Data were subjected to statistical evaluation using student's "t" test.

Results and Discussion

The results obtained in the present investigation are summarized in the Table 16.1.

Effluent discharged from sugar mill constitute number of chemical pollutants. When these effluents are discharged into the environment, they disturb the ecological cycle of living organisms (Behra and Missra, 1985). The effluent alters the biochemical parameters like carbohydrate, total protein, cholesterol, enzymes like GOT, GPT, ACP, and ALP of the fish. This is amply supported by many observations.

Table 16.1: A Study on Biochemical Changes in Liver Due to Sugar Mill Effluent in Freshwater Fish *Cirrhinus mrigala*

Parameters	*Concentration %*		*Exposure Period*		
			24	*48*	*72*
Glycogen (mg/g)	2.6	Con	81.51±1.161	81.33±1.175	81.44±1.015
		Exp	78.24±1.125	74.20±1.134	70.40±1.304
		t-test	3.48106*	8.43964*	10.164*
	3.6	Con	81.51±1.161	81.33±1.175	81.44±1.015
		Exp	67.29±1.075	65.30±1.175	63.20±1.150
		t-test	15.5545*	16.709*	18.7952*
	4.6	Con	81.51±1.161	81.33±1.175	81.44±1.015
		Exp	61.24±1.125	59.23±1.130	57.77±1.902
		t-test	21.702*	23.4728*	25.8832*
Total Protein (mg/g)	2.6	Con	83.27±1.100	82.26±1.200	82.08±1.51
		Exp	78.34±1.263	73.20±1.250	68.60±1.250
		t-test	3.15554*	9.02479*	13.2276*
	3.6	Con	83.27±1.100	82.26±1.200	82.08±1.51
		Exp	64.27±1.125	61.36±1.200	59.18±1.160
		t-test	20.9155*	21.325*	23.7115*
	4.6	Con	83.27±1.100	82.26±1.200	82.08±1.51
		Exp	53.22±1.140	46.25±1.060	39.28±1.135
		t-test	32.851*	38.9436*	45.0809*
Cholesterol (mg/g)	2.6	Con	78.21±1.115	77.27±1.100	77.13±1.095
		Exp	75.66±1.340	72.47±1.280	69.44±1.341
		t-test	2.63104*	49.9255*	8.59148*
	3.6	Con	78.21±1.115	77.27±1.100	77.13±1.095
		Exp	67.17±1.100	64.26±1.212	62.8±1.924
		t-test	13.9446*	13.7687*	17.0674*
	4.6	Con	78.21±1.115	77.27±1.100	77.13±1.095
		Exp	58.15±1.010	55.20±1.215	52.74±1.297
		t-test	22.8919*	23.3219*	24.876*
GOT (IU/L)	2.6	Con	41.24±1.110	41.42±1.015	41.64±1.105
		Exp	44.34+1.185	48.39±1.266	52.25±1.250
		t-test	3.41634*	7.44159*	51.2509*
	3.6	Con	41.24±1.110	41.42±1.015	41.64±1.105
		Exp	57.33±1.165	60.55±1.195	64.15±1.135
		t-test	17.3248*	21.1298*	24.7095*
	4.6	Con	41.24±1.110	41.42±1.015	41.64±1.105
		Exp	69.43±1.298	73.85±1.060	77.42±1.197
		t-test	7.5892*	38.2737*	38.2882

Contd...

Table 16.1–Contd...

Parameters	*Concentration %*		*Exposure Period*		
			24	*48*	*72*
GPT (IU/L)	2.6	Con	63.25±1.120	63.50±1.090	64.55±1.073
		Exp	68.48±1.145	72.44±1.235	76.20±1.195
		t-test	5.66252*	9.50667*	12.5604*
	3.6	Con	63.25±1.120	63.50±1.090	64.55±1.073
		Exp	80.27±1.180	84.24±1.145	88.46±1.218
		t-test	18.1149*	22.9796*	24.0346*
	4.6	Con	63.25±1.120	63.50±1.090	64.55±1.073
		Exp	94.38±1.105	98.58±1.205	102.90±1.080
		t-test	26.9789*	37.7855*	48.2285*
ACP (IU/L)	2.6	Con	163.42±1.220	163.10±1.053	162.13±1.101
		Exp	159.95±1.110	155.24±1.105	151.21±1.105
		t-test	4.86098*	9.386698	5.64565*
	3.6	Con	163.42±1.220	163.10±1.053	162.13±1.101
		Exp	147.42±1.3170	143.76±1.274	139.66±1.200
		t-test	15.1029*	18.5242*	23.8999*
	4.6	Con	163.42±1.220	163.10±1.053	16213±1.101
		Exp	135.78±1.100	131.41±1.270	127.35±1.260
		t-test	29.1434*	35.9616*	48.2314*
ALP (IU/L)	2.6	Con	103.34±1.197	103.12±1.110	103.27±1.146
		Exp	102.21±1.176	98.37±1.215	100.13±1.201
		t-test	1.16792*	4.92388*	3.43452*
	3.6	Con	103.34±1.197	103.12±1.110	103.27±1.146
		Exp	91.62±1.205	87.53±1.165	89.15±1.130
		t-test	3.59044*	16.4215*	15.3411*
	4.6	Con	103.34±1.197	103.12±1.110	103.27±1.146
		Exp	80.33±1.180	75.22±1.120	77.32±1.200
		t-test	3.7112*	29.8682*	22.2666*

The liver has vital physiological role to perform under any stress conditions. The liver has numerous specialized biochemical functions. Firstly the toxicant in the system should be metabolished, degraded and eliminated from the organism. Secondly, there is a necessity for increased energy production for physical activities manifested under stress effect. The process of glycogenolysis and glyconeogenesis are utilized. In the later process amino acids from the precursors in fish liver (Premakumari, 1998). Therefore the liver is important. The liver tissues possess enzymatic machinery to carry out the energy production and detoxification. Glycogen is the form in which carbohydrate is in the organism especially in the liver.

In the present investigation, sharp decline in glycogen content of liver was noticed, when the fishes were exposed to the effluent for 24, 48 and 72 hrs exposure at 2.6, 3.6 and 4.6 per cent concentration.

The significant decrease in glycogen content in the liver indicated the excess utilization of the reserve glycogen to withstand heavy metal toxicosis. Similar observation were made by Amudha *et al.* (2002).

Proteins are highly sensitive to pollutants and are one of the earliest indicators of heavy metals poisoning (Jacob *et al.*, 1977). The results of the present investigation showed significant decrease in total protein in the liver. This decrease may be due to lowered protein synthesis indicating a lowered metabolism in these fishes. The findings of Anitha Kumari and Sree Ram Kumar (1996) also supporting the above view.

The liver plays a central role in the regulation of the cholesterol balance of the body (Champe and Harvey, 1995). In the present investigation the cholesterol was found to be decreased in the liver may be due to depressed *de novo* synthesis along with greater mobilization of liver cholesterol for the synthesis for corticosteroids or their conversion to bile acids that form the main route for the ecretion of cholesterol. Similar observation was made by Kondal *et al.*, 1988.

GOT and GPT activity in the present study showed significant increase in liver. GOT and GPT increased during liver damage. Increased activity of GOT and GPT indicates an increased rate of proteolysis in the tissues. Vineeta Shukla and Sastry, 1998 supported the above reason.

The result of the present study has reflected decreased in activity of both acid and alkaline phosphatases significantly in liver. Decreased activity of ACP could be due to inhibition of enzymes. Similar observations were made by Roux and Parker *et al.* (1984). The inhibition of alkaline phosphatase activity in liver was to facilitate the breakdown of glycogen to meet energy demand under effluent stress or other physiological process like transphosphorylation or uncoupling of oxidative phosphorylation as also supported by Sarabadhikary and Sur (1992).

Hence in the present investigation the alteration of glycogen, total protein, cholesterol and enzymes like GOT, GPT, ACP and ALP content can be taken as a meaningful biochemical indices of sugar mill effluent toxicity to assess pollution of the aquatic environment.

References

Amudha, P., G. Sangeetha and S. Mahalingan. 2002. Dairy effiuent induced alterations in the protein, carbohydrate and lipid metabolism of a fresh water teleost fish *Oreochromis mossambicus*. *Poll Res.*, 21(1): 51–53.

Anitha Kumari, S. and N. Sree Ram Kumar. 1996. Effect of water pollution on the biochemical parameters of selected tissues of *Channa striatus*. *J. Ecobiol.*, 8(3): 221–227.

Bebra, B.K. and B.N. Missra. 1985. The effect of sugar mill effluent on enzyme activities of rice seedling. *Industrial Research*, 37: 390–398.

Champe, P. and Havey, R.A. 1994. *Illustrated Reviews-Biochemistry*, 2nd edn. Pamela C. Champe and Richard A. Harvey, J.B. Lippincott Company. Philadelphia.

Chawdury, N. 1977. Worst polluted areas India. *Ind. J. Environ. Hlth.*, 21: 73.

Haniffa, M.A. and S. Arulselvan. 1991. Relative toxicity of bleaching dyeing and mixed textile mill effluent to the fresh water fish *Oreochromis mossambicus* (Trawavesp). *J. Environ. Biol.*, 12: 359–362.

Jacob, J.M., N. Carmichael and J.P. Cavanagh.1977. Ultra structural changes in the nervous system of rabbits poisoned with methl mercury. *Toxicol. Appl. Pharmacol.*, 39: 249–261.

Kemp, A. and Kits Vaheijnigen, A.J.M. 1945. A colorimetric micro method for the determination of glycogen in tissues. *J. Biochem.*, 56(4): 646–648.

Kind, P.R.N. and King, F.J. 1954. Estimation of plasma phosphatase by determination of hydrolysed phenol with amino antipyrine. *J. Clin. Pathol.*, 7: 372.

King, J. 1959. *J. Med. Lab. Tech.*, 16: 265.

Kondal, J.K., Saxena, P.K. and Soni, G.L. 1988. Effect of vegetable oil factory effluent on the liver of a fresh water teleost *Heteropneuates fossilis* (Bloch). A biochemical study. *Biol. Wastes*, 25: 17–24.

Lowry, O.H., Rose Brough, A.L. Farr and R.J. Randall. 1951. Protein measurements with folin-phenol reagent. *J. Biol. Chem.*, 193: 265–275.

Neena Vyas. 1993. Survey of the environment. Govt. stance of hoped and reality. *The Hindu*, p. 1–196.

Palanichamy, S., Baskaran, P. and Balasubramanian, M.P. 1986.

Parker, C.L., Patterson, D.R., Van Gelder, G.A., Valerio, M.G. and Hall, W.C. 1984. Chronic toxicity and carcinogenicity evaluation of fenvalerate in rats. *J. Toxicol. Environ. Hlth.*, 13: 83–97.

Prema Kumari, R. 1988. Some aspects of physiology of fish under nutritional stress. *Ph.D. Thesis*, Osmania University, Hyderabad, A.P., India.

Reitman, S. and Frankel, S. 1957. A colorimetric method for the determination of glutamic oxalo acetic and glutamic pyruvic transaminases. *Amer. J. Clin. Pathol.*, 28: 56–63.

Richmond, W. 1973. Preparation and properties of a cholesterol oxidase from *Nocardia* Sp. and its application to the enzymatic assay to total cholesterol in serum. *Clin. Chem.*, 19: 1350–1356.

Roux, F. Bescol-liver sac J. Guillam, C. and Foornier, E. 1976. Studies on the toxic action of Lindane: Biochemical and ultrastructural modifications of the lysosomal system in cultured hepatocytes. *Eur. J. Toxicol. Environ. Hyg.*, 9: 357–372.

Sarabadikary, A. and R.K. Sur. 1992. Effect of short duration exposure to methyl parathion followed by recovery of activities of some enzymes of fish *Oreochromis niloticus*. *Environ. Ecol.*, 10: 333–340.

Vineeta Shukla and K.V. Sastry. 1998. Acute toxic effects of cadmium on various parameters in the freshwater fish. *Proc. Acad. Environ. Biol.*, 7(2): 155–160.

Weister, W. and S. Hinterleitler. 1980. Serum enzymes in rainbow trout as tools in the diagnosis of water quality. *Bull. Environ. Contam. Toxicol.*, 25: 188–193.

Chapter 17

Impact of Fungicide Validacin-3L on Bioenergetics of the Freshwater Fish Silver Carp *Hypophthalmichthys molitrix*

S. Athikesavan, S. Vincent and B. Velmurugan

Unit of Environmental Health and Biotechnology, PG and Research Department of Zoology, LIFE, Loyola College, Chennai – 600 034, India

ABSTRACT

Fungicide Validacin-3L was investigated in the present study. Silver carp (*Hypophthalmichthys molitrix*) were selected for the bioassay experiments. Fishes were exposed to four sublethal concentrations of validacian-3L for 21 days, and bioenergetics aspects such as food consumption, absorption, convertion, metabolized, efficiency and net conversion efficiency were analysed. The bioenergetics values are reduced compared to control fishes in Validacian-3L treated fishes.

Keywords: *Fungicide Validacin-3L, bioenergetics, Hypophthalmichthys molitrix.*

Introduction

Pollution of aquatic system by chemicals used in industries including those of textiles, tanneries, rayon, paper, synthetic drugs, oil refineries, photofilms, insecticides, fungicides, agricultural drainage water upon entering the water body have a fatal effect on fish (Metelev *et al.*, 1983). Used in fungi control, these fungicides have contaminated the ecosystem and non target species. Also, the fungicides stay in the environment for long period and remain a permanent threat to human health and safety (Revelle and Revelle, 1974).

Aquatic pollutants such as pesticides and metals reduce the food intake in fishes (Arunachalam *et al.*, 1980). Industrial effluents have shown to affect food intake and growth in many fishes (Stones and Livington, 1978). The food consumption was decreased with increasing concentration of superphosphate and the feeding rate was also lower at the higher concentration (Khillare and Wagh, 1987). Feeding rate in *Macrones keletiusi* when exposed to sublethal concentration was observed to decrease with increasing concentration of Rogor (Hameed and Vadamailai, 1986).

Growth is a broad index of various physiological functions and may be affected by decrease in food consumption or absorption and conservation efficiency: an increase in metabolism may also influence the growth rate. Growth directly depends on food consumption. The inverse relationship between the feeding rate and concentration suggests that the dye stuff effluents bring about growth reduction through lowered food consumption (Sakthivel and Sampath, 1989).

This present study was conducted to determine the sublethal effect of bioenergetics status of Validacin-3L in silver carp *Hypophthalmichthys molitrix.* The observation of food utilization parameter *viz.*, food consumption efficiency, absorption efficiency, conversion efficiency, with exposed different concentration of Validacin-3L.

Materials and Methods

The freshwater fish *Hypophthalmichthys molitrix* were obtained from the Tamil Nadu fish seed farm, near Poondy Reservoir. Fish fingerlings (6±1 gm/wt and measuring 5.5±0.5 cm) were selected for the experiment. In the laboratory fishes were acclimated in tap water in aquaria with 100 l capacity for 10 days before the commencement of the experiment. Fish were fed *ad libitum* with fed with commercial pellet. The fungicide Validacin-3L used in static bioassay stock solution was prepared using in double distilled water. For the bioenergetics studies fishes exposed to sublethal concentrations of 0.01, 0.02, 0.03 and 0.04 ppm for 21 days.

The scheme of energy budget followed is IBP formula (Petrusewicz and Macfudyen, 1970), represented as:

Average food consumed = Food given – Unfed

Average food absorption was estimated as the difference between the average food consumed and that of faeces.

Average food absorbed = Average food consumed – Faeces

Growth or food converted was calculated by substracting final weight of the fish and initial weight of the fish.

Average food converted = Final weight of the fish – Initial weight of the fish

Metabolism was calculated as the difference between average food absorbed and average food converted.

Average food metabolized = Average food absorbed – Average food converted

$$\textbf{Absorption Efficiency} = \frac{\text{Absorption rate}}{\text{Consumption rate}} \times 100$$

$$\textbf{Net conversion efficiency} = \frac{\text{Absorption rate}}{\text{Consumption rate}} \times 100$$

Results and Discussion

Food utilization parameters and efficiencies of the fish *Hypophthalmichthys molitrix* alteration in food utilization under the stress of sublethal concentration of the fungicide Validacin-3L was recorded in, Table 17.1. Minimum food consumption was observed in *H. molitrix* treated with at 0.04 ppm, 1.246 g/dry weights after 21 days for exposure of Validacin-3L. Food consumption was maximum found in control fish (1.324mg/g/dry weight) after 21 days. Consumption rate was decreased with increasing concentrations of the fungicide. The decrease in food absorption with increase in concentration of the fungicide, the quantity of faeces egested by the fish decreased as 1.057 to 1.051 mg/g/21 days. Food conversion also decreases with increase in the concentration of the fungicide from 0.112 to 0.080 mg/g/21 days. Similarly with increase in concentration of the fungicide the energy spent on metabolism decreased from 0.963 to 0.943 mg/g/21 days for 0.01 to 0.04 ppm fungicide. The absorption efficiency of the fish decreased with the increase in fungicide concentration from 83.62 to 80.07 for fishes treated with 0.01 to 0.04 ppm respectively. The net conversion efficiency shows decreased in fishes treated with Validacin-3L, the values varies from 8.494 to 6.42 for 0.01 to 0.04 ppm respectively.

Table 17.1: Effect of Validacin-3L on Food Utilization in *Hypophthalmichthys molitrix* Food Consumed, Food Absorbed, Food Converted, Food Metabolized, Absorption Efficiency and Net Conversion Efficiency Expressed in mg/g/21 Days

Different Concentration	*Food Consumed*	*Food Absorbed*	*Food Converted*	*Food Metabolized*	*Absorption Efficiency*	*Net Conversion Efficiency*
Control	1.324±0.003	1.064±0.002	0.123±0.034	0.980±0.01	85.38±0.32	9.07±0.008
0.01	1.296±0.001	1.057±0.094	0.112±0.011	0.963±0.078	83.62±0.310	8.494±0.12
0.02	1.276±0.012	1.054±0.02	0.10±0.01	0.95±0.002	82.28±0.210	7.83±0.521
0.03	1.257±0.01	1.052±0.021	0.090±0.003	0.947±0.02	81.77±0.42	7.15±0.01
0.04	1.246±0.001	1.051±0.02	0.080±0.01	0.943±0.003	80.07±0.062	6.42±0.04

Values are means of three replication ± S.E.

Feeding behavior is often a sensitive indicator of toxicity (Little *et al.*, 1985). Feeding initiation, the major behavioral changes, as observed in the present investigation a reduction in consumption from 1.296 to 1.246 gm/dry weight/21 days of Validacin-3L concentration is comparable to the observations of Bull and McNerney (1974) in the *Coho salmon* exposed to fenitrothion. In the present study lowered food intake which was inversely proportional to toxicants similar to the findings of (Palanisamy and Ravichandran, 1990).

A loss of appetite might have reduced the glucose level and in turn, the feeding (Colgat, 1973). According to Sarkar (1988), fertilizers could cause damage to the buccal epithelium, rendering it unable to detect the palatability of the food material. According to Arunachalam *et al.*, (1980) aquatic pollutants such as fungicides and metals reduce the food intake in fishes, with the increase in the concentration and decreasing body weight food consumption rate for the fish decreased gradually.

The variation in feeding efficiency with difference in concentration of fungicide in accordance with the results of earlier studies (Haniffa and Venkatachalam, 1980), food type is known to significantly affect food utilization. In conclusion the present study reveals that the fungicide Validacin-3L affects the bioenergetics aspects on dose dependent manner in fish silver carp *Hypophthalmichthys molitrix*.

References

Arunachalam, S.K., Jayalakshmi and Abobucker, S. 1980. Toxic and sublethal effects of Carbaryl on fish. *Arch. Environ. Contam. Toxicol.*, 9: 307–316.

Bull, C.J. and McNerney. 1974. Behaviour of Juvenile Coho Salmon *Oncorhynchus mykiss* exposed to sumithion (Fenitrothion) an organophasphate insecticide. *J. Fish. Res. Bd. Can.*, 31: 1872–1876.

Colgat, P. 1973. Motivational analysis of fish feeding behaviour. *J. Fish. Res. Bd. Can.*, 45: 38–66.

Hameed, P.S. and Vadamalai., P. 1986. Effect of sublethal concentration of dimethoate EC 30 on feeding, growth, O_2 consumption and activity in *Macrones kelatins* (Dumeril). *J. Environ. Biol.*, p. 277–284.

Haniffa, M.N. and Venkatachalam. 1980. Effect of food quality on energy budget and chemical composition of the grass carp *Ctenopharyngodon idella. Proceeding of the International Symposium on Conservation: Inputs from Life Sciences.* University Kebangsaan, Malayasia, p. 163–173.

Khillarre, Y.K. and Wagh, S.B. 1987. Effect of Agricultural fertilizer superphosphate on food and growth in fish, *Barbus stigma. Poll. Res.*, 6(1): 25–28.

Little, B.E., Flerou, B.A. and Ruzhinskaya, N.V. 1985. Behavioural approaches in aquatic toxicity investigation, a review. In: *Toxic Substances in the Aquatic Environment: An International Aspect*, (Eds.) Jr. P.M Mehrle, R.H. Gray and R.C. Kendal. Amer. Fish. Soc., Water Quality Section, Bethesda, M.D., p. 72–98.

Metelev, V.V., Kansev, A.K. and Dzasokhova, N.G. 1983. *Water Toxicology*, Amerind, New Delhi, pp. 216.

Palanisamy, R. and Ravichandran, P. 1990. Sublethal effects of Diammonium phosphate on food utilization in *Labeo rohita. Proc. Sump. Trends in Ectoxical.*, p. 279–284.

Petrusewicz, K. and Macfadyen, 1970. *Productivity of Terrestrial Animals: Principles and Methods.* Black Well, Oxford.

Revelle and Revelle. 1974. *Sourcebook on the Environment.* Houghton Miffin Company, London, p. 227–228.

Sakthivel, M. and Sampath, K.1989. Effect of textile dye stuff effluent on food utilization of *Cyprinus carpio* (Linn). *Ind. J. Exp. Biol.*, 27: 1032–1034.

Sarkar, S.K. 1988. Influence of Ammonium sulphate on the feeding rate of fish under multivariate temperature. *Comp. Physiol. Ecol.*, 13(1): 30–33.

Stoner Allam and Livingston, V. 1978. Respiration, growth and food conversion efficiency of Pin Fish (*Lageodon rhomboids*) exposed to bleaked kraft mill effluent. *Environ. Poll.*, p. 207.

Chapter 18

Evaluation of Sensory and Proximate Composition of the Fish *Cyprinus carpio* at Refrigeration Storage (4°C)

V. Indira, A. Kumaresan** and P. Dhasarathan****

**Department of Biochemistry, Sri Parasakthi College, Courtallam – 627 802*
***Head and Professor, ICAR Centre, Sri Parasakthi College, Courtallam*
****Department of Biotechnology, Sri Kaliswari College, Sivakasi – 627 432*

ABSTRACT

Proximate composition and sensory evaluation of *Cyprinus carpio* var. communis (Linn.) during refrigeration storage at 4°C was studied for a period of 21 days at an interval of 3 days. Proximate composition study showed a slight decrease in carbohydrate, protein and lipid content. Total volatile base nitrogen (TVB–N) increased from 18 to 69.0 mg per cent (r + 0.9978) and free fatty acid value was 0.0766 per cent per g fat to 0.1469 per cent per g fat from the beginning and completion of refrigeration storage. These values indicate that the quality of refrigerator stored *C. carpio* is quite good and the fat oxidation of this species during refrigerator storage is very minimum *i.e.*, in initiation phase. The results of the whole study indicate that despite some changes in the proximate composition and sensory properties during prolonged refrigeration storage at 4°C, the *C. carpio* appeared to be suitable for refrigerator storage preservation.

***Keywords**: Total volatile base nitrogen, Fatty acid, Proximate composition, Cyprinus caprio and Sensory evaluation.*

Introduction

Cyprinus carpio (Cypriniformes, Cyprinidae) is a popular and delicious fish species in Tamil Nadu, India. Anglers readily appreciate it. It occupies third place in aquaculture production. Its total yield in 1992 was over 1200,435 metric tones (FAO, 1992). *C. carpio* production was 13705 metric tones in 1996 (FAO, 1998). The major parts of carps are consumed locally in the producing countries mainly China and India. Common carp can be reared in polyculture systems that make optimum use of the natural productivity of ponds and water bodies where they are stocked. They have good markets in Asian countries due to traditions and relatively low prices.

Major portion of the catch is sold as iced fish. During harvesting season when the supply exceeds the demand in the market there are reports of post harvest loss. Some time the catch is so high that the ice making plants fail to fulfill the requirement of ice necessary for short-term preservation of *C. carpio* fish to transport the catch in distant places. As a result a sizeable quality of the total catch losses much of its taste, flavor and nutritional quality by the time it reaches the consumers (Kim *et al.*, 1995). The present study was undertaken to have information on the suitability of refrigeration storage of *C. carpio* on commercial scale.

Materials and Methods

Cyprinus carpio samples were collected from the fish culture pond of Government fisheries Department Manimuthar, Tamil Nadu state, India. The fish were brought to the laboratory using sterile plastic buckets containing pond water where the fish live. On arrival to the laboratory, the fish were killed by shock treatment. The fish samples were divided into three lots as shown below

1. Raw whole fish were taken in sterile plastic trays and stored in the refrigerator (4±1°C).
2. The second lot was processed as fillets, packed and stored at refrigerator (4±1°C).

Three lots of samples were evaluated for proximate composition analysis at 0, 1, 2, 3, 6, 12 and 21 days.

Initial freshness of the raw fish was evaluated organoleptically using grade and score system and refrigerator stored samples by inspecting the general appearance, colour, taste and flavor. Proximate composition (protein, lipid, ash and moisture), fat/lipid quality (peroxide value and free fatty acids value) was determined according to the method of AOAC (1975) and the results of FFA were expressed as percentage of Oleic acid. Total volatile base nitrogen (TVB–N) was determined according to the Micro diffusion technique of Conway (1950).

Results and Discussion

Organoleptic evaluation of refrigerated raw whole fish showed good condition until three days of storage. The fish appeared acceptable and remained edible for three days. Under this environmental condition, spoilage occurred within four to six days. Generally, the fish fillets are considered as acceptable to human consumption until the sensory score reached 3 with the following characteristics: Slight loss of bright colour of skin, moderately soft texture, curdy/fruity and red colouration in a portion of fillet. In the present study, organoleptically the shelf life of fillets stored at room and refrigeration temperatures are eight hours and six to eight days respectively (Table 18.1).

Changes in the protein, carbohydrate and fat contents of raw whole fish *C. carpio* during storage at refrigeration temperature are presented in Table 18.2. The protein, carbohydrate and fat contents of raw whole fish decreased from 16.75 to 14.50 per cent, 2.30 to 1.90 per cent and 3.13 to 2.32 per cent respectively. The result of the present study showed that there was a steady decrease in protein, carbohydrate and fat throughout the storage period at refrigeration temperature, negatively correlated with the storage period and the respective values are r = –0.89, –0.85 and –0.80. Similar to raw whole fish stored at refrigeration temperature. The protein, carbohydrate and fat contents of fish fillets stored at refrigeration temperature got lowered from 16.75 to 14.50 per cent, 2.30 to 1.90 per cent and 3.13 to 2.32 per cent respectively and negatively correlated with the period of storage. The correlation co-efficient are r = –0.90, –0.82 and –0.98 respectively. Reduction of these values during storage of raw whole fish was higher than the fish fillets stored at refrigeration temperature.

Table 18.1: Sensory Score of *Cyprinus carpio* Fillets During Storage at Refrigeration Temperature

Storage Period (Days)	*Skin Appearance*	*Texture*	*Odour*	*Discolouration*	*Overall Sensory Score*
0	4.50	5.00	5.00	5.00	4.88
3	4.00	4.50	4.00	4.50	4.25
6	4.00	4.00	3.50	3.50	3.75
9	3.50	3.00	2.50	2.50	2.88
12	2.50	2.50	2.00	2.00	2.25
15	2.00	2.00	1.50	1.50	1.75
18	2.00	1.50	1.00	1.00	1.38
21	1.50	1.00	1.00	0.50	1.00

Table 18.2: Proximate Composition of *Cyprinus carpio* Muscle Tissue During Refrigeration Storage

Storage Period (Days)	*Proximate Composition (%)*			
	Moisture	*Carbohydrate*	*Protein*	*Fat*
0	79.24±0.20	2.30±0.03	16.75±0.14	3.13±0.10
1	80.13±0.10	2.30±0.02	16.62±0.12	3.13±0.06
2	80.73±0.12	2.15±0.04	15.96±0.11	3.10±0.04
3	81.25±0.04	2.08±0.07	15.78±0.09	3.02±0.02
6	81.65±0.13	1.96±0.06	15.24±0.18	2.95±0.01
12	81.07±0.09	1.94±0.02	15.12±0.14	2.50±0.14
21	82.21±0.12	1.90±0.01	14.50±0.12	2.32±0.12

Changes in the total volatile base nitrogen content of the fish fillets during storage at refrigeration temperature are presented in Table 18.3. The total volatile base nitrogen content of fish fillets of *C. carpio* was varied from 18.00 to 69.50 mg per cent for the samples stored in refrigeration temperature. The result of the present study showed that there was a constant increase in TVB-N values throughout the period of observation in refrigeration temperature. There was an increase in the TVB-N values were found to be the highest in fillets stored at refrigeration temperature storage. Positive correlation was established (r = + 0.95, 1.00) at refrigeration temperature.

Table 18.3: Changes in Total Volatile Base Nitrogen (TVB-N) and Free Fatty Acids (FFA) in Fillets of *Cyprinus carpio* During Storage at Refrigeration Temperature

Storage Period (Days)	*Storage at Refrigeration Temperature/4°C*	
	TVB–N (mg%)	*FFA % per g Fat*
0	18.00±0.65	0.0766±0.01
3	24.00±0.62	0.0998±0.01
6	30.00±0.88	0.1014±0.02
9	38.50±0.68	0.1190±0.03
12	48.00±1.05	0.1232±0.01
15	56.50±0.88	0.1330±0.04
18	62.00±0.94	0.1414±0.02
21	69.50±1.01	0.1469±0.01

Oxidation and hydrolysis of lipid in fish during storage cause quality deterioration. Lipid hydrolysis results in the formation of free fatty acids. Changes in free fatty acid values of fish fillets of *C. carpio* stored at refrigeration temperature are presented in Table 18.3. In the present study, the FFA values increased significantly with storage period of refrigeration temperature. The rise in FFA was found to be more (0.1336 per cent per g fat at fifteenth day) and statistically these values were positively correlated ($r = +0.78, +0.98$) with storage period. The result of the present study revealed that minimum FFA on initial day (0.0766 per cent per g fat) with steady rise (0.1469 per cent per g fat) till 21st day of storage. Statistically, this rise in FFA levels between various periods of storage is highly significant. The quality of fish will be very poor if the fish is not stored properly. Fish is usually subjected to a variety of processing procedures before consumed. The effects of these processes on nutritive value are important. Consumers are reluctant about some aspects of handling fish.

In the present study, the fish fillets were stored at refrigeration temperature. Changes in the average organoleptic scores of the fish fillets stored at refrigeration temperature are shown in Table 18.1. From these scores, the fish fillets were acceptable for six to eight days at refrigeration temperature. Similar observation was made with aerobically packed fillets of tilapia (Reddy *et al.*, 1994) and *Lethrinus* sp. (Shanmugam *et al.*, 2000) that were spoiled after nine, seven and eight to nine days respectively when stored at 4°C. Odour and appearance on untreated catfish fillets resembled as that of fresh fillets upto three days and became unacceptable after this period (Kim *et al.*, 1995). Ramachandran *et al.* (1990) reported that the results of the visual and olfactory assessment of *Hilsa toli* had a correlation organoleptic changes and the microbiological quality. It is generally agreed that spoilage of catfish fillets occurs within five to eight days during refrigerated storage (Kim *et al.*, 1995). However, the shelf life of same fillets increased when they were acid dipped and stored under refrigeration (Farid *et al.*, 1998).

Proximate composition changes in the fish samples during storage and spoilage occur predominantly on the major constituents like moisture, carbohydrate, protein and fat. Moisture level of fish plays an important role in bacterial spoilage. Lowering of moisture retards the spoilage. The result of the present investigation indicated that the moisture content of the raw whole fish and fillets stored at refrigeration temperature were increased whereas carbohydrate, protein and fat were decreased during the storage period.

Increase in moisture and decrease in protein content were observed in spotted seer during ice storage by Shenoy and James (1974). Devadasan and Nair (1977) also reported decrease in protein fraction in fish muscle during ice storage. Devadasan and Nair (1977) observed lipid breakdown in oil sardine stored at refrigeration temperature. Oxidation and hydrolysis of lipid presenting fish cause quality deterioration during storage, which leads to the formation of free fatty acids (Huss, 1971). Olley and Dunean (1965) studied the lipid hydrolysis by fish muscle tissue lipases and bacterial lipases. In the present investigation, the FFA value increased significantly with storage period. However, the changes in FFA correspond to the bacterial population as well as organoleptic scores. Removal of reserve fat from the flesh resulted in a product that developed less off flavor and odour during cold storage. The present investigation also emphasized that further depletion of flesh fat reserves resulted an increase in the water content and free fatty acids.

Acknowledgement

The authors are thankful to the Secretary, Management and Principal, Sri Kaliswari College and Principal, Sri Parasakthi College for providing facilities to carry out the work.

References

AOAC. 1975. *Official Methods of Analysis*, 12th edn. (Ed.) Horwitz, W. Association of Official Analytical Chemists, Washington, p. 1–1094.

Conway, E.J. 1980. *Microdiffusion: Analysis and Volumetric Error.* Crossby, Loockwood and Sons, London.

Devadasan, K. and Nair, M.R. 1977. Studies on changes in protein fractions on fish muscle during storage in ice. *Fish Technol.*, 14(2): 127–130.

FAO. 1992. *FAO Yearbook of Fishery Statistics* Vol. 74. Food and Agriculture Organization of the United Nations, Rome, pp. 667.

FAO. 1998. *FAO Yearbook of Fishery Statistics Capture Production*, 86/1.

Farid, M., Bala, A and Douglas L. Marshall. 1998. Organic acids dipping of catfish fillets; Effect on colour, microbial load and Listeria monocytogenes. *J. Food Prot.*, 61(11): 1470–1474.

Huss, H.H. 1971. Prepacked fresh fish. In: *Fish Inspection and Quality Control,* (Ed.) Kreyzere, R. Fishing News (Books) Ltd., London, p. 60–65.

Kim, C.R., Hearnsberger, J.O., Vickery, A.P., White, C.H and Marshal, D.L. 1995. Sodium acetate and bifidobacteria increase the shelf life of refrigerated catfish fillets. *J. Food Science*, 60(1): 25–27.

Olley, J. and Duncan, R.H. 1965. Lipids and protein denaturation in fish muscle. *J. Sci. Food. Agri.*, 16: 99.

Ramachandran, A., Bedonia, R. and Viswanathan Nair, P.G. 1990. Effect of delayed icing on the microbiological quality of fish *Hilsa toil. Fish. Technol.*, 27: 31–35.

Reddy, N.R., Schreiber, K.S., Bazard, K.S., Skinner, G.E and Armstrong, D.J. 1994. Shelf life of fresh tilapia fillets packaged in high barrier film with modified atmospheres. *J. Food. Sci.*, 59(2): 260–264.

Shanmugam, S.A., Shalini, R. and Indra Jasmine, G. 2000. Sensory and proximate composition characteristics of sodium acetate and potassium sorbate treated vacuum-packed *Lethrinus lentjan* fillets. *Indian J. Microbiol.*, 40: 113–117.

Shenoy, A.V. and James, M.A. 1974. Spoilage of spotted seer *Scomberomorous guttatus* during ice storage. *Fish. Technol.*, 11: 67–72.

Chapter 19

Pesticidal Stress Influenced Respiratory Alterations in the Freshwater Fish, *Mystus vittatus*

R. Sonaraj, A.J.A. Ranjit Singh** and A. Pushparaj**

**Reader, Research Department of Zoology, Kamaraj College, Thoothukudi*

***Reader, Department of Biology, S.P.K. College, Alwarkurichi*

ABSTRACT

The fish *Mystus vittatus* exposed to two sublethal concentrations of parrysulfan (0.1672 ppm and 0.3345 ppm) and organochlorine and two sub-lethal concentrations of sicocil (0.335 ppm and 0.6717 ppm) and organophosphate pesticide showed an elevated consumption of oxygen on the very first day. The elevated change in parrysulfan medium was 15.85 per cent and 18.23 per cent in 0.1672 ppm and 0.3345 ppm respectively and 11.52 per cent and 15.19 per cent in the 0.3358 ppm and 0.6717 ppm in sicocil medium respectively. As the duration of exposure extends beyond 24 hrs, the rate of oxygen consumption showed a decreasing trend. The percentage decline in oxygen consumption in the parrysulfan exposed fish was –8.74, –24.91, and –30.28 after 10, 30 and 60 days of exposure in the concentration of 0.1672 ppm and –10.44, –26.06 and –36.35 after 10, 30 and 60 days of exposure in the 0.3345 ppm respectively. The same trend was noticed in sicocil exposed fish also and the percent reduction in oxygen consumption was –7.52, –22.34 and –26.14 after 10, 30 and 60 days of exposure in 0.3358 ppm and –9.52, –23.28 and –28.45 after 10, 30 and 60 days of exposure in 0.6717 ppm concentration. The decline in oxygen consumption was higher in the fish exposed to parrysulfan medium than to sicocil added medium.

Keywords: *Parrysulfan, Sicocil, Oxygen consumption, Mystus vittatus.*

Introduction

Among different environments aquatic environment is the serious sufferer of the damage caused by manmade and natural chemicals. The fauna of aquatic environment affected by pesticides is a subject of great concern. The pesticides are drained into the nearby pond and riverine system during monsoon and the land run off. The indiscriminate and widespread use of pesticides and their subsequent passing into the aquatic system question the very survival and the behavioural, physiological and biochemical functioning of aquatic animals, particularly fishes of commercial import. The extent and degree of their harmful effect on fish can be gauged to a greater extent by experimental studies in the laboratory.

Studies on oxygen consumption form a useful tool in the assessment of toxicant stress on the aquatic organisms and give an index of energy expenditure mechanisms for environmental variations. Hence, in the present study, an attempt is made to study the oxygen consumption of a catfish, *Mystus vittatus,* exposed to two pesticides commercially called Parrysulfan (Organochlorine) and Sicocil (Organophosphate).

Materials and Methods

The fish, *Mystus vittatus.* of the size 5–6 gms were collected from the laboratory acclimatized stock. The fish were fed on *ad libitum* with pellets prepared from groundnut oil cake and rice bran. The excreta of fishes were removed periodically (every one hour) without disturbing the fish.

Respiratory measurements were made by the method of Ganapati and Prasada Rao (1960) and dissolved oxygen was estimated by Winkler's method (Welsh and Smith, 1953). The fishes exposed to two different sublethal concentrations of Parrysulfan (0.1672 ppm and 0.3345 ppm) and Sicocil (0.3358 and 0.6717 ppm). The oxygen consumption of the exposed fish to different sublethal medium was analysed on 1st, 10th, 30th and 60th days of exposure.

The oxygen consumption was measured initially in the pesticide free medium and test concentrations separately. Prior to each experiment, the medium was renewed and then the oxygen consumption of the fish was estimated. The fish were not disturbed during the course of the experiment. A total of 5 fish was exposed separately in each of the different concentrations to obtain the average oxygen consumption. Triplicates were maintained. All the experiments were carried out at temperature of 29±1°C. The results were presented as ml of oxygen consumed/g/hr. A test of significance on the difference in the mean values between the experimental fish and control fish, using correlation coefficient and regression equation was made.

Results

The effect of oxygen consumption on the fish, *Mystus vittatus,* exposed to various sublethal concentrations of Parrysulfan and Sicocil is presented in Table 19.1. The rate of oxygen consumption got elevated on the 1st day of release of fish into the pesticide medium. The elevated change in Parrysulfan medium was 15.85 per cent and 18.23 per cent in 0.1672 ppm and 0.3345 ppm respectively and 11.52 per cent and 15.19 per cent in the 0.3358 ppm and 0.6717 ppm of Sicocil medium respectively. As the duration of exposure extends beyond 24 hrs a negative trend was observed in oxygen consumption. The percentage decline in oxygen consumption in the Parrysulfan exposed fish was –8.74, –24.91 and –30.28 after 10th, 30th and 60th days of exposure in the concentration of 0.1672 ppm and –10.44, –26.06 and –36.35 after 10th, 30th and 60th days of exposure in the 0.3345 ppm respectively. In the Sicocil exposed medium the percentage reduction was –7.52, –22.34 and –26.14 after 10th, 30th and 60th days of exposure in 0.3358 ppm and –9.52, –23.28 and –28.45 after 10, 30 and 60 days of exposure in the 0.6717 ppm concentration respectively. The decline in oxygen consumption was higher in the fish exposed to Parrysulfan that to Sicocil.

Table 19.1: Changes in the Level of Oxygen Consumption (mg O_2 consumed/g/h) of the Fish *Mystus vittatus* Exposed to Sublethal Concentrations of *Parrysulfan* and *Sicocil.* Values in the parentheses indicate percentage of change over cont. Correlation coefficient 'r' and regression equation are also indicated.

Pesticide	*Concentration ppm*	*Oxygen Consumption/ml/g/h*			
		1 Day	*10 Days*	*30 Days*	*60 Days*
Parrysulfan	Control	0.3555±0.0053	0.3499±0.0024	0.3415±0.0048	0.3352±0.0038
	0.1672	0.4119±0.0062	0.3193±0.0042	0.2564±0.0050	0.2323±0.0057
		(+15.85)	(–8.74)	(–24.91)	(–30.28)
		r=0.7692	r=0.7957	r=0.7633	r=0.3716
		Y=0.0944+0.8930X	Y=0.1665–1.3884X	Y=0.0165–0.7992X	Y=0.0437+0.5627X
		t= 13.8024, p < 0.05 S	t=12.6558, p < 0.05 S	t=24.6581, p < 0.005 S	t=30.22422, p < 0.005 S
	0.3345	0.4203±0.0106	0.3134±0.0022	0.2524±0.0043	0.2120±0.0064
		(+18.23)	(–10.44)	(–26.06)	(–36.35)
		r=0.7643	r=0.8276	r=0.7112	r=–0.4385
		Y=0.1223–1.5262X	Y=0.5816–0.7665X	Y=0.0311+0.6481X	Y=0.8533–1.9131X
		t=10.881, p < 0.05 S	t=22.3054, p < 0.05 S	t=27.5981, p < 0.05 S	t=14.6767, p < 0.05 S
Sicocil	0.3358	0.3964±0.0068	0.3236±0.0033	0.2652±0.0075	0.2461±0.0063
		(+11.52)	(–7.52)	(–22.34)	(–26.14)
		r=0.3988	r=–0.0354	r=–0.4065	r=0.1656
		Y=0.2143+0.5124X	Y=0.3395–0.0455X	Y=0.0484+0.6350X	Y=0.1522+0.2801X
		t =9.4423, p < 0.05 S	t=12.9981, p < 0.05 S	t=17.2457, p < 0.05 S	t=24.2022, p < 0.02 S
	0.6717	0.4095±0.0067	0.3166±0.0041	0.2620±0.0041	0.2384±0.0052
		(+15.19)	(–9.52)	(–23.28)	(–28.45)
		r=0.7080	r=0.6757	r=–0.5260	r=–0.5494
		Y=0.0951+0.8824X	Y=0.0818–1.1384X	Y=0.4160–0.4507X	Y=0.4945–0.7640X
		t=12.6386, p < 0.05S	t=14.1558, p < 0.05S	t=25.3132, p < 0.05S	t=30.1476, p <0.05S

The correlation coefficient 'r' estimated for oxygen consumption as a function of duration of exposure confirmed that the decrease in oxygen consumption was statistically significant, in all cases except for the 1st day of exposure. The 't' value estimated confirmed that there was a significant difference ($p < 0.05$) in the intake of oxygen with that of the control value in all concentrations and duration of exposure. The regression equations derived indicated that there was an inverse relationship between the oxygen consumption and duration of exposure to the pesticide irrespective of the concentration (Table 19.1).

Discussion

Oxygen intake by an organism is an important physiological event. The quantum of intake of oxygen determines the physiological status and metabolic activity of the animal. The oxygen consumption not only indicates the metabolic rate, but it also provides an index of stress condition (Mount, 1962; Ferguson and Bingham, 1966; and Hutner *et al.*, 1968). The intimate contact of the gills with the polluted water may lead to alteration in normal respiratory mechanism, lower the diffusion mechanism through gill and thereby oxygen consumption is reduced, which creates a physiological imbalance to the organism (Finney and Berman, 1976). Accelerated rate of oxygen consumption of *Anabas scandens* after exposure to sumithion was reported by Natarajan (1980). Rao *et al.* (1980) showed that in case of *L. rohita* oxygen consumption increased progressively with increase in the concentration upto certain level and decreased with further increase of toxicant until death ensured. Roy and Datta (1988) reported that the *Cirrhinus mrigala* exposed to malathion showed a decreasing trend in oxygen intake in relation to concentration and duration of exposure. Rafia and Uma (1995) stated that the oxygen consumption of *Mystus gulio* was altered due to exposure to different concentrations of copper sulphate and zinc sulphate. According to Saxena and Chauhan (1996) the reduction in oxygen consumption in *Labeo rohita* exposed to effluent was becanse of coagulation of gill mucus, which has resulted in asphyxiation as well as inhibition of enzymes at mitochondrial level due to the toxic components in the effluent Devi (2000) reported that the oxygen consumption in *O. mossambicus* exposed to endosulfan showed an initial increase in the sublethal and medium lethal concentrations followed by a decline in the subsequent hours.

In the present study, the oxygen consumption of the fish, *Mystus vittatus*, exposed to two sublethal concentrations of Parrysulfan and Sicocil showed an increasing trend during the first day of exposure and a declining trend as a function of the duration of the exposure. The increase in oxygen intake during the first day of exposure in the present study might be due to behavioural manifestation like hyperactivity and escape behaviour, which was observed after exposure to Parrysulfan and Sicocil stress. The decrease in the rate of oxygen consumption after exposure to Parrysulfan and Sicocil for 10, 30 and 60 days might be due to the sluggishness of the fish as a result of the pesticide stress and also the secretion of excessive mucus which formed a thin film over the gill thereby preventing absorption of oxygen during the process of gaseous exchange.

References

Devi Swetharanyam. 2000. The effects of endosulfan on oxygen consumption of the fish *O. mossambicus*. *J. Ecotoxicol. Environ. Monit.*, 10(1): 21–24.

Ferguson, D.E. and C.R. Bingham. 1966. The effect of combination of pesticides on susceptible and resistant mosquito fish. *Bull. Environ. Contam. Toxicol.*, 1: 97.

Finney, L. and E.R. Berman. 1976. Studies in oxygen uptake. *Invest. Opthalmol.*, 15: 789.

Ganapati, P.N. and D.G.V. Prasada Rao. 1960. Studies on the respiration of barnacles. Oxygen uptake and metabolic rate in relation to body size in *Balanus amphitrite communis* (Darwin). *J. Anim. Morph. Physiol.m* 7: 27–31.

Hutner, J.Y., B.F. Dowden and H.J. Benett. 1968. The effect of endrin on the oxygen consumption of the blue gill sun fish *Lepomis macrocharus*. *Proc. Ind. Acad. Sci.*, 90: 80–86.

Mount, D.I. 1962. Chronic effect of endrin on blunt nose minnows and guppies. *U.S. Fish Wild. Ser. Res. Resp.*, 58: 1–38.

Natarajan, G.W. 1980. Effect of sumithion on oxygen consuption of *Anabas scandens*. *Geobios*, 7: 128–130.

Rafia Sultana and V. Umadevi. 1995. Oxyen consumption in a cat fish, *Mystus gulio* (Ham.) exposed to heavy metals. *J. Environ Bio.*, 16(3): 207–210.

Rao, D.M.R., A.P.Devi and A.S. Murthy. 1980. Relative toxicity of endosulfan, its isomers and formulated production to the fresh water fish *Labeo rohita*. *J. Toxicol. Environ. Hlth.*, 6: 323.

Roy, P.K. and J.S. Dattamunshi. 1988. Oxygen consumption and ventilation rate of a freshwater major carp *Cirrhinus mrigala* (Ham) in fresh and malathion treated waters. *J. Environ. Biol.*, 9(1): 5–13.

Sexana, K.K. and R.R.S. Chauhan. 1996. Oxygen consumption in fish *Labeo rohita* (Ham) caused by distillery effluent. *Poll. Res.*, 15(3): 277–280.

Welsh, J.H and R.I. Smith. 1953. *Laboratory Exercises in the Invertebrate Physiology*. Burgess Publishing Co., Minneapolis, pp. 126.

Chapter 20

Effect of Stocking Density on the Blood Parameters of Goldfish *Carassius auratus*

A. Elezabeth Mary and M. Sakthivel***

**St. Mary's College, Tuticorin – 628 001, Tamil Nadu, India*

***Kamaraj College, Tuticorin – 628 003, Tamil Nadu, India*

ABSTRACT

The impact of stocking density on the blood parameters was studied in *Carassius auratus*. The group reared in the density of 10 individuals/aquarium showed maximum values of all parameters studied. The other groups 1, 5, 20 and 40 individuals/aquarium displayed lower blood values in relation to the density and exposure period. The change in the total RBC of the group with 20 individuals/aquarium was not significant. The RBC reduction was significant in the group with 40 individuals/aquarium. Haemoglobin content, Haemotocrit Thrombocytes and Coagulation time also showed similar trend. But MCV, MCH, MCHC values showed increase. Total WBC count of the higher density group was reduced. Relatively the lymphocytes showed a significant increase in higher density groups.

Keywords: *Haematological parameters, Stocking density, Carassius auratus.*

Introduction

In fish farming practice, high density culture is often opted to maximize the available water resources to enhance the yield. The interaction of stocking density and water turnover is studied by many authors (Andrews *et al.*, 1971; Sampath, 1985; Jena *et al.*, 2001). Reports on the impact of water quality parameters and other toxic products of metabolism on growth of fish in high density culture are also available (Armstrong and Boyd, 1982; Soderberg *et al.*, 1983; James and Sampath, 2003). Sakthivel (1993) reported the density effect on the blood parameters of *Cyprinus carpio*. Similar studies in ornamental fish is almost lacking. Moreover, the effect of environmental factors on blood will be expressed in advance of reduced growth (Wedemeyer, 1996a). Considering these, the present study was undertaken to ascertain the effect of stocking density on the blood parameters of gold fish, *Carassius auratus*.

Materials and Methods

Carassius auratus (10±2g live weight) were procured from an ornamental fish farm at Sawyerpuram, Tuticorin District, Tamil Nadu and acclimated for 60 days on a pelleted diet with 35 per cent crude protein prepared according to the square method (Hardy, 1980). Test fish were divided into 5 density groups with 1, 5, 10, 20 and 40 individuals/aquarium of 120 L capacity. Triplicates were maintained for each group. Experiments were conducted in glass aquaria. The test fish were fed twice daily at 08.00 and 17.00 hours at 5 per cent body weight every time. The following experimental parameters were maintained throughout the experiment; Temperature 26±1°C; pH 7.1; D.O > 8 mg/l and photoperiod (12LL : 12DD).

Haematological parameters were estimated at a 30 days interval for a period of 120 days. Blood samples were collected in capillary tubes by severing the caudal peduncle. Wintrobe's method (1978) was adopted to estimate the chosen blood parameters. Red cell indices were calculated according to Johansson-sjobeck and Larson (1979). Differential count of WBC was made by classical dry smear method. The relative percentage of differential Leucocyte count was made based on the count of 1000 cells from two imprints from each fish.

Results and Discussion

The changes observed in RBC and related parameters of *C. auratus* are given in Table 20.1. The group reared in the density of 10 individuals/aquarium showed maximum values of all blood parameters tested. But the other groups displayed gradual reduction in relation to the density and exposure period. In groups with the density of 20 or 40 individuals/aquarium the RBC count decreased from 1.64 to 0.56 × 10^6/mm^3 or 0.34 × 10^6/mm^3 during 120 days exposure. The change in RBC of the group with 20 individuals/aquarium was not statistically significant ($t = 2.9$; $P > 0.05$). But the RBC reduction was statistically significant in the groups with 40 individuals/aquarium ($t = 3.68$; $P < 0.05$). Similar trend was observed in Hb content and haematocrit values. On the other hand, MCV, MCH and MCHC values showed significant increase in higher density groups.

Total WBC decreased from 52.6 to 32.6 × 10^3/mm^3 ($t = 7.21$; $P < 0.01$) at the end of 90 days in the highest density of group of 40 individuals/aquarium. Similar decrease was observed in thrombocytes and coagulation time. Such significant variations occurred only after 120 days in the other groups with 5 or 20 individuals per aquarium whereas, the lymphocytes displayed an opposite trend, for example, in higher density groups (20 or 40 individuals/aquarium) lymphocyte percentage was increased from 37.7 to 67.6 per cent and 37.9 to 71.6 per cent respectively during 120 days exposure (Table 20.2).

The present observations clearly indicate that the increase in density of *C. auratus* has a significant impact on blood parameters. The blood parameters in the smaller groups (1 or 5 individuals) are not much affected. The intermediate group (10 individuals/aquarium) shows fairly good results when compared with other groups of 10 individuals/aquarium. But when the stocking density is further increased from 10 to 40 the RBC is decreased from 1.64 to 0.34 × 10^6/mm^3 and the total WBC is reduced from 52.6 to 32.6 × $10^3/mm^3$ during 90 days exposure. This shows that at higher densities the crowding stress over rides the benefits of optimum density.

Table 20.1: Effects of Various Stocking Density on RBC and Related Values in *Carassius auratus*. Each value is the average ($\overline{X}$ ± SD) of 6 individual observations.

Blood Parameters	*Exposure Period (days)*	*Density (No. of individuals/aquarium)*				
		1	*5*	*10*	*20*	*40*
RBC count	0	1.63±0.11	162±0.18	1.64±0.15	1.64±0.19	1.64±0.25
(X 10 ^{6}/mm^3)	30	1.58±0.18	1.50±0.26	1.62±0.22	1.31±0.21	1.14±0.32
	60	1.51±0.31	1.32±0.36	1.63±0.18	0.99±0.25	0.84±0.28
	90	1.29±0.25	1.18±0.28	1.64±0.26	0.66±0.31	0.45±0.21
	120	1.18±0.32	1.06±0.32	1.62±0.32	0.56±0.34	0.34±0.18
HB content	0	6.81±0.21	6.78±0.42	6.84±0.32	6.83±0.12	6.82±0.31
(g %)	30	6.67±0.32	6.48±0.21	6.83±0.42	6.10±0.36	6.06±0.18
	60	6.52±0.28	5.81±0.18	6.81±0.15	5.26±0.19	4.81±0.21
	90	5.58±0.18	5.33±0.26	6.82±0.28	3.63±0.16	2.78±0.28
	120	5.26±0.16	4.62±0.12	6.79±0.19	3.17±0.20	2.30±0.33
Haematocrit	0	26.23±1.23	26.42±1.24	27.36±1.18	27.64±1.29	27.68±1.18
(%)	30	25.48±1.34	24.86±1.28	27.62±1.15	22.73±1.26	21.84±1.21
	60	24.52±1.39	22.41±1.38	27.51±1.27	18.85±1.23	16.31±1.24
	90	21.18±1.14	19.18±1.35	27.49±1.33	11.80±1.22	9.05±1.32
	120	19.86±1.15	16.25±1.36	27.52±1.35	10.22±1.11	6.94±1.36
MCV (fl)	0	160.92±1.31	163.09±1.14	166.83±1.11	168.54±1.36	168.78±1.61
	30	161.27±1.62	165.73±1.31	170.49±1.21	173.51±1.21	191.58±1.82
	60	162.38±1.91	169.77±1.76	168.77±1.36	190.40±1.19	194.17±1.73
	90	165.47±1.21	171.02±1.69	167.62±1.55	178.79±1.53	201.11±1.32
	120	168.31±1.43	177.36±1.48	169.88±1.90	182.5±1.28	204.12±1.18
MCH (pg)	0	41.78±0.84	41.85±0.72	41.71±0.63	41.65±0.92	41.59±0.68
	30	42.22±0.64	43.20±0.68	42.16±0.72	46.56±0.61	53.16±0.71
	60	43.18±0.38	44.02±0.32	41.78±0.81	53.13±0.72	57.26±0.36
	90	43.59±0.09	45.17±0.52	41.59±0.58	55.00±0.31	61.78±0.63
	120	44.58±0.92	48.87±0.19	41.91±0.29	56.60±0.42	67.65±0.32
MCHC (g/l)	0	25.96±0.28	25.66±0.78	25.01±0.72	24.71±0.82	24.64±0.94
	30	26.18±0.72	26.07±0.62	24.73±0.66	26.84±0.64	24.75±0.62
	60	25.59±0.66	27.26±0.58	24.76±0.51	27.90±0.58	29.49±0.78
	90	27.46±0.65	28.35±0.79	24.81±0.65	30.76±0.38	30.72±0.42
	120	28.49±0.73	29.46±0.82	24.67±0.74	31.02±0.73	33.14±0.38

The decline in leucocrit value obtained in the present study corroborates with the findings of Sinha. Ranjana *et al.,* (1999) in *Labeo rohita* under crowded conditions. Leucocrit is regarded as a sensitive indicator of physiological stress. The stress response of salmonids and fishes normally include a moderate to severe leucopenia. (Srivastava and Agarwal, 1977; McLeay and Gordon, 1977).

Table 20.2: Effects of Various Stocking Density on Total WBC, DLC and Coagulation Time in *Carassius auratus*. Each value is the average ($\overline{X}$ ± SD) of 6 individual observations.

Parameters	*Exposure Period (days)*	*Density (No. of individuals/aquarium)*				
		1	*5*	*10*	*20*	*40*
Total WBC	0	52.8±3.32	52.4±2.19	52.8±3.14	52.6±2.06	52.6±3.055
(X 10^3/mm^3)	30	51.3±1.80	50.1±2.23	52.8±2.05	45.6±3.29	42.5±2.10
	60	50.7±2.32	49.4±3.19	53.0±1.18	41.3±4.34	39.4±1.82
	90	49.1±3.46	47.9±2.08	52.7±3.19	38.6±3.14	35.2±1.19
	120	48.2±1.28	46.5±2.36	52.5±2.19	36.3±4.80	32.6±1.28
Lymphocyte	0	37.6±1.21	36.9±1.32	37.8±1.24	37.7±1.05	37.9±1.02
(%)	30	39.0±1.05	39.8±1.22	37.6±1.16	41.3±1.34	43.1±1.32
	60	41.3±1.14	46.8±1.08	37.9±1.08	49.4±1.14	48.2±1.21
	90	46.0±1.62	53.9±1.36	36.9±1.34	57.2±1.46	66.5±1.43
	120	53.9±1.44	57.1±1.44	37.6±1.42	67.6±1.06	71.6±1.36
Neutrophil	0	22.5±1.12	22.6±1.34	22.6±1.32	22.8±1.19	22.6±1.26
(%)	30	21.7±1.18	21.3±1.05	22.0±1.36	19.7±1.14	19.1±1.16
	60	20.5±1.02	19.6±1.15	22.5±1.24	16.5±1.05	16.3±1.22
	90	20.3±1.36	17.7±1.28	22.3±1.08	13.7±1.32	13.3±1.19
	120	17.2±1.08	15.2±1.19	22.5±1.06	10.2±1.28	10.7±1.08
Basophil	0	24.9±1.02	24.7±1.05	24.8±1.08	24.8±1.24	24.9±1.25
(%)	30	22.2±1.21	20.6±1.15	25.0±1.32	18.3±1.28	18.1±1.31
	60	19.3±1.36	16.7±1.23	24.7±1.43	13.5±1.32	13.2±1.18
	90	16.0±1.26	12.8±1.41	24.8±1.03	10.2±1.36	8.6±1.08
	120	12.8±1.44	8.6±1.29	25.2±1.15	6.8±1.31	6.1±1.19
Thrombocyte	0	18.9±1.15	18.9±1.08	19.2±1.05	18.8±1.36	19.0±1.11
(X 10^3/mm^3)	30	16.5±1.09	15.3±1.24	18.8±1.41	13.7±1.05	16.3±1.44
	60	14.8±1.12	12.4±1.44	18.8±1.36	10.9±1.11	9.2±1.08
	90	12.5±1.42	11.5±1.32	18.9±1.26	8.2±1.24	7.6±1.25
	120	10.6±1.06	8.6±1.08	19.1±1.28	6.4±1.32	5.1±1.34
Coagulation	0	22.61±1.05	22.63±1.08	22.58±1.28	22.68±1.06	22.72±1.07
time (sec)	30	22.38±1.09	20.41±1.22	22.62±1.09	19.74±1.03	29.01±1.04
	60	20.51±1.22	19.91±1.06	22.47±o.09	17.33±1.08	16.44±1.05
	90	20.18±1.08	17.34±1.08	21.92±1.11	16.15±1.26	15.95±1.21
	120	19.62±1.26	15.24±1.12	22.48±0.05	15.63±1.18	15.04±1.40

In the higher density groups while the lymphocytes a significant reduction in total RBC or WBC is observed in higher density groups (20 or 40 individuals/aquarium). This adverse effect on blood parameters of higher density groups may be due to the physiological stress caused by overcrowding.

It is well established that the intensity of competition for food is directly related to the population density (Weatherley, 1966). Though overcrowding may affect the health and growth of fish, an optimum degree of grouping exists among fish populations (Seng-Keh Teng and Thia-Engchua, 1978). Such a grouping effect is reflected in the present study also. The blood parameters of the fish at the lower stockjng densities of 1 or 5 individuals/aquarium do not show significantly better values than those of the fish at a comparatively higher density and basophils are increased, the thrombocytes and neutrophils are significantly reduced. This finding is supported by similar haematological effects due to physiological stress caused by dietary nutrient deficiency (Mahajan and Agarwal, 1979) or by industrial effluents (McLeay, 1973) or Crowding (Sakthivel, 1993) reported earlier.

The increase in lymphocytes observed in higher density group may be due to the stimulated lymphopoiesis or due to an increased release of lymphocytes from lymphomycloid tissues (Ellis, 1976) under physiological stress caused by overcrowding.

References

Armstrong, M.S. and Boyd, C.E. 1982. Oxygen transfer calculation for a tractor powered wheel aerator. *Trans. Amer. Fish. Soc.,* 111: 363–366.

Andrews, J.W., Knight, L.H., Page, J.W., Matsuda, Y. and Brown, E. 1971. Interactions of Stocking density and water turnover on growth and food conversion of channel catfish reared in intensively stocked tanks. *Prog. Fish Cult.,* 33: 197–203.

Ellis, A.E. 1976. Leucocytes and related cells in the place *Pleuronectus platessa. J. Fish. Biol.,* 8: 143–156.

Hardy, R. 1980. Fish feed formulation. In: *Fish Feed Technology, Agriculture Development and Coordination Programme,* FAO/ADCP/ReP/80/11: 233–239.

James, R. and Sampath, K. 2003. Effect of stocking density on growth and fertility in the ornamental fish, *Xiphophorus helleri. Indian J. Fish,* 50(3): 339–347.

Jena, J.K., Ayyappan, S. and Aravinda Krishnan, P.K. and Hick Mudul. 2001. Comparative evaluation of growth, survival and production of carp species at different stocking densities under polyculture. *Indian J. Fish,* 48(1): 17–25.

Johansson-Sjobeck and Larson, A. 1979. Effect of inorganic lead on d-aminolevulinic acid dehydratase activity and haematological variables in the rainbow trout, *Salmo gairdneri. Arch. Environ. Contam. Toxicol.,* 8: 419–431.

McLeay, D.J. and Gordon, M.R. 1977. Leucocrit: A simple haematological technique for measuring acute stress in salmonid fish, including stressful concentration of pulp mill effluent. *J. Fish. Res. Bd. Can.,* 34: 2164–2175.

McLeay, D.J. 1973. Effects of 12h and 25 day exposure to kraft pulp mill effluent on the blood and tissues of juvenile Coho salmon. *J. Fish. Res. Bd. Can.,* 30: 395–400.

Sakthivel. 1994. Effect of stocking density on the blood parameters of the fish *Cyprinus crapio* L. *Environment and Ecology,* 12(2): 424–428.

Sampath, K. 1985. Food intake, conversion and surfacing activity as a function of density and water change frequency in an air breathing fish, *Channa striatus. Aquaculture,* 46: 201–213.

Seng-Keh Teng and Thia-Eng Chua. 1978. Length-weight relationship and growth of young estuary grouper, (*Epinephelus salmonides maxwell*) caught from the middle Bank, Penang straits, Sch. Biol. Sci. Univ. Sains Malasia Penang, Tech. Rep. USM/IFM/CTE No. 26: 15.

Soderberg, R.W., Flynn, J.B., Schmittou, H.R. 1983. Effects of ammonia on growth and survival of rainbow trout in intensive static water culture. *Trans. Amer. Fish. Soc.*, 112: 448–451.

Sinha, Ranjana, Dutta, Tansuree and Das, M.K. 1999. Impact of stress on blood parameters and interrenals of *Labeo rohita.* In: Eco-friendly management of resources for doubling fish production: Strategies for 21st century. *Proceedings of the National Seminar,* December 22–23. Inland Fisheries Society of India, Barrackpore.

Srivastava, A.K. and U. Agarwal. 1997. Involvement of pituitary internal axis and cholenergic mechanism during cold shock leucocyte sequence in a freshwater tropical teleost, *Colisa fasciatus. Archs. Anat. Microsc.*, 66: 97–108.

Weatherley, A.K. 1966. Ecology of fish growth. *Nature,* 212: 1321–1324.

Wedemeyer, G. 1996a. Rearing conditions effects on fish in intensive culture. In: The Physiology of Fish (Edn.) Chapman Hall, New York.

Wintrobe, M.M. 1978. Cliniacl Haematology, 7th edn. Philadelphis Lea and Feluger Publ.

Chapter 21

Biochemical Studies of Insulin Effect on Freshwater Fish, *Oreochromis mossambicus*: 2. On Muscle Tissue

M.R. Delphine Rose*, W. Premasugantha, A. Jeyaseeli*, S. Anitakannagi* and M. Narayanan*****

**Department of Zoology, Jayaraj Annapackiam College (Autonomous), Periyakulam – 625 601, India*
***Department of Zoology, Holy Cross College (Autonomous), Trichy – 620 002, India*
****Department of Advanced Zoology and Animal Biotechnology, St. Xavier's College (Autonomous), Palayamkottai – 627 002, India*

ABSTRACT

Changes in the protein, glycogen and RNA contents were studied in the muscle of *Oreochromis mossambicus* exposed to different concentrations of insulin (1, 5 and 10 µg) at different time intervals (0, 6, 9, 12, 15, 18, 24 and 48 hrs). The amount of glycogen was decreased at 18 and 48 hrs in all the concentrations when compared to the control. The protein and RNA were observed ranging from 168.80 ± 1.486–57.28 ± 2.073 mg/g and 0.12 ± 0.009–0.48 ± 0.018 mg/g of the fish during the investigation.

***Keywords**: Glycogen, Protein, RNA, Insulin.*

Introduction

Biochemicals that give energy for all life activities are classified into three major groups, *viz.*, carbohydrates, fats and proteins. The utilization, transformation and storage of these biochemicals to a considerable extent are regulated by hormones. Among the hormones, insulin facilitates the utilization of glucose by the cells, promotes glycogen synthesis, storage and prevents the excessive breakdown of glycogen stored in the liver and muscle tissues. The initial action of insulin is stimulation of cyclic AMP phosphodiesterase activity which lowers the tissue concentration of cyclic AMP (Janssens and Maher, 1986). The hypoglycemic action of mammalian insulin, has been demonstrated in fishes (Thrope and Ince, 1974, Lewander *et al.*, 1976 and deRoos and deRoos, 1979), but the doses necessary to produce hypoglycemia vary greatly among species, being generally much higher in fishes than mammals. The effect of insulin on carbohydrate and protein metabolism on the fish, *Ophiocephalus striatus* has been studied, following injection of insulin (Seshadri, 1964). Various experiments has performed on fish species with the concept that the main role of insulin is the regulation of protein metabolism in muscle (Cowey *et al.*, 1977, and Ince, 1983). The present investigation is aimed to estimate the levels of glycogen, proteins and RNA in the untreated and treated muscle tissue of *Oreochromis mossambicus* and to find out the optimal concentration of insulin for enhancing the levels of glycogen, proteins and RNA.

Materials and Methods

The experimental fishes *O. mossambicus* weighing 15.00±1.00 g were collected from a local pond in Trichy. These fishes were acclimated for ten days in the laboratory conditions at room temperature (30±1°C) before carrying out the experiments. During this entire period of acclimation, they were fed *ad libitum* with algae and were maintained in plastic containers of 20 litres capacity. Water was changed every day to get rid of nitrogenous and non-nitrogenous wastes.

The experiments were carried out in four sets, one set served as control, whereas the other sets were treated with insulin at three levels *viz.*, 1 µg, 5 µg and 10 µg. Insulin was injected intraperitenially using a syringe. Control samples were analysed at 0, 6, 9, 12, 15, 18, 24 and 48 hrs, whereas the treated, were sampled at 6, 9, 12, 15, 18, 24 and 48 hrs after insulin injection. In all, five replicates were used. The muscle of the experimental fishes were dissected out, weighed (wet wt.) and stored in the freezer for estimation of glycogen, protein and RNA content.

Glycogen, protein and RNA were estimated respectively by another method (Hassid and Abraham, 1957), Biuret method (Gomall *et al.*, 1949) and Orcinol method (Ceriotti, 1955). The obtained results were analysed statistically (Zar, 1984).

Results and Discussion

Fishes administered with different doses of insulin showed an increase in glycogen content at 6, 9, 12, 15 and 24 hrs when compared to the control except at 9 hr for 5 µg. The glycogen content showed a drastic decrease at 18 and 48 hrs in all treated. In the untreated muscle tissue glycogen showed a gradual increase up to 15 hr, after which, there is a drastic increase at 18hr, followed by a sudden decrease (Table 21.1). This reveals that the glycogen synthesis and its utilization is a gradual process. For the synthesis of glycogen, D-glucose units are utilized in a sequential manner with the involvement of two enzymes *viz.*, glycogen phosphorylase and phospho glucomutase. Besides, another enzyme, glycogen synthetase is responsible for the formation of glycogen from glucose phosphate (Lehninger, 1970). It has also suggested that glycogen phosphorylase is responsible for glycogen breakdown in the cell. In the present study this is shown by the gradual increase followed by decrease of glycogen in muscle tissue. The decrease in muscle glycogen may also be due to balanced glucose consumption by oxidative processes on its entering the muscle tissue (Cameiro and Amaral, 1983). The calculated ANOVA value within the treatment was insignificant ($F = 1.33$; $P > 0.05$) and between the treatment was higher ($F = 185.6$) indicating a significant dose dependent effect of insulin ($P < 0.05$).

Table 21.1: Glycogen Content in Muscle Tissue (mg/g) of *O. mossambicus* Exposed to Insulin at Different Time Intervals. Each value represents mean ± SD of five replications.

Duration (hrs)	*Control*	*Insulin Concentration*		
		1 μg	*5 μg*	*10 μg*
0	0.26±0.007	–	–	–
6	0.08±0.006	0.28±0.10	0.16±0.007	0.10±0.009
9	0.08±0.004	0.21±0.013	0.02±0.001	0.17±0.027
12	0.09±0.010	0.25±0.029	0.13±0.008	0.16±0.009
15	0.10±0.007	0.17±0.012	0.14±0.010	0.24±0.008
18	0.24±0.009	0.16±0.007	0.19±0.007	0.24±0.008
24	0.06±0.002	0.14±0.010	0.26±0.025	0.34±0.019
48	0.08±0.004	0.07±0.005	0.06±0.012	0.075±0.002

Fishes treated with different quantities of insulin failed to show any clear trend as a function of either dose or time interval. However, a peak protein content both in control experimental animals at 12 h after treatment was obvious (Table 21.2). But for this peak value, the overall protein content in fish muscles in control in experimental fishes showed a lower value. Subjecting the data to ANOVA showed that the insulin effect within the treatment was insignificant ($F = 0.43$; $P > 0.05$) and between the treatment was significant ($F = 227.59$; $P < 0.05$). The aforesaid overall decrease is paradoxical to the finding that insulin enhanced the incorporation of C^{14}–glycine into the muscle protein as seen in the opercular muscle of eel (Inui and Ishioka, 1983) and northern pike (Ince and Thorpe, 1976).

Table 21.2: Protein Content in Muscle Tissue (mg/g) of *O. mossambicus* Exposed to Insulin at Different Time Intervals. Each value represents mean ± SD of five replications.

Duration (hrs)	*Control*	*Insulin Concentration*		
		1 μg	*5 μg*	*10 μg*
0	124.26±5.112	–	–	–
6	77.92±2.752	71.74±3.067	74.42±2.373	71.11±3.431
9	83.631±2.995	59.62±2.944	80.15±7.840	78.05±1.315
12	168.80±1.486	126.92±4.895	111.75±2.096	105.07±1.326
15	70.68±1.560	84.34±0.918	117.85±4.693	94.75±2.047
18	70.33±1.652	65.35±0.936	62.56±3.286	65.46±1.852
24	72.31±5.111	79.10±3.497	68.63±2.723	57.28±2.073
48	125.7±3.952	73.03±0.902	104.57±6.348	97.16±3.962

The amount of RNA in the untreated muscle tissue at the commencement of the experiment (0 hr) was 0.48±0.018 mg/g. The amount of RNA content got reduced after the beginning of the experiment. Though the values of RNA obtain in the muscle tissues of experimental animals showed a low value, a peak RNA content after 12 h of insulin treatment was noted (Table 21.3). In general, the RNA content in all experimental animals up to 12 hr showed a letter value than control irrespective of insulin dose and as the time progressed, the total RNA content got decreased when compared to the control fishes.

(Table 21.3) the untreated. However at 48 hr an increase in RNA level was noticed for 5µ (0.31±0.038 mg/g; Table 21.3).

Table 21.3: RNA Content in Muscle Tissue (mg/g) of *O. mossambicus* Exposed to Insulin at Different Time Intervals. Each value represents mean±SD of five replications.

Duration (hrs)	*Control*	*Insulin Concentration*		
		1 µg	*5 µg*	*10 µg*
0	0.48±0.018	–	–	–
6	0.16±0.022	0.19±0.029	0.24±0.013	0.28±0.007
9	0.22±0.030	0.26±0.050	0.24±0.008	0.15±0.008
12	0.29±0.007	0.38±0.004	0.26±0.008	0.32±0.014
15	0.36±0.005	0.26±0.13	0.22±0.009	0.30±0.007
18	0.27±0.007	0.15±0.005	0.12±0.009	0.26±0.007
24	0.20±0.033	0.16±0.006	0.18±0.041	0.17±0.004
48	0.25±0.067	0.18±0.030	0.31±0.038	0.21±0.005

The ANOVA value within the treatment was insignificant ($F = 3.55$, $P > 0.05$) and between the treatment was significant ($F = 49.2$; $P < 0.05$). When recalculated the Table 21.3 dates into percentage, an overall decrease in the RNA content (1 µ = 92, 5 µ = 83 and 10 µ = 99.5 per cent) was noticed for the different levels of insulin treated. However, there was an overall time related decrease in RNA content in the treated muscle tissue could be attributed to the utilization of the available transcripts in protein synthesis, as shown by the increase in the protein content (Rose *et al.*, 2005). Thus the decrease in RNA in the muscle tissue of *O. mossambicus* could be due to the regulatory role of insulin on rRNA transcription as suggested by Granner (1988) and Kleinsmith and Kish (1988).

References

Carneiro, N.M. and Amaral, A.D. 1983. Effects of insulin and glucagons on plasma glucose levels and glycogen content in organs of freshwater teleost, *Pimelodes maculates. Gen. Comp. Endocrinol.*, 49: 115–121.

Ceriotti, G. 1955. Determination of nucleic acids in animal tissues. *J. Biol. Chem.*, 214: 59–70.

Cowey, C.B., Higuera dela M. and Adron, J.W. 1977. The effect of dietary composition and of insulin on gluconeogenesis in rainbow trout, *Salmo gairdneri. Br. J. Nutr.*, 38: 385–395.

Delphine Rose, M.P., Prema sugantha, W., Jeyaseeli, A., Anitakannagi, S. and Narayanan, M. 2005. Biochemical studies of insulin effect on freshwater fish, *Oreochromis mossambicus:* 1. On liver tissue. *J. of Current Sciences.* (Communicated).

DeRoos, R. and DeRoos, C.C. 1979. Severe insulin–induced hypoglycaemia in the spiny dog fish shark, *Squalus acanthias. Gen. Comp. Endocrinol.*, 37: 186–191.

Gornall, A.G., Bardawill. C.J., and David, M.M. 1949. Determination of total serum protein by means of biuret reaction. *J. Biol. Chem.*, 177: 751 –766.

Hassid and Abraham, 1957. Chemical procedures for analysis of polysaccharides. In: *Methods in Enzymology*, (Eds.) Colowick, P. and Kaplan. Academic Press Inc., New York, 3: 34.

Ince, B.W. 1983. Insulin Secretion from the *in situ* perfused increase of the European silver eel, *Anguilla anguilla. Gen. Comp. Endocrinol.,* 37: 533.

Inui, Y. and Ishioka, H. 1983. Effects of insulin and glycogen on amino acid transport into the liver and opercular muscle of the eel *in vitro. Gen. Comp. Endocrinol.,* 51: 213–218.

Janssens, A.P. and Maher, F. 1986. Glucogon and insulin regulate *in vitro* hepatic glycogenolysis in the axolotl, *Ambystoma mexicanum* via changes in tissue cyclic AMP concentration. *Gen. Comp. Endocrinol.,* 61: 64–70.

Lehninger, A.L., 1970. *Biochemistry: The Molecular Basis of Cell Structure and Function.* Worth Publishers, Inc., New York.

Lewander, K., Dave, G., Johenson-Sjobeck, M.L., Larsson, A. and Lidman, V. 1976. Metabolic affects of insulin in the European eel, *Anguilla anguilla* L. *Gen. Comp. Endorinol.,* 29: 455–465.

Mittermayer, C., Braun, R. and Rusch, H.P. 1964. RNA synthesis in the mitotic cycle of *Physarum polycephalim. Biochem. Biophys. Acta,* 91: 399–405.

Sashadri, B. 1964. Effect of insulin on blood glucose level in the fish, *Ophiocephalus straitus* (Bloch). *J. Anim. Morph. Physiol.,* 11: 302–308.

Thorpe, A. and Ince, B.W. 1974. The affects of pancreatic hormones, catecholamines, and glucose loading on blood metabolites in the northern pike, *Esox lucius* L. *Gen. Com. Endocrinol.,* 23: 29–44.

Zar, J.H. 1984. *Biostatistical analysis.* Prentice Hall, London, UK.

Chapter 22

Observations on a Trypanosome *Trypanosoma notopterae* n.sp. from a Freshwater Teleost Fish, *Notopterus notopterus* (Pallas)

T.K. Kundu* and Monali Chatterjee**#

**Reader, Ranaghat College, Ranaghat, West Bengal, India*

***Guest Lecturer, Ranaghat College, Ranaghat, West Bengal, India*

ABSTRACT

The communication records the description of a haemoflagellate belonging to the genus *Trypanosoma* Gruby 1843 from the blood of the teleost fish *Notopterus notopterus* (Pallas), collected at Ranaghat and from various places of Nadia district in West Bengal.

Keywords: *Teleost, Freshwater, Notopterus notopterus, Protozoa: Haemoflagellate: Trypanosoma: New species.*

Introduction

Parasites affect fish population in a variety of ways. Large fish mortality or fish kill frequently occurs in rivers, ponds and tanks due to environmental stress and heavy parasitic infections. Among

---

Corresponding Author: E-mail: dr_monali02@hotmail.com.

the ichthyoparasites, protozoans constitute a major group and include representatives from haematozoans, sporozoans, myxozoans and ciliates. Lom (1979), Lom and Dykova (1992) and Woo (1994) have described haemoflagellates in fishes from other countries. However the study of the haematozoan infection in fishes has been left far behind in India in comparison to other protozoan parasitic groups. Kundu and Haldar (1981) reported the presence of a trypanosome from the same host fish without assigning any specific status to it. The present species under discussion resembles fully both morphologically and morphometrically, the flagellate as described by Kundu and Haldar.

Materials and Methods

The host fishes were collected from various ponds and reservoirs of Ranaghat and other places in Nadia district of West Bengal and brought alive to the laboratory. Usually the hosts were examined immediately afterwards. Some of the host fishes were kept in glass aquaria (90 cms × 45 cms × 45 cms) fitted with aerators and artificial food for study on subsequent days.

Morphological and Morphometric Study

Study of Fresh Preparations

Method of Collection

The caudal peduncle of the fish was wiped with free of water and mucous with clean towel and also with cotton and alcohol and a drop of uncontaminated blood from the caudal artery was streaked on a clean slide and examined under a cover slip.

Study of Permanent Preparations

Method of Preparation of Giemsa Stock Solution

About 0.6 gms of certified powder of Giemsa stain was dissolved in 50 ml of glycerine in a flask and the flask was heated in a water bath at 55°C–60°C for 2 hours or longer with occasional shaking till all the stain was dissolved. Now the flask was allowed to cool to room temperature and 50 ml of methyl alcohol was added to it. The flask was then stoppered tightly and kept in dark place for 10–15 days for maturation. Finally the stain was filtered and stored in black coloured bottles and kept in a dark place.

Preparation of Permanent Slides

Smears were made, on the slides at the center, from the drop of plasma taken with the help of a needle and allowed to dry. For the preparation of blood films, grease-free slides were taken. A drop of fresh blood was taken at one end of the slide from the severed peduncle of the fish. Immediately, a second slide, one corner of which had been cut away, was placed at the inner margin of the blood drop and the blood was allowed to spread along the edge of the slide The second slide was pushed along the surface of the first at an angle of about 45° towards the other end. Thus a thin film of blood was prepared and allowed to dry. The dried smears were fixed in absolute acetone-free methyl alcohol for 5 minutes and allowed to dry. Now the stock solution of Giemsa was diluted at the rate of 1 drop/ml of water buffered at ph 7.2. The slides were then placed on a staining rack and covered with dilute stain for 40–50 minutes. Finally, the slides were washed by pouring neutral distilled water or buffer solution until the colour did not change to noticeable extent and the slides were dried in air. A synthetic resin DPX had been used as mounting medium.

Measurements and Figures

The measurements of the nucleus and the kinetoplast have been taken using a calibrated ocular micrometer. Figures have been drawn with the aid of a camera lucida and the measurements of trypanosomes were made by the Thread overlay method of Minchin (1909).

Etymology

The specific epithet has been derived from the generic name of the host.

Results

Taxonomic Diagnosis

Trypanosoma notopterae sp.n.

The trypanosome shows pleomorphism *i.e.*, is found in three different morphological forms, *viz.*,

1. A short slender form having incipient undulating membrane with a very short free flagellum.
2. An intermediate form with 5–6 undulating membranes and a moderately long free flagellum, and
3. A long slender form with more than 8 undulating membranes and a very long free flagellum.

In all the three forms the nucleus is ovoidal in shape and situated at the middle of the body and it takes purple stain with Giemsa. The kinetoplast is small and ovoidal, located at the posterior tip of the body. The cytoplasm stains light purple with Giemsa and is granulated in nature. A few vacuoles are also observed in the cytoplasm of the 'short slender' and 'long slender' forms. We could not encounter the divisional stages so far. The flagellate is apparently non-pathogenic to its host.

Type Specimen

Hapantotype

Trophozoite on slide no. FOL/TR/I 1 prepared from a drop of uncontaminated blood collected from the caudal artery of the host fish *Notopterus notopterus* collected from the ponds and reservoirs of Ranaghat, Nadia district, West Bengal, India by Dr. Tarun K. Kundu on March 27, 1999. Slides containing the hapantotype and syntype material have been deposited in the collection museum of the Protozoology Laboratory, Department of Zoology, University of Kalyani.

Table 22.1: Showing Measurements (in microns) of the Present Species from the Freshwater Teleost Fish *Notopterus notopterus*

Characters	*Short Slender Form*		*Intermediate Form*		*Long Slender Form*	
	Range	*Mean*	*Range*	*Mean*	*Range*	*Mean*
Total length of the parasite (including free flagellum)	24.4–29.9	27.1	41.0–44.4	42.7	49.9–55.5	52.7
Length of the cell body	22.2–26.6	24.4	31.0–32.1	31.6	36.6–38.8	37.7
Breadth of the cell body	1.1–1.6	1.3	13.19	1.6	1.9–2.7	2.3
Length of the free flagellum	2.2–3.3	2.7	8.8–13.3	11.1	13.3–16.6	14.9
Length of the nucleus	2.2–3.3	2.7	2.2–3.6	2.9	3.3–4.4	3.8
Breadth of the nucleus	1.1–1.3	1.2	1.3–1.6	1.5	1.6–1.9	1.8
Length of the kinetoplast	0.83–1.1	0.97	0.83–1.1	0.97	1.1–1.1	1.1

Figure 22.1: Short Slender Form

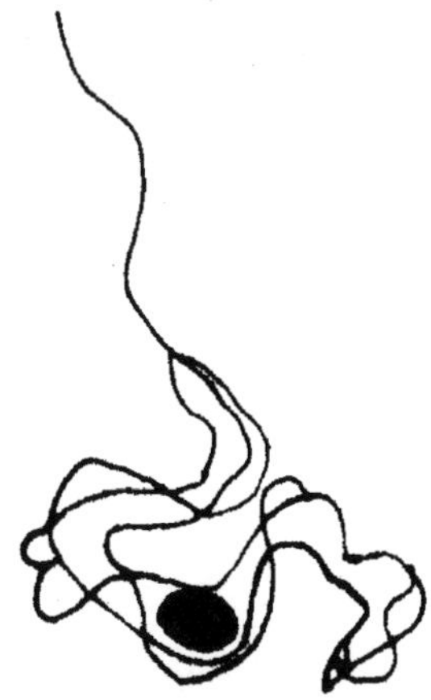

Figure 22.2: Intermediate Form

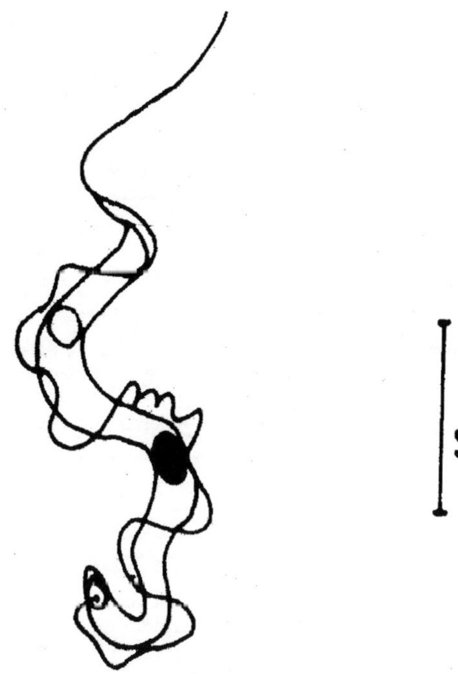

Figure 22.3: Long Slender Form

Figures 22.1–22.3: Line Drawings of the Trimorphic Forms of the Newly Described Species *Trypanosoma notopterae* sp.n. from *Notopterus notopterus*

Syntype

Many on the above numbered slide and on other slides; other particulars are same as for the hapantotype material. One slide containing some syntype materials has been deposited at the Zoological Survey of India, Kolkata.

Ecological Consideration

Roy and Haldar (1997) observed that macro-environmental factors like temperature, dissolved oxygen and alkalinity have got immense role in modulating the course of infection in fishes by haematozoans. Temperature tolerance of different morphs in the host fishes shows a wide range of variations. Maximum incidence of infection was observed in the months of March and April, which gradually decreased in the months of July and August. There was minimum infection in the months of September and October, which gradually decreased to a nil in the months of December to January with very low incidence of infection. The course of infection has also been found to be definitely correlated with the age, body weight and sex of the examined host fishes.

Comparative Discussion

Trypanosoma striati (Qadri 1955) from *Ophicephalus striatus* is marked polymorphic, having namely large, intermediate and small forms in the length/breadth ratio of 54.7/8.0; 39.2/3.0 and 32.0/2.0 respectively.

Trypanosoma channai from *Channa punctata* (Narasimhamurti and Saratchandra, 1980) is pleomorphic having a slender, medium and short forms and measures 38.4/62.0 µm (including free flagellum). The length/breadth ratio of the slender, medium and short forms ranges 14.5/17.3; 7.5/12.0 and 3.5/5.6 respectively.

Trypanosoma attii (Gupta and Jairajpuri, 1982) was described from *Wallago attu* and is polymorphic in nature having large, intermediate and small forms with length/breadth ratio of 28.1/1.4; 20.5/1.5 and 15.0/1.2 respectively.

Trypanosoma colisi from *Colisi fasciata* (Gupta, 1986) show three distinct forms; long and narrow slender forms; intermediate forms and small slender forms in the ratio of 23.3/1.5; 19.5/1.6 and 14.2/1.8 respectively.

Trypanosoma mastacembeli (Wahul, 1987) from *Mastacembelus armatus* showed trimorphism of large, medium and small forms with a length/breadth ratio of 38.6/5.5; 34.3/4.1 and 31.1/2.8 respectively. The body of the trypanosome is variable in width but distinctly elongated in all forms. *Trypanosoma murtii* (Wahul, 1987) from the same host is pleomorphic being significantly long but with distinct variations in the body width. The length/breadth ratio in long, medium and small forms is 38.8/4.6; 36.2/3.5 and 31.1/2.1 respectively.

The presently described species *Trypanosoma notopterae* sp.n. from the host *Notopterus notopterus* resembles the above-mentioned species in having pleomorphic forms, short slender, intermediate and long slender forms. But differences lie in the morphometric ratios (length/breadth), which truly justify its designation as a new species.

Sporozoeans belonging to the genus *Dactylosoma* Labbé 1894 were found in the erythrocytes of the freshwater teleost *Notopterus notopterus* by Kundu and Haldar in 1984. Thus this genus in no way resembles the newly described species *Trypanosoma notopterae* sp. n. and is characterized by the presence of schizogony without producing pigment and with more than four merozoites in the red cells.

Comparative account of the length/breadth ratio of *T. notopterae* sp.n. and its related trimorphic forms are given in Table 22.2.

Table 22.2: Comparative Account of the Length/Breadth Ratio of *T. notopterae* sp.n. and its Related Trimorphic Forms

Previously Described Related Species	*Short Slender Form*	*Intermediate Form*	*Long Slender Form*	*Host*	*Reference*
Trypanosoma channai	3.5 : 5.6	7.5 : 12.0	14.5 : 17.3	*Channa punctata*	Narasimhamurti and Saratchandra, 1980
Trypanosoma colisi	14.2 : 1.8	19.5 : 1.6	23.3 : 1.5	*Colisa fasciata*	Gupta, 1986
Trypanosoma striati	32.0 : 2.0	39.2 : 3.0	54.7 : 8.0	*Ophicephalus striatus*	Qadrii, 1955
Trypanosoma mastacembeli	31.1 : 2.8	34.3 : 4.1	38.6 : 5.5	*Mastacembelus armatus*	Wahul, 1987
Trypanosoma murtii	31.1 : 2.1	36.2 : 3.5	38.8 : 4.6	*Mastacembelus armatus*	Wahul, 1987
Trypanosoma attii	15.0 : 1.2	20.5 : 1.5	28.1 : 1.4	*Wallago attu*	Gupta and Jairajpuri, 1982
Trypanosoma notopterae sp.n.	24.4 : 1.3	31.6 : 1.6	37.7 : 2.3	*Notopterus notopterus*	Present study

References

Gupta, N. 1986. *Trypanosoma colisi* sp.n. from a freshwater fish, *Colisa fasciata* (Bloch). *Fiev. Iber. Parasitol.*, 46: 1–6.

Jairajpuri, D.S. and Gupta, N. 1982. A new polymorphic trypanosome from an Indian freshwater fish *Wallago attu* (Bl. and Schn.). *Riv. Parasitol.*, 63: 287–292.

Kundu, T.K. and Haldar, D.P. 1981. Preliminary observations on a trypanosome from the freshwater teleost fish, *Notopterus notopterus* (Pallas). *All. Ind. Symp. Perspectives in Research on Protozoa*, (Abstract), pp. 20.

Kundu, T.K. and Haldar, D.P. 1984. *Dactylosoma notopterae* n.sp. from the freshwater teleost *Notopterus notopterus* (Pallas). *Riv. Di. Parassit.*, 1: 391–398.

Lom, J. 1979. Biology of the trypanosomes and trypanoplasms of fish. In: *Biology of Kinetoplastida Vol. II*, (Eds.) Lumsden, W.H.R. and Evans, D.A. Academic Press, London and New Zealand.

Lom, J. and Dykova, I. 1992. *An Introduction to the Study of Fish Parasites*. Elesiver Publications, Netherlands.

Minchin, E.A. 1909. Observations on the flagellate parasitic in the blood of freshwater fishes. *Proc. Zool. Soc.*, London, 1: 2–31.

Narasimhamurti, C.C. and Saratchandra, B. 1980. Two new species of trypanosomes, *Trypanosoma channai* n.sp. from *Channa punctata* and *Trypanosoma qadri* n.sp. from *Clarias batrachus*. *Proc. Indian Acad. Sci.*, 89: 371–376.

Qadrii, S.S. 1955. The morphology of *Trypanosoma striati* n.sp. from an Indian freshwater fish. *Parasitology*, 45: 79–85.

Roy, S. and Haldar, D.P. 1997. Protozoan haemoflagellates in common air-breathing fishes of West Bengal. *Trans. Zool. Soc.*, India, 1: 28–33.

Wahul, M.A. 1987. Studies on the morphology of three new species of trypanosomes from the freshwater fish *Mastacembelus armatus* in Maharashtra, India. *Arch. Protistenkd.*, p. 111–124.

Woo, P.T.K. 1994. Haemoflagellates of fishes. In: *Parasitic Protozoa*, Vol. IV (Ed.) J.P. Kreier. Academic Press. Inc., New York.

Chapter 23

A Study on the Selection of Habitat by the Post Larval and Juvenile Banana Shrimp, *Penaeus merguiensis* in and Around Kali Estuary, Karwar

Sampath Kumar S. and Kusuma Neelakantan

School of Ocean Sciences, Kodibag, Karwar – 581 303

ABSTRACT

The distribution of post larval and juvenile stages of *Penaeus merguiensis* in four different habitats adjoining the Kali estuary having different abiotic characteristics were exarnined for a period of one lunar month during the pre-monsoon season, when maximum availability of this species occurred. The results indicate that the variation in the density was due to the difference in substrate features such as pH, Organic content, sediment fraction type and vegetation cover. The study shows that slightly low pH might make the immigrating post larval forms to move out of a particular habitat. The present investigation provides evidence to suggest that though muddy substratum is the most preferred habitat, it can tolerate other substrate types, if the habitat provides sufficient amount of food and protection. The occurrence and density of post larvae and juvenile of *P. merguiensis* in a specific habitat may be dependent on the particular environmental factor like muddy substratum, optimal pH of water and sediments, water temperature, oxygen content, vegetation covers etc.

Keywords: *Habitat selection, Banana shrimp, Penaeus merguiensis.*

Introduction

The influence of environmental factors on the process of habitat selection by penaeids in general and *Penaeus merguiensis* in particular is not well understood. Most species utilize habitats, which encompass ecologically hospitable abiotic and biotic parameters, guided by their physiological limits of tolerance. The habitat selection is essentially the relationship between behavior and environment which determines the local distribution (Meadows and Campbell, 1972).

P. merguiensis like most penaeids make use of mangrove lined estuary as the nursery feeding ground (Gunter, 1967; Staples and Vance, 1979; Staples *et al.*, 1985; Dall *et al.*, 1990; Vance *et al.*, 1990). Studies have been carried out in the Kali estuary Karwar to assess the seasonality in the distribution and abundance of penaeid prawns in general (Gunaga *et al.*, 1989) and *P. merguiensis* in particular (Sampath Kumar, 1996). Some information is available on the habitat requirement of *P. merguiensis* from Australian waters (Staples *et al.*, 1985; Vance *et al.*, 1990; Robertson, 1988) and also from the waters of Goa (Achuthankutty, 1988).

The present study was an extension of a work on the seed abundance of *P. merguiensis* in relation to some abiotic and biotic factors in Kali estuary for a period of two years (Sampath Kumar, 1996). During the study it was noted that even though immigration was occurring into all the creeks during the flood tides, the density of post larvae and juveniles in different creeks adjoining the estuary during low tide were different. As a result of this observation, this particular study was planned and carried out, during the peak immigration period, to know the environmental conditions prevailing in the adjoining locations and the factors limiting the survival in these areas, which in turn might help in elucidating the habitat preference of *P. merguiensis* and to identify the natural causes of variation in density of post larval and juveniles of *P. merguiensis* in the Kali estuary.

Materials and Methods

The habitat preference and selection were calculated on the basis of the number of individuals obtained per unit area and is expressed as no/m^2.

The study was carried out during the pre-monsoon season, when the immigration was frequent and the abundance of the post larvae and juvenile stages of *P. merguiensis* in the estuary was at its peak (Sampath Kumar, 1996) at four study stations, referred to as Habitat 1, Habitat 2, Habitat 3 and Habitat 4.

Weekly sampling was carried out at all the four habitats, during low tide for a period of one lunar month. The sampling was done using a standard velon screen hand held drag net, having an area of 3 × 1 m and 2 mm mesh size. The sampling was done at each sampling area, by dragging the net over a distance of 10 mts, near the edge of the water, by holding the two ends of the net by two people and counting the number of post larvae and juveniles obtained per drag. The net was dragged 3 times and the average number was calculated. Concomitantly, surface water and sediment (core) samples were collected for analysis.

Temperature was measured using a centigrade thermometer. Salinity was determined by Mohrknudson's method (Strickland and Parson, 1956), dissolved oxygen was estimated by Winkler's method (Strickland and Parson, 1956), and nutrients like Nitrate, Nitrite, Phosphate and silicate were determined by following the methods of Strickland and Parson (1956). Organic matter in sediments was estimated by the method of ElWakeel and Riley (1956) and pH was measured using pH meter (Systronic make, model 7024). Size fraction was determined by sieve analysis and pipette analysis (Holme and Intyre, 1984). Suspended load was determined by filtering 1 liter of water sample collected from the sampling site, through preheated weighed G.F/C filter paper and weighing it after drying till constant weight was attained and is expressed as mg/l.

Study Stations

Habitat 1, Kodibag Creek

This creek joins the Kali estuary along the southern bank at a distance of 0.3 km from the mouth. Even though only about 1 km long, the link with the estuary is lost during low tide as major portion of the creek bottom lies exposed during low tide, except the far end where the water depth is around 1.5m. It appears like a pond, when the connection with the estuary is lost, during low tide. The area is bordered by a few mangrove shrubs. The initial survey conducted revealed that immigration of post larvae of *P. merguiensis* does occur during the flood tide, but sometimes no shrimps were caught during the ebb tide.

Habitat 2, Mavinahole Creek

This creek joins the estuary along the northern side, nearly opposite to Kodibag creek and is approximately 6 kms long. The creek near the study area is bordered by fringing mangroves. The creek bottom is covered by sea grass and water is shallow and sediment is sandy.

Habitat 3, Sandy Backwater Area

This is located on the southern side of the Kharland bund near the mangrove free area of the Kadwad backwaters of the Kali estuary. It is approximately 6 kms from the mouth of the estuary and has a sandy bottom with no vegetation cover, which gets exposed during the low tide.

Habitat 4, Muddy Mangrove Area

This is located in the backwater area of Kadwad, near to Habitat 3 on the northern side of the Kharland bund. This area includes a steep intertidal bank adjoining the fringing mangrove forest.

Statistical Analysis

The data acquired from the investigation was subjected to correlation analysis (Snedecor and Cochran, 1967) to test the significance of the relationship.

Results and Discussion

From Table 23.1, it can be seen that only post larvae and early stages of juveniles were seen at Habitat 1 and Table 23.2 shows that during 2nd and 4th collection they were totally absent at Habitat 1 thus suggesting that even though post larvae entered the creek, they did not remain or survive there as some of them were seen to move out during the following ebb tide along with the water and were collected in the drag net tide against the receding waters. This may be attributed to the slightly low pH of the water.

Table 23.1: Percentage Availability of Different Size Groups Collected from Each Habitat

	Size Group	*Habitat 1*	*Habitat 2*	*Habitat 3*	*Habitat 4*
Percentage	60	0	0	0	21
availability of	50 mm	0	4	0	42
different size groups	40 mm	0	33	0	92
collected from	30 mm	0	21	0	105
each habitat	20 mm	18	40	0	147
	10 mm	7	6	0	12

Table 23.2: Variation in the Density of Post Larval and Juveniles of *P. merguiensis* at Different Habitats

	Collections	*Habitat 1*	*Habitat 2*	*Habitat 3*	*Habitat 4*
	1	19	83	0	234
Post larval and	2	0	121	0	583
Juvenile Density	3	84	144	0	302
(no/m^2)	4	0	117	0	148
	5	21	56	0	827

William's had indicated that hydrogen ion concentration of the substrate may influence attraction to a substrate, but the present observation suggests that slightly low pH can induce repulsion from a particular substrate type, when it is not tolerable. Pedini (1981) has shown that acidic sediments can decrease growth rates and increase mortality in aquaculture ponds.

The results of hydrological analysis shows that, the pH of water. pH of the sediment and dissolved oxygen was less in comparison to all other habitats (Table 22.3). Even though there was no significant relationship existing between seed abundance and any particular environmental factor statistically, at Habitat 1 there existed significant relationship between water temperature and sediment temperature (at $P < 0.05$), water temperature and dissolved oxygen (at $P < 0.0\ 1$), which might suggest that as the soil is exposed to the sun during day time at low tide it gets heated up and in turn increasing the temperature of the inflowing flood tide waters, which might result in decreasing the oxygen content of water.

Table 23.3: Variation in the Abiotic Factors and Seed Abundance at Different Habitats During the Study Period

Parameters	*Habitat 1*	*Habitat 2*	*Habitat 3*	*Habitat 4*
Seed Abundance (no/m^2)	124	521	0	2094
Water Temperature (°C)	30.1	31.3	30.4	30.4
Salinity (%)	27.5	28.6	19.6	20.2
pH	6.4	7.3	7.8	7.8
Suspended Load (mg/g)	0.009	0.009	0.007	0.007
Dissolved Oxygen (ml/l)	2.9	3.9	4.6	5.2
Sediment Temperature (°C)	29.7	31.2	31.1	30
pH	6.6	7.6	7.7	7.7
Organic Carbon (mg/g)	4.9	5.4	0.4	6.3
Sand (%)	99.5	86.3	99.5	11.8
Silt (%)	0.5	10.5	0.5	33.7
Clay (%)	0	3.2	0	54.5

At Habitat 2 the density was (56–144 no/m^2) greater than Habitat 1 (0–84 no/m^2). This habitat recorded almost all the size groups except the larger size group (Table 23.1), even though the density was low in comparison to Habitat 4 (148–827 no/m^2). The continued presence of early stages suggests regular immigration during flood tides (Staples and Vance, 1979). The amount of organic content of the sediment implies richness in terms of food availability for the shrimps. The post larvae appear to survive even though the substrate was predominated by sand (Table 23.3). The lack of muddy substrate, which is its preferred habitat (Robertson and Duke, 1987) may be outweighed by the protection rendered by vegetation cover and the organic richness as a food source of food.

Habitat 3 was sandy and without any vegetation cover and no post larvae were collected. This might be attributed to the lack of food (absence of organic matter) and lack of protection by way of the absence of vegetation cover and no soft muddy substratum. The presence of coarse sand makes it difficult to bury quickly, unlike muddy substrate. Somers *et al.* (1987) inferred from their study that low organic carbon becomes a limiting factor in the distribution of the commercial species, during their estuarine phase. Variation in shrimp abundance has been linked to the organic carbon content of the sediment both in Gulf of Mexico (Gardy, 1971) and in the Sudanese Red Sea (Branford, 1981).

Habitat 4 recorded very high densities (148–827 no/m^2) and all the size groups were represented. This may be due to factors like, increased oxygen content, high organic detritus (Table 23.3), floral and faunal abundance in the water and sediments (Sampath Kumar, 1996) and protection offered by the soft muddy substratum.

Correlation analysis did not reveal any significant relationship between seed abundance and environmental factors at any Habitats and hence the data is not presented.

Sick *et al.* (1972) have inferred that a suitable substratum is not only important for prawn species to maintain their normal burrowing habits but also in promoting high survival and growth rates in culture systems. The present result apparently agrees with that of Young and Carpenter (1977) who determined that the final selection of a nursery ground probably depends on habitat features such as depth, substratum and types of plant or other cover for early juveniles of various species.

The greater densities in muddy mangrove area are mainly due to the aggregation of immigrating population (Sampath Kumar, 1996). Vance *et al.* (1996) observed during their study that large catch of *P. merguiensis* post larvae were only recorded from mangrove lined banks of the estuary, although small numbers of post larvae settled in all parts of the estuary.

The present observation at Habitat 4 provides further evidence for the hypothesis suggested by Robertson and Duke (1987) that mangrove habitat is more productive and has greater food resource for consumers. However they further added that the structural heterogeneity provided by mangrove tree, roots and mangrove snags in creeks afford greater protection from predators. In the present study it was observed that during the low tide the water recedes to such an extent that it reaches below the level of any growing mangrove trees, and has no prop roots or stilt roots to provide protection to the inhabiting organism. The only way of self defense in this habitat of high density seems to be by burying themselves in the muddy substratum and disappearing from its predators. Hence it shows that the degree of protection that the shrimps gains from the mangrove roots is less significant in comparison to the muddy substratum.

The observation recorded at Habitat 1 and Habitat 2 is similar to the influence drawn by Vance *et al.*, 1990. From their study in the Embley estuary they concluded that, if the habitat in which the post larvae and small juveniles settles, bears less than optimal environmental conditions, then either they move back into the main estuary to be carried by subsequent flood tides (Habitat 1) or remain where they first settle and be subjected to increased predation and decreased growth (Habitat 2). The difference in the densities of post larvae and juveniles of *P. merguiensis* recorded at Habitat 2 and Habitat 4 might help us in suggesting that, Habitat 2 does not bear very optimal environmental condition for the species, but remains there as it gets food and protection. In addition the temperature recorded is relatively higher than all other habitats.

Table 23.4: Probable Abiotic Factors that might be having Some Influence on Seed Abundance at Different Study Sites

Factors	*Habitat 1*	*Habitat 2*	*Habitat 3*	*Habitat 4*
Seed Abundance (no/m^2)	124	521	0	2094
pH of Water	6.4	7.3	7.8	7.8
Dissolved Oxygen (ml/l)	2.9	3.9	4.6	5.2
Sediment pH	6.6	7.6	7.7	7.7
Organic Carbon (mg/g)	4.9	5.4	0.4	6.3
Sand (%)	99.5	86.3	99.5	11.8
Clay (%)	0	3.2	0	54.5

As the environmental condition of the estuary during the study in general was appropriate for immigration and proper growth (Gunaga *et al.*, 1989; Sampath Kumar, 1996), the differences in densities at each habitat was due to the differences in the prevailing environmental conditions (Table 23.4). Thus it is possible to conclude that, muddy areas provide the most protection to these prawns and it may be their most preferred habitat and if they land up in a sandy habitat, they might remain there if the area provides protection by means of vegetation cover. If the inundating waters take the post larvae to an area with unfavorable pH, then they do not remain there and move out along with the receding waters.

References

Achuthankutty, C.T. 1988. Biology of the banana prawn, *Penaeus merguiensis* de Man from Goa waters. *Journal of Aquatic Organisms*, 21: 1–10.

Branford, J.R. 1981. Sediment and distribution of penaeid shrimp in Sudanese Red Sea. *Estuar. Coast Shelf Sci.*, 13: 349–354.

Dall, W., Hill, B.J., Rothlishberg, P.C. and Staples, D. 1990. The biology of the penaeidae. *Adv. Mar. Biol.*, 27: 1–489.

El-Wakeel, S.K. and Riley, J.P. 1956. Determination of organic carbon in marine muds. *J. Cons. Perm. Int. Explor. Mer.*, 22: 180–183.

Gardy, J.R. 1971. The distribution of sediment properties and shrimp catch on two shrimp shrimping grounds on the continental shelf of Gulf of Mexico. *Proc. Gulf and Caribbean Fish. Inst.*, 23 (Ann. Sess.): 139–148.

Gunaga, V.R., Neelakantan, K. and Neelakantan, B. 1989. Seasonal abundance of penaeid post larvae in the Kali estuary, Karwar. *Mahasagar*, 22(4): 193–197.

Gunter, G. 1967. Some relationships of estuaries to the fisheries of the Gulf of Mexico. In: *Estuaries*, (Ed.) by G.H. Lauff. American Association for the Advancement of Science, Washington, D.C., p. 621.

Holme, N.A. and McIntyre, A.D. 1984. *Methods for the Study of Marine Benthos*. Blackwell Scientific Publication, Oxford, pp. 172.

Meadows, P.S. and Campbell, J.I. 1981. Habitat selection by aquatic invertebrates. In: *Adv. Mar Biol.*, (Eds.) Russel, F.S. and M. Young. Academic Press, London, 10: 271.

Pedni, M. 1981. Penaeid shrimp culture in tropical developing countries. *FAO Fish Cir.*, pp. 732.

Robertson, A.I. 1988. Abundance, diet and predators of juvenile banana prawn, *Penaeus merguiensis* in a tropical mangrove estuary. *Aust. J. Mar. Freshwat. Res.*, 39: 467–478.

Robertson, A.I. and Duke, N.C. 1987. Mangroves are nursery sites: Comparison of the abundance and species composition of fish and crustaceans in mangroves and other near shore habitats in tropical Australia. *Mar. Biol.* (*Berl.*), 96: 193–206.

Sampath Kumar, S. 1996. Some eco-physiological aspects of *Penaeus merguiensis* from Karwar waters. *Ph.D. Thesis,* Karnatak University, pp. 265.

Sick, L.V., Andrews, J.W. and White, D.B. 1972. Preliminary studies of selected environmental and nutritional requirements for the culture of penaeid shrimps. *Fish. Bull. US Fish Wildlife Ser.*, 70: 101.

Sivabalan, A. 1993. Studies on the ecology of the mangrove of Kali estuary, Karwar. *Ph.D. Thesis,* Karnatak University, pp. 390.

Snedecor, G.W. and Cochran, W.C. 1967. *Statistical Methods*, 6th edition. Oxford and IBH Publishing Co., New Delhi, pp. 172.

Somers, I.F., Croaces, P.J. and Hill, B.J. 1987. Distribution and abundance of tiger prawns *Penaeus esculentus* and *Penaeus semisulcatus. Aust. J. Freshwat. Res.*, 38: 63.

Staples, D.J. and Vance, D.J. 1979. Effects of changes in catchability on sampling of juvenile and adolescent banana prawns *Penaeus merguiensis* de Man. *Aust. J. Mar. Freshwat. Res.*, 30: 511–519.

Staples, D.J., Vance, D.J. and Heals, D.S. 1985. Habitat requirement of juvenile penaeid prawn and their relationship to offshore fisheries. In: *Sec. Aust. Prawn Sem.*, (Eds.) Rothlisberg, P.C., Hill, B.J. and Staples, D.J. NPS_2 Clevland, Australia, pp. 47.

Strickland, J.D.H. and Parson, T.R. 1956. A practical handbook of seawater analysis. *Bull. Fish. Res. Bd., Canada,* 167: 310.

Vance, D.J., Haywood, M.D.E. and Staples, D.J. 1990. Use of a mangrove/estuary as a nursery area by post larval prawns, *Peneus merguiensis* de Man, in northern Australia. *Estuar. Coastal Shelf Sci.*, 31: 689–701.

Williams, A.B. 1958. Substrate as a factor in shrimp distribution. *Limnol. Oceanogr.*, 3: 283–291.

Young, P.C. and Carpenter, S.M. 1977. Recruitment of post larval prawns to nursery areas in Moreton Bay, Qld. Australia. *Aust. J. Mar. Freshwat. Res.*, 28: 745–773.

Chapter 24

Impact of Flyash of a Thermal Power Station on Biochemical Parameters of a Shrimp, *Penaeus monodon* Inhabiting Ennore Brackishwater

E. Ekambaram and D. Sudarsanam
School of Genomics and Bioinformatics, P.G. and Research Department of Zoology, Loyola College, Chennai – 34, India

ABSTRACT

Brackish water shrimp, *Penaeus monodon*, were exposed to flyash for a short-term and bioassayed for acid phosphatases (ACP) and alkaline phosphatases (ALP). The targeted tissues such as gill, hepatopancreas and muscle were selected for the assay at 24, 48, 72 and 96 hours and it revealed fluctuations.

Keywords: *Penaeus monodon, Thermal flyash, ACP, ALP.*

Introduction

Thermal pollution (calefaction) is defined as the output of heat by a population or industry by which the ambient temperature of water is increased. Heated effluents of thermal power plants are supersaturated with nitrogen and it is being taken by shrimps and fish into the blood stream through gills. Direct temperature effect of thermal effluents has also been responsible for fish kills near thermal power plants. One such case was reported in 1968 in Maryland, USA where 40,000 blue crabs were killed (Clark and Brownell, 1973). Flyash of thermal power station contains a wide range of harmful metals like As, Cd, Pb, Ni, Cr, Cu etc., (Majumudar and Mukharjee, 1983; Sikka and Kansal, 1994). The release of total flyash from thermal plants is estimated more than three times in India than other countries (Gupta and Birendra, 1998). Effect of pollutants on the respiration of aquatic invertebrates have received wide spread attention and are reviewed respectively by Hughe (1976) and Wright (1978).

In the present investigation the impact of flyash of Thermal Power Station (NCTPS) (Ennore, Chennai, Tamil Nadu) on the enzyme profile (acid and alkaline phosphatases) of a brackishwater shrimp, *Penaeus monodon* was assayed.

Materials and Methods

Penaeus monodon were collected from Pulicat lake (Pazhaverkadu) 60 km away from North Chennai, Tamil Nadu and they were acclimated in the laboratory for a week in well aerated sea water and fed every day with boiled egg (white) and small pieces of fish muscle. The feeding was arrested one day prior to experimentations to avoid interference. After acclimation, LC_{50} values were estimated for 24, 48, 72 and 96 hrs. Subsequently the shrimps were exposed to the ash water at sublethal concentrations 2 per cent and 3 per cent along with a control set up. Eventually the acid and alkaline phosphatases were assayed spectrophotometrically (Bergmeyer, 1974).

Results and Discussion

Acid and alkaline phosphatases being hydrolytic lysosomal enzymes bring about hydrolysis of foreign material. The enzymes are the indicators of various aspects of metabolism and are used to evaluate the physiological, biochemical, genetical and metabolic defects in gill, hepatopancreas and muscle of shrimps. These enzymes are deemed 'inducible enzymes' and their activity increases when there is a toxic impact and to counteract the toxic effect (Leland, 1983). Whitt and Savage (1981) analysed in germ freemice showed increased in ALP levels when compared to the mice infested with germs. A remarkable increase in the alkaline phosphatase level in the gill and hepatopancreas tissues at both the sublethal concentrations over the control during 24 hours of exposure was recorded. However, a decline in the enzyme was noted at 48, 72 and 96 hours of exposure (Tables 24.1–24.4) and the ANOVA revealed significant at $p < 0.05$ between the exposure period in all the three tissue samples and insignificant between the concentrations. Addition of heavy metal cadmium to crustaceans showed increase in acid and alkaline phosphatases (Geetha Ramesh and Govindan, 1992). Friberg *et al.* (1979) indicated that majority of toxins alter biochemical profiles by acting at the molecular level as inhibitors of enzymes system, and thereby an alteration in the level of enzyme and specificity and their permeability into the biological membranes.

Table 24.1: Acid Phosphatase Activity in Tissues of *Penaeus monodon* During Exposure Periods at 2 per cent Sublethal Concentration of Flyash

Tissues	*Control*	*24 hours*	*48 hours*	*72 hours*	*96 hours*
Gill	5.61 ± 0.11	8.25 ± 0.06	5.77 ± 0.03	4.31 ± 0.02	2.71 ± 0.06
Hepatopancreas	4.64 ± 0.09	4.29 ± 0.04	2.82 ± 0.05	1.38 ± 0.04	1.13 ± 0.12
Muscle	10.07 ± 0.16	8.72 ± 0.07	6.82 ± 0.04	4.59 ± 0.08	2.85 ± 0.04

Table 24.2: Acid Phosphatase Activity in Tissues of *Penaeus monodon* During Exposure Periods at 3 per cent Sublethal Concentration of Flyash

Tissues	*Control*	*24 hours*	*48 hours*	*72 hours*	*96 hours*
Gill	5.61 ± 0.11	14.31 ± 0.04	10.31 ± 0.04	8.79 ± 0.05	5.78 ± 0.05
Hepatopancreas	4.64 ± 0.09	6.28 ± 0.06	4.74 ± 0.4	3.82 ± 0.06	1.86 ± 0.05
Muscle	10.07 ± 0.16	12.80 ± 0.05	9.83 ± 0.03	5.77 ± 0.05	3.76 ± 0.04

Table 24.3: Alkaline Phosphatase Activity in the Tissues of *Penaeus monodon* During Exposure Periods at 2 per cent Sublethal Concentration of Flyash

Tissues	*Control*	*24 hours*	*48 hours*	*72 hours*	*96 hours*
Gill	3.41 ± 0.15	5.25 ± 0.27	4.49 ± 0.13	3.55 ± 0.11	3.53 ± 0.10
Hepatopancreas	5.88 ± 0.10	6.72 ± 0.02	6.77 ± 0.17	5.40 ± 0.09	4.86 ± 0.08
Muscle	21.92 ± 0.60	23.79 ± 0.13	23.99 ± 0.31	22.62 ± 0.20	21.88 ± 0.03

Table 24.4: Alkaline Phosphatase Activity in the Tissues of *Penaeus monodon* During Exposure Periods at 3 per cent Sublethal Concentration of Flyash

Tissues	*Control*	*24 hours*	*48 hours*	*72 hours*	*96 hours*
Gill	3.41 ± 0.15	5.68 ± 0.15	7.34 ± 0.09	5.74 ± 0.02	2.58 ± 0.12
Hepatopancreas	5.88 ± 0.10	8.54 ± 0.15	7.59 ± 0.06	4.44 ± 0.12	3.47 ± 0.12
Muscle	21.92 ± 0.60	26.50 ± 0.12	24.56 ± 0.12	22.45 ± 0.09	21.80 ± 0.10

The alkaline phosphatase of the control gill sample revealed a mean value of 3.41, while the experimental hepatopancreas showed 5.87, and muscle 21.92. In the exposure period of 24 and 48 hours, a declining trend and at 72 and 96 hours, an increasing activity were observed (Tables 24.3 and 24.4). Besides, the ANOVA revealed significant relationship between the sublethal concentrations and the exposure periods. The increased activity of acid and alkaline phosphatases is due to the labilization of cell membrane, including lysosomes leading to hepatosis.

Acknowledgement

We are grateful to Rev. Fr. V. Joseph Xavier, Principal, Loyola College for providing facilities, help and encouragement.

References

Bergmeyer, H.V. (1974. *Methods of Enzymatic Analysis*. Academic Press. New York.

Clark, J. and Brownell. 1973. *Electric Power Plants in the Coastal Zone: Environmental Issues*. American Littoral Society. Sp. Pub. No. & Highlands, N.J.

Friberg, L. Nordberg, G.F. and Vonk, V.B. 1979. *Handbook on the Toxicology of the Metals*. Elsevier, North Holland Biomedical Press, New York.

Geetha, R. and Govindan, V.S. 1992. Cadmium induced metabolic changes in zooplankton. *Indian J. Env. Pro.*, 12: 727–732.

Gupta, A. and Birendra, K. 1998. Environmental impact and leaching characteristics of fly ash obtained from thermal power station. *Res. J. Chem. Env.*, 2: 23–28.

Hughes, G.M. 1976. Polluted fish respiratory physiology. In: *Effects of Pollutants on Aquatic Organism*, (Ed.) Lock-Wood, A.P.M. Cambridge University Press, Cambridge, p. 163–183.

Majumudar, A. and Mukharjee, S.N. 1983. Trace element in Indian fly ash. *Fuel Sci. Tech.*, 2: 80–83.

Sikka, R. and Kansal, B.D. 1994. Characteristics of thermal power plant fly ash for agronomic purposes and to identify pollution hazards. *Bioresour. Tech.*, 50: 269–273.

Whitt, D.D. and Savage, D.C. 1981. Influence of indigenous microbiota on amount of protein and activities of ALP and disaccharides in extracts of intestinal mucosa in mice. *App. Env. Mitcobiol.*, 42: 513–520.

Wright, Z.W. 1978. Heavy metal accumulation by aquatic vertebrates. *Appl. Biol.*, 3: 331–394.

Chapter 25

Studies on the Correlation Between the Neurosecretory Activity and Ovarian Cycle of a Freshwater Prawn *Macrobrachium dayanum* (Henderson, 1893)

K. Borana and Arvind Kumar***

**Department of Applied Aquaculture, Barkatullah University Bhopal – 462 026, M.P.*
***Environmental Science Research Unit, P.G. Department of Zoology, S.K.M. University, Dumka – 814 101, Jharkhand, India*

ABSTRACT

The present work establishes that the type 'A', 'B' and 'C' NSCS of the thoracic ganglion and the 'B' and 'D' of the brain release the ovary stimulating hormones, during the breeding period, accelerating the reproductive activity, and the 'C' type NSCS of the eye-stalk produce the hormones which inhibit the ovarian development during the non-breeding period in *M. dayanum.*

Keywords: *Thoracic ganglion, Stimulating hormones, Breeding, Eye-stalk.*

Introduction

Reproduction in crustacea is under the control of neuroendocrine factors (Adiyodi and Adiyodi, 1971). The role of neurosecretory material (NSM) produced by the eye-stalks, brain and the thoracic ganglion, in reproduction and several other physiological functions in Crustacea, is sufficiently worked out but the exact source (cell types) of the production of particular hormone is not yet ascertained in any of the crustacean member. Hanstrom (1937) for the first time carried out the pioneering experiments on the physiological changes taking place in the neuroendocrine centers of crustaceans. Otsu (1963) working on the freshwater crab, *Potamon dehaani* confirmed the hypothesis that the sinus gland and/or the eye-stalks contain an ovary inhibiting hormone.

Few reports are available regarding the correlation between the reproductive cycle and the neurosecretion in crustaceans (Parameswaran, 1955; Diwan and Nagabhushanam, 1975; Joshi, 1980 and Bhanudas Rao, 1989). Nagabhushanam *et al.* (1989) explained the pheromonal and hormonal control of reproduction in the freshwater prawn, *Macrobrachium kistnensis.* Rajender Rao (1989) demonstrated the relation among neuroendocrine and its control over reproduction in *Macrobrachium lamarrei.* Han and Kim (1993) advocated the effect of neurosecretory cells and X-organ on gonadal maturation in freshwater prawn, *Macrobrachium nipponense.* Wilder (1998) demonstrated that in decapod crustacea, the physiological process of molting and reproduction are linked together and governed by the hormonal activities. Upadhyay (2001) demonstrated the correlation between neurosecretory cells and the gonads in freshwater prawn, *Macrobrachium malcolmsonii,* but our knowledge is limited in relation to freshwater prawns in this regard. It is, therefore, the present investigation has been undertaken to find out whether there exists any inter-relationship among these two systems of *M. dayanum.*

Materials and Methods

For the purpose of ovarian cycle, the prawns were seggregated depending upon their ovarian condition *viz.*, Previtellogenic, Vitellogenic-I, Vitellogenic-II and Spent (prawns spawned in the laboratory were used). They were sacrificed and their eye-stalks, brains and the thoracic ganglia were dissected out and fixed in Bouin's fluid. The material was then processed for histological preparation, sectioned at 4–6 µm thickness, stained with Gomori's chromealum hematoxylin phloxine (Gomori, 1941). Histological observations on the neurosecretory profile in five prawns of each group were made and the results averaged. The neurosecretory profile of the pre-vitellogenic individuals was considered as standard with which the neurosecretory profile of the other groups was compared.

Results

Tables 25.1–25.3 exhibit the changes taken place in the amount of NSM and the nuclear diameter of the neurosecretory cells (NSCS) of the eye-stalk, brain and the thoracic ganglion of *M. dayanum* at various stages of maturity.

The type 'C' NSCS of the eye-stalk show the decrease in the nuclear diameter in the vitellogenic-I, vitellogenic-II and spent conditions of the ovarian cycle. However, the type 'D' and type 'E' cells do not show any correlative change in their nuclear diameter in relation to the ovarian cycle.

In case of brain, the type 'B' NSCS show the increase in their nuclear diameter during vitellogenic-I stage. However, the increase in the nuclear diameter of these cells during vitellogenic-II and spent stages is comparatively less. The type 'C' and 'E' NSCS do not show any significant correlative change in them in relation to the ovarian cycle. The type 'D' NSCS exhibit marked increase in their nuclear diameter and the increase in the quantity of NSM in their cytoplasm during vitellogenic-II and spent stages.

The type' A ' NSCS of the thoracic ganglion show the increase in their nuclear diameter and NSM in the cytoplasm during the vitellogenic-I, vitellogenic-II and the spent stages. However, the increase

Table 25.1: Alteration in the Neurosecretory Profile of the Eye-stalk of *M. dayanum* in Relation to the Ovarian Maturation

Ovarian Stages Used for Study	*Pre-vitellogenic*			*Vitellogenic-I*			*Vitellogenic-II*			*Spent*		
NSC Types	*C*	*D*	*E*	*C*	*D*	*E*	*C*	*D*	*E*	*C*	*D*	*E*
NSM	*3*	*2*	*1*	*2*	*1*	*1*	*0*	*2*	*3*	*1*	*2*	*1*
Nuclear diameter (μm)±SD	9.5± 0.50	7.2± 00.57	6.22± 0.63	9.1± 0.65	7.12± 0.60	6.36± 0.50	**8.24± 0.17	6.96± 0.65	6.43± 0.42	**8.48± 0.09	6.83± 0.59	6.38± 0.36
% Change in Nuclear Diameter	–	–	–	– 4.21	– 1.11	2.25	– 13.26	– 3.33	3.37	– 10.73	– 5.13	2.57

** P ∠ 0.001 Significant increase over control.

0: NSM absent; 1: NSM less; 2: NSM moderate; 3: NSM more.

Table 25.2: Alteration in the Neurosecretory Profile of the Brain of *M. dayanum* in Relation to the Ovarian Maturation

Ovarian Stages Use for Study	*Pre-vitellogenic*				*Vitellogenic-I*				*Vitellogenic-II*				*Spent*			
NSC Types	*B*	*C*	*D*	*E*	*B*	*C*	*D*	*E*	*B*	*C*	*D*	*E*	*B*	*C*	*D*	*E*
NSM	*1*	*2*	*2*	*1*	*3*	*2*	*2*	*2*	*2*	*2*	*3*	*2*	*3*	*2*	*2*	*1*
Nuclear diameter (μm)±SD	15.33± 0.54	9.38± 0.53	8.09± 0.48	6.28± 0.57	***16.43± 0.45	9.70± 0.40	8.25± 0.50	6.40± 0.57	15.71± 0.51	9.80± 0.31	**8.88± 0.49	6.60± 0.51	15.60± 0.52	9.56± 0.42	*8.73± 0.49	6.48± 0.58
% Change in nuclear diameter	–	–	–	—	7.17	3.41	1.97	1.91	2.47	4.47	9.76	5.06	1.76	1.91	7.91	3.18

* P < 0.02 Significant increase over control.

** P < 0.01 Significant increase over control.

*** P < 0.001 Significant increase over control.

0: NSM absent; 1: NSM less; 2: NSM moderate; 3: NSM more.

Table 25.2: Alteration in the Neurosecretory Profile of the Thoracic Ganglion of *M. dayanum* in Relation to the Ovarian Maturation

Ovarian Stages Used for Study	Pre-vitellogenic					Vitellogenic-I					Vitellogenic-II					Spent				
NSC Types	A	B	C	D	E	A	B	C	D	E	A	B	C	D	E	A	B	C	D	E
NSM	2	2	2	2	1	3	2	2	2	1	3	3	3	2	2	2	2	3	1	1
Nuclear diameter (µm)±SD	34.00± 0.35	15.50± 0.47	10.18± 0.51	8.23± 0.51	6.28± 0.57	**** 33.42± 0.55	15.96± 0.40	10.46± 0.69	8.35± 0.53	6.42± 0.49	*** 35.22± 0.65	* 16.06± 0.40	** 11.10± 0.58	8.43± 0.49	6.56± 0.45	** 34.56± 0.37	15.84± 0.27	* 10.83± 0.58	8.31± 0.48	6.34± 0.43
% Change in nuclear diameter	–	–	–	–	–	4.17	2.96	2.75	1.45	2.22	3.58	3.61	9.03	2.43	4.45	1.64	2.19	6.38	0.97	0.95

* $P < 0.05$ Significant increase over control.

** $P < 0.02$ Significant increase over control.

*** $P < 0.01$ Significant increase over control.

**** $P < 0.001$ Significant increase over control.

0: NSM absent; 1: NSM less; 2: NSM moderate; 3: NSM more.

in the NSM in the spent stage is comparatively less. The type 'B' NSCS show pronounced increase in their nuclear diameter and the NSM during the vitellogenic-II stage. The Type 'C' NSCS exhibit significant increase in their nuclear diameter and the NSM during the vitellogenic-II and the spent stages. Type 'D' and 'E' cells do not exhibit any change in their structure in relation to the cyclic activity of the ovary.

Discussion

Five types of NSCS ('A', 'B', 'C', 'D' and 'E') are observed to occur in the neurosecretory system of *M. dayanum* (Rajesh Biswas, 1991). Out of them, only 'C', 'D' and 'E' type NSCS are found in the eye-stalk; 'B', 'C', 'D' and 'E' in the brain and all the five types are found in the thoracic ganglion. Changes in the NSCS profile, related to the reproductive activity have been noticed.

Alterations have been noticed in the NSCS of the eye-stalk, in relation to the ovarian cycle. Type 'C' NSCS exhibited decrease in their neurosecretory activity with the advancement of ovarian maturation, while type 'D' and 'E' NSCS did not react to this situation. This observation conforms with that of Joshi (1980) who has also registered the decreased neurosecretory activity in type IV NSCS commensurating with the advancement of the ovarian maturation in *P. stylifera.* Though he observed the increased neurosecretory activity in type VII NSCS in fully mature on recently spawned animals, but could not find any neurosecretory activity in type V NSCS.

The NSCS profile of the brain also exhibits the changes in relation to the ovarian maturation. Type 'B' NSCS become more active at the advent of vitellogenic-I stage while they grow less active during the vitellogenic-II stage. The same was recorded by Nagabhushanam *et al.* (1984) in type 'IV' NSCS (comparable to type 'B' NSCS of the present investigation) of *Parapenaeopsis stylifera.* The type 'C' and 'E' NSCS of the brain showed the same activity as has been shown by the type 'D' and 'E' NSCS of the eye-stalk. Type 'D' NSCS of the brain have been found to be more active in the vitellogenic-II stage as compared to the vitellogenic-I and the spent period of the ovarian cycle. In *P. stylifera* (Nagabhushanam *et al.*, 1984), the type VII and V NSCS exhibited the same features as the type 'C' and 'E' NSCS exhibited in case of *M. dayanum.*

The NSCS of the thoracic ganglion are no exception in exhibiting the changes in relation to the ovarian maturation. 'A' type NSCS showed gradually increasing activity corresponding to the advancement of vitellogenic-I and vitellogenic-II stages and during the spent period. These NSCS became hypo-active after the extrusion of the oocytes from the ovaries. Parameswaran (1955) correlated the high secretory activity of the type 'A' NSCS of the thoracic ganglion of the freshwater crab, *Paratelphusa hydrodromous* with the reproductive activity of the females (gravid and ovigerous). Joshi (1980) observed that the type 'I' NSCS of *P.stylifera* display maximum neurosecretory activity during the vitellogenic-I and vitellogenic-II stages which decreases during the spent period. The type 'B' and 'C' NSCS are found active during the vitellogenic-II stage and the spent period of ovarian activity, respectively. The type 'D' and 'E' NSCS of the thoracic ganglion show the features, parallel to those of the type 'D' and 'E' NSCS of the eyestalk and the 'C' and 'E' cells of the brain.

The above observations are suggestive that the type 'C' NSCS of the eye-stalk, type 'B' and 'D' of the brain and the type 'A', 'B' and 'C' of the thoracic ganglion exhibit alteration in their activity in relation to the ovarian maturation in *M. dayanum.* Matsumoto (1958) has established the relationship between 'A', 'A^{1}' and 'E' cells of the central nervous system and the sexual cycle in five species of crabs, he studied. Nagabhushanam and Diwan (1975) have observed that the type 'A' NSCS, in the brain and the thoracic ganglion of the crab, *B. cunicularis,* were active during the pre–spawning and spawning periods. Joshi (1980) and Nagabhushanam *et al.* (1984) observed that the types IV, VI and

VII NSCS of the eye-stalk, type IV and VII of the brain and the type IV and VII of the thoracic ganglion exhibit alterations in their activity in correlation with the ovarian maturation in the prawn, *P. stylifera.*

The above observations suggest that the neurosecretory hormones are synthesized by the 'A', 'B' and 'C' NSCS of the thoracic ganglion and the 'B' and 'D' NSCS of the brain and that the hormones, synthesized by these cells, stimulate the process of oogenesis. On the other hand, the 'C' type NSCS of the eye-stalk, which are more active during the previtellogenic period, might be producing the hormone exerting inhibitory action on the ovarian development. Adiyodi (1969) suggested the possibility, in the crab, *Paratelphusa hydrodromous* that the 'C' type NSCS of the eye-stalk might be producing the gonad inhibitory hormone. Parryman (1969), working on *Procambarus simulans*, proposed that the X-organ (type one cells) produces the ovulation inhibitor, and the type three cells of the brain produce accelerator substances. Bomirski and Klek (1974) and Klek and Bomirski (1975) reported that the ovary inhibitory hormone activity, of the eye-stalk, is responsible for the occurrence of the resting phase in the ovary of *Crangon crangon.* Upadhyay (2001), while working on the freshwater prawn, *M. malcolmsonii*, proposed that the process of gametogenesis accelerated by the neurosecretory hormones released from type 'A' 'B' and 'C' NSCS of thoracic ganglion and type 'B' and 'C' NSCS of the brain during breeding period and type 'C' NSCS of eye-stalk secrete the ovary inhibitory hormones.

The biological changes observed in different neurosecretory cell types from different ganglion of *M. dayanum* in relation to ovarian cycle lend enough support to postulate tentatively the following facts:

1. Type 'C' NSCS of eye-stalk are involved in ovarian inhibition *i.e.*, the source of gonad inhibiting hormone (GIH).
2. Type 'D' and type 'E' cells from eye stalk and type 'C' and 'E' of brain have no role to play in reproduction.
3. Type 'B' and type 'D' NSCS from brain are source of ovarian stimulating hormone (OSH). They probably stimulate different vitellogenic stages *i.e.*, vitellogenic-I and vitellogenic-II, respectively.
4. Type 'A', 'B', 'C' and 'D' NSCS from thoracic ganglion are also the source of ovarian stimulating hormone (OSH). They probably act on different stages of ovarian cycle.

References

Adiyodi, R.G. 1969. Protein metabolism in relation to reproduction and moulting in the crab, *Paratelphusa hydrodromous* (Herbst) III: RNA activity and protein yolk biosynthesis during normal vitellogenesis and under condition of acute insanition. *Indian J. Exp. Biol.*, 7: 13–16.

Adiyodi, R.G. and Adiyodi, K.G. 1971. Endocrine control of reproduction in decapod Crustacea. *Biol. Rev., Cambridge Phil. Soc.*, 45: 121–165.

Bhanudas Rao, S.K. 1989. Reproductive biology of Indian marine prawn and its commercial importance. *Ph.D. Thesis*, Marathwada University, Aurangabad, India.

Bomirski, A. and Klek-Kawinsxa, E. 1974. Action of the eye-stalk on ovary in *Rhithropanopeus harrisii* and *Crangon crangon* (Crustacea: Decapoda). *Mar. Biol.*, 24: 329–337.

Diwan, A.D. and Nagabhushanam, R. 1975. Neurosecretory cells of the central nervous system of the freshwater crab, *Barytelphusa cunicularis* (Westwood, 1836). *Rivista di Biologia LXVIII-FASc 1/2 Gennaio-Giugno Nuova Starte*, 28: 79–115.

Gornez, R. and Nayar, K.K. 1965. Certain influence in the reproduction of the crab, *Paratelphusa hydrodromous. Zool. Jb. Abt. Physiol.*, 71: 699–701.

Han, C. and Kim, D.J. 1993. Studies on the X-organ of eye-stalk and the photoperiod for the control of gonadal maturation in a freshwater prawn, *Macrobrachium nipponense* (De Haan). *Bull. Korean Fish. Soc.*, 26(1): 76–90.

Hanstorm, B. 1937. Die sinus druse under hormonal bedingte Furbwechsel der crustacean King. *Svenska. Vetenskapsakod, Handl.*, 16: 1–99.

Joshi, P.K. 1980. Reproductive physiology and neurosecretion in some Indian marine prawn. *Ph.D. Thesis,* Marathwada University, Aurangabad, India.

Klek-Kawinska, E. and Bomirski, A. 1975. Ovary inhibition hormone activity in shrimp *Crangon crangon* eye-stalk. *Gen. Comp. Endocr.*, 25: 9–13.

Matsumoto, K. 1958. Morphological studies on the neurosecretion in crabs. *Biol. J.*, Okayflma University, 4: 176.

Nagabhushanam, R. and Diwan, A.D. 1975. Neuroendocrine control of reproduction in the male crab, *Barytelphusa cunicularis. Proc. V. Ann. of Oceania. Comp. Endocrinol.*, Chandigarh, 5(1): 375–385.

Nagabhushanam, R., Joshi, P.K. and Sarojini, R. 1984. Changes in the activity of neurosecretory cells during ovarian maturation in *Parapenaeopsis stylifera. Uttar Pradesh J. Zool.*, 4(2): 185–192.

Otsu, T. 1963. Bihormonal control of sexual cycle in the freshwater crab, *Potamon dehaani. Embriologica*, 8: 1–20.

Parameswaran, R. 1955. Neurosecretory cells in *Paratelphusa hydrodromous* (Herbst). *Curr. Sci.*, 24: 23–24.

Perryman, E.K. 1969. *Procambarus simulans.* Light induced changes in neurosecretory cells in ovarian cycle. *Trans. Am. Microsc. Soc.*, 88: 518–524.

Rajendra Rao, K. 1989. Reproductive biology of the freshwater prawn *Macrobrachium lamarrei. Ph.D. Thesis,* Marathwada University, Aurangabad, India.

Rajesh Biswas. 1991. Developmental Biology and correlative cyclic changes in the neurosecretory cells and gonads of freshwater prawn, *Macrobrachium dayanum. Ph.D. Thesis,* Barkatullah University, Bhopal, India.

Upadhyay, R.K. 2001. Co-relative cyclic changes in neurosecretory cells and gonads of a freshwater prawn, *Macrobrachium malcolmsoni* (H. Milne Edwards, 1844) and its fishery in Weinganga River at Balaghat (M.P.).

Wilder, M.N. 1998. Reproductive mechanisam in *Macrobrachum rosenbergii* and *Penaeus japonicus* Endocrinological research and potential applications in aquaculture. New–Hampshire Univ. Durham USA Sea Grant Program 1998: 125–136.

Chapter 26

Effect of Monocrotophos on Biochemical Markers DNA and RNA in *Cyprinus carpio* During Exposure and Recovery Period

C. Maruthanayagam and G. Sharmila
A.V.C. College (Autonomous), Department of Zoology, Mannampandal, Mayiladuthurai – 609 305

ABSTRACT

The present study was designed for 15 days Monocrotophos exposure period followed by 15 days recovery period to understand the toxic effects of toxicants on the nucleic acid aspect of *Cyprinus carpio.* The pesticide leads to several changes in the biochemical markers like DNA and RNA may be due to the increased activity of the enzyme DNAase and the inhibition of RNA polymerase function. But during recovery period, the DNA and RNA level both progressively increased indicating a probable recovery from the disruption of internal organs.

Introduction

The indiscriminate use of various types of pesticides in the modern world has led to much greater emphasis on the probability of serious environmental contamination to the survival of mankind (Cremlyn, 1978). Such indiscriminate use of pesticides in agriculture and public health operations has increased the scope of disruption of ecological balance as many of the 'non-target' organisms which include important members of the food chain that are perishing (Saxena, 1989). Ahana Lakshmi

(1992) has reported that pesticidal contamination of water bodies causes changes in the hydrobiological characters of water, ultimately affecting the aquatic biota in general and fishes in particular. In fact fishes are very sensitive to a wide variety of toxicants in water (Holden, 1972). The non-target species such as fishes from marine and freshwaters have been monitored for pesticide contamination (Couch, 1978).

Another toxic effect of the pesticide is seen in the alterations it brings in the nucleic acid content of the animal. The role of nucleic acids particularly RNA/DNA and protein/DNA rations, which are used as an index of protein synthesis and cell size, are considered to be important and form an integral part of the growth and toxicity studies in animals. But such studies in fishes are scanty. The treatment with the pesticides causes variability in the nucleic acid content in different tissues and the degree of variability or extent of alterations caused by the pesticides is found to be dose dependent (Malla Reddy and Bashamohideen, 1988).

Materials and Methods

The fishes were collected from the control and the experimental tanks at the end of 15 days of exposure to Monocrotophos and at the end of 15 days of recovery period and all the fishes were subjected to the nucleic acid analysis individually.

The nucleic acids were extracted from the tissues by the method of Schneider (1957). Subsequently, the DNA and RNA concentrations of muscle, liver and intestine were estimated by the method of Burdon (1957) and Ceriotti (1965) respectively.

Results

When the fishes were exposed to presumable harmless concentration of 51.57 ppm of Monocrotophos, the tissues of muscle, liver and intestine showed significant amount of decreased in the nucleic acid content (RNA and DNA) during exposure period. However during recovery period, both the RNA and DNA level progressively increased (Tables 26.1–26.2).

Table 26.1: Sublethal Effect of Pesticide on RNA Content (mg/g) in Different Tissues of *Cyprinus carpio* During Exposure and Recovery Periods. Each value is Mean ± S.E (n = 5).

Period of Study	*Muscle (mg/g)*	*Liver (mg/g)*	*Intestine (mg/g)*
Control (0 day)	5.4±0.33	6.3±0.41	4.8±0.26
Exposure (15 days)	3.4±0.24*	4.2±0.31**	3.1±0.18**
	(– 37.03)	(– 33.3)	(– 35.41)
Recovery Control	5.5±0.53	6.8±0.63	4.9±0.16
Recovery (15 days)	5.0±0.32**	5.9±0.33**	4.8±0.09*
	(– 9.09)	(– 13.2)	(– 2.04)

(–) Values in parenthesis indicates percent decrease over control.

** Values are statistically significant at $p < 0.5$ (Student's 't' test)

* Values are statistically significant at $p < 5$.

Discussion

The pesticides lead to several changes in the biochemical markers such as RNA and DNA non-target agents of aquatic pollution resulting in secondary responses (Mazeaud *et al.*, 1977). In the present study, the exposure of fishes to Monocrotophos resulted in the depletion of RNA and DNA. This decrease in nucleic acid suggests the decrease in protein synthesis and further damage to the liver, which is the major metabolic organ of drug detoxification (Ramalingam *et al.*, 2000). The decrease of RNA may be suggested that the daily addition of pesticides results in the swelling and chromatolysis of Nissle bodies which are rich in RNA (Watson. 1969). According to Mukhopadhyay and Dehadrai (1980), the decrease of RNA may also be due to interference in the incorporation of precursor in the nucleic acid synthesis or inhibiting the function of RNA polymerase. Dawood (1986) and Benjamin (1990) have suggested that the decrement of RNA may also be due to the non-coding for the process of protein synthesis, thereby decrease in the mRNA content, which in turn would have reduced the concentration of RNA. Organophosphorus compounds exhibit strong mutagenic and clastogenic potentiality (Patankar and Vaidya, 1980), which may be responsible for the alteration of DNA level. However the decrease of DNA is not very prominent when compared to RNA. The decrease may be attributed to the increased activity of DNAase as suggested by Tayyaba *et al.* (1981). The decrement of DNA may also be due to the degeneration of hepatic cells due to the effect of Organophosphate compounds, which is in agreement with the work of Bachelard (1974).

Table 26.2: Sublethal Effect of Pesticide on DNA Content (mg/g) in Different Tissues of *Cyprinus carpio* During Exposure and Recovery Periods. Each value is Mean ± S.E. (n = 5).

Period of Study	*Muscle (mg/g)*	*Liver (mg/g)*	*Intestine (mg/g)*
Control (0 day)	1.8±0.07	2.9±0.07	1.4±0.15
Exposure	1.4±0.04**	1.9±0.13**	1.0±0.12**
(15 days)	(– 22.2)	(– 34.4)	(– 28.5)
Recovery Control	1.9±0.09	2.4±0.03	1.5±0.11
Recovery (15 days)	1.5±0.08**	1.98±0.13**	1.3±0.13**
	(– 21.0)	(– 17.5)	(– 13.3)

(–) Values in parenthesis indicates percent decrease over control.

** Values are statistically significant at $p < 0.5$ (Student's 't' test).

The observed persistent significant decrease in both protein and nucleic acids would suggest that pollutant impair the process of protein synthesis in the tissues of fishes exposed to pesticides. Since RNA is the biochemical mid wife in the formation of proteins, the diminished RNA content also affects the cellular protein content. Eichorn (1995) have also suggested that the depression in DNA synthesis is not energy dependent and may be due to the disruption of the replication process. In the recovery period, the RNA and DNA level progressively increased indicating a probable recovery from the disruption of internal organ and decrease of hypoxic condition to the fish, as evidenced by the recovery of whole animal activity after their shift from pesticide water to normal water.

Hence in the present investigation, the reduction of nucleic acids can be taken as meaningful biochemical indices of pesticides toxicity to assess the extent of pollution of the aquatic environment.

Acknowledgement

We express our sincere thanks to Mr. S. Swetharanyam, Head of the Department, Department of Zoology, Dr. V. Radhakrishnan, Principal, A.V.C. College for giving constant support and encouragement and providing us the necessary laboratory facilities in doing this research work.

References

Ahana Lakshmi, P. 1992. Pesticides in India. Risk assessment and policy. Abstract No. SI-24. 2nd *European Conference on Ecotox.*, Amsterdam., Netherlands.

Bachelard, H.S. 1974. Outline studies in biology. *Brain Biochem.*, Chapman and Hall, London, pp. 8.

Benjamin, N. 1990. Biochemical changes due to learning in the brain of rat exposed to an Organophosphate pesticide. *M.Phil. Thesis*, University of Madras.

Burdon, R.M. 1957. In: *Methods in Enzymology*, (Eds.) S.P. Colowick and N.O. Kaplam. Academic Press, New York, 1: 267.

Ceriotti, G. 1965. A microchemical determination of deoxyribonucleic acid. *J. Biol. Chem.*, 198: 297–303.

Couch, A.J. 1978. Histopathological effects of pesticides and related chemicals on the liver of fishes. *Nature*, 273: 559–586.

Cremlyn, R. 1978. *Pesticides: Preparation and Mode of Action*. John Wiley and Sons, New York, pp. 240.

Dawood, N. 1986. Studies on the functional organization of the brain of rabbit, *Oryctolagus cuniculus*. *Ph.D. Thesis*, University of Madras.

Eichorn, G.L. 1995. Organohalogen compound: Active sites of biological micromolecules and their interaction with heavy metals. In: *Ecological Toxicology Research: Effects of Heavy Metals*, (Eds.) A.D. McIn tyre and C.F. Mills. Plenum press, New York, p. 123–143.

Holden, A.V. 1972. The effects of pesticides on life in freshwater. *Proc. R. Soc. Cond.*, 180: 383–394.

Malla Reddy, P. and M.D. Bashamohideen. 1988. Toxic impact of Fenvelerate on the Protein metabolism in the branchial tissues of fish *Cyprinus carpio*. *Curr. Sci.*, 57: 211–212.

Mazeaud, M.M., Mazeaud, F. and E.M. Donaldson. 1977. Primary and secondary effects of stress in fishes: Some new data with a general review. *Trans. Am. Fish. Soc.*, 106: 201–212.

Mukhopadhyay, P.K. and P.V. Dehadrai. 1980. Biochemical changes in the air-breathing cat fish, *Clarias batrachus* (Linn.) exposed to Malathion. *Environ. Pollut.* 22(A): 149–158.

Patankar, N. and V.G. Vaidya. 1980. Evaluation of genetic toxicity of the insecticide Phosphamidon using *in vitro* mammalian test systems. *Ind. J. Exp. Biol.* 18: 1145–1147.

Ramalingam, V., Vimaladevi, Narmadaraji, R. and P. Prabakaran. 2000. Effect of the lead on the haematological and biochemical changes in the freshwater fish, *Cirrhina mrigala*. *Poll. Res.*, 19(1): 81–84.

Saxena, R.C. 1989. Detection of insecticides residues on fruits and vegetables sold in local market, Project report: 82: 8.

Schneider, R.P. 1957. Mechanism of inhibition of rat brain (Na^{++})K^+ Adenosine triphosphate by 2,2-bis (P-chlorophyll)-1,1,1-Trichloroethane (DDT). *Bio. Chem. Pharmacol.*, 24: 939–946.

Tayyaba, K., Hasan, M.F. and N.H. Khan. 1981. Organophosphate Metasustox-R, Induced regional alterations in brain nucleic acid metabolism. *Indian J. Exp. Biol.*, 19: 688–690.

Watson, W.E. 1969. The response of motor neurons to intramuscular injection of botulinum toxin. *J. Physiol.*, 202: 611–630.

Chapter 27

Length-Weight Relationship of the Gobiid Fish *Pseudapocryptes lanceolatus* (Bloch and Schneider, 1801) of the Vasista Godavari Estuary, East Coast of India

K.V.C.S. Appa Rao and K. Sreeramulu*

Department of Zoology, Andhra University, Visakhapatnam

ABSTRACT

The length-weight relationship of *Pseudapocryptes lanceolatus* (Bloch and Schneider) was computed. The data of 493 specimens including 226 females and 267 males was separately identified. The regression coefficient was found to depart significantly from the cubic value. The values of 'b' for males was 2.1365 and females was 2.2865. The females showed better condition factor (0.3666) than males (0.3281).

Keywords: *Pseudapocryptes lanceolatus, Vasista Godavari estuary, Length weight relationship, Condition factor.*

Introduction

Pseudapocryptes lanceolatus (Bloch and Schneider), is commonly called as a lanceolate tailed goby or Amphibious goby, is an important food fish occurring in the east coast estuaries of India. It is well distributed in almost all the riverine estuaries from Ganges in the north to Cauvery in the South. The fish is found inhabiting the tide pools as well as mud bottoms of the estuaries, especially in the mangrove ecosystems.

---

* Lecturer in Zoology, Sri Y.N. College, Narsapur – 534 275, West Godavari District.

The length-weight relationship is an important parameter in the fishery biology. The relationship indicates the taxonomic differences and events in the life history and also denotes the fatness and general well being of a fish or groups of fishes. The length weight relationship is also necessary to set yield equations (Beverton and Holt, 1957) and at times useful as a character for differentiation of "small taxonomic units" (La Cren, 1951). Length-weight relationships studies include determination of body condition factor that indicates the relative well being of a fish and conversion of length-growth data to weight-growth models (Kubbicki *et al.*, 1993 and King 1996). The present investigation concerned with the length-weight relationship of *P. lanceolatus* of the Vasista Godavari estuary, East coast of India.

Materials and Methods

A total of 493 specimens, 226 females measuring 10.1 cm to 20.0 cm in length and 267 males measuring 9.1 cm to 21.0 cm in length were collected from the local fish markets of Narsapur during December, 2000 to November 2001. Immediately after bringing to the laboratory, the total length (T.L.) was measured from the tip of the snout to the tip of the tail in centimeters and weight was noted in grams.

Length-weight relationship of either sex of the fish *P. lanceolatus* was calculated separately by using the formula of Tesch (1971).

$$W = aL^b$$

The expressions were later on transferred to logarithmic form using the expression $\text{Log } W = \log a + b \log L$

Subsequently, the identity of regression between the males and females was found out from the student 't' test. Coefficient of correlation (r) between the length and weight were also determined. The regression coefficient of each group was tested for significance of difference from 'Cube' relation Beverton and Holt (1957) employing students 't' test (Bailey, 1959).

The following equation was adapted to calculate the Relative Condition factor (Kn)

$$\left[Kn = \frac{W}{aL^n} \right]$$

where,

W: Weight of fish

L: Length of fish;

a: Constant and

n: Exponent.

Results and Discussion

Scatter diagrams plotted separately for one length and weight of males and females of *P. lanceolatus*, suggested a curvilinear relationship between the two variables (Figures 27.1–27.3). The derived parabolic equations for males and females were:

Male $W = 0.233\ L^{2.1365}$

Female $W = 0.204\ L^{2.2865}$

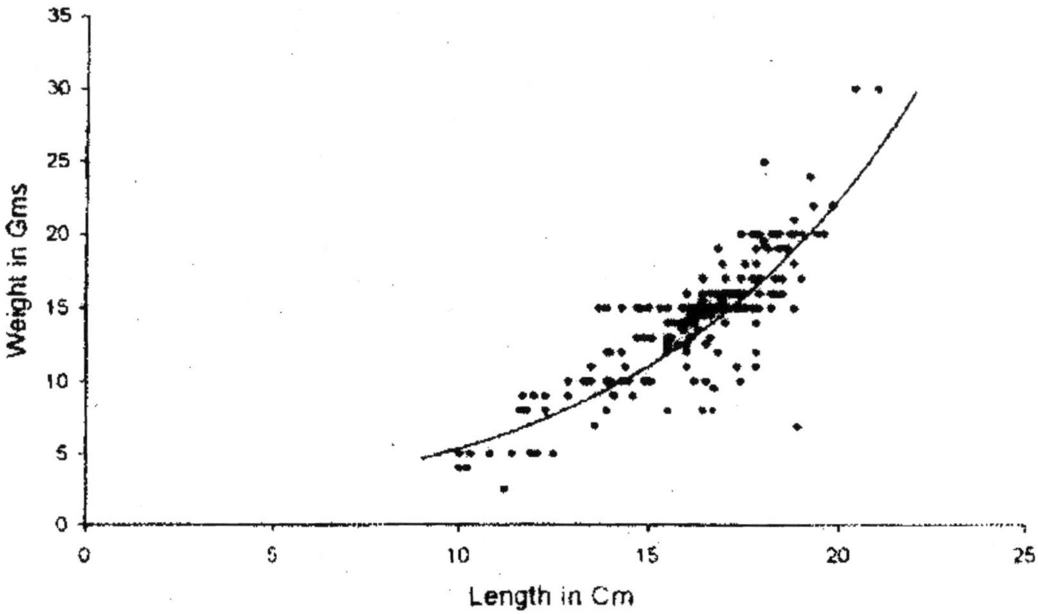

Figure 27.1: Length-weight Relationship of Observed Values in Males of *P. lanceolatus*

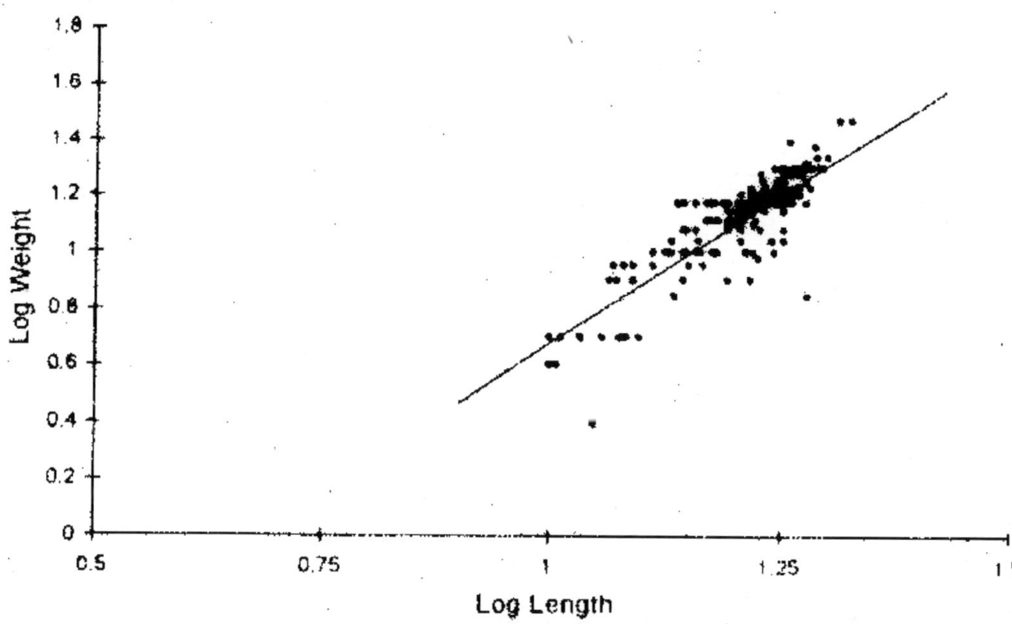

Figure 27.2: Log Length and Log Weight Relationship in Males of *P. lanceolatus*

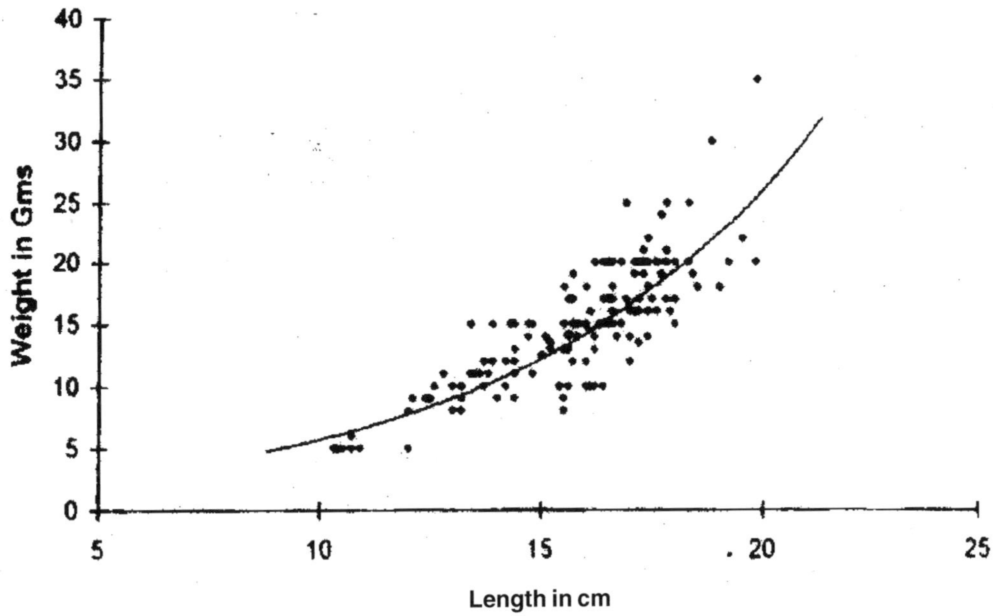

Figure 27.3: Relationship of Observed Values of Length and Weight in Females of *P. lanceolatus*

The regression lines of linear relationship arrived at on employing the method of least squares for the two categories were:

Male Log W = 2.1365 log L-1.4559 (Figure 27.2)

Female Log W = 2.2865 log L-1.5903 (Figure 27.4)

Test of significance revealed that regression coefficients of both the sexes differed significantly at 5 per cent level (t = 1.96).

The condition factor of the females was 0.3666 and the males was 0.3281. The condition factor was least, 0.2127 for males and 0.2566 for females during January and highest, 0.4302 for males and 0.4623 for females during June and February respectively. The peak factor of 0.4302 for males almost matched with the second highest peak of 0.4325 for females in June while the second highest peak factor of 0.4623 for females (Figure 27.5).

A change in the value of exponent is due to a change in specific gravity or shape and contour of the body or condition of fish or environmental conditions or a combination, of some or all of these factors. Morphological changes due to age may also cause substantial changes in the exponent of length on weight (Le Cren, 1951). In such cases the cube law need not always hold good as pointed out by Rounsfell and Everhaurt (1953).

The graphic representation of length-weight data of *P. lanceolatus* indicates a curvilinear relationship in the case of observed values and a straight line relationship in respect of the logarithmic transformations.

Analysis of the length-weight data revealed that the 'b' values were 2.1365 and 2.2865 for males and females respectively. The results indicated that both the sexes of the fish do not fall in line with the

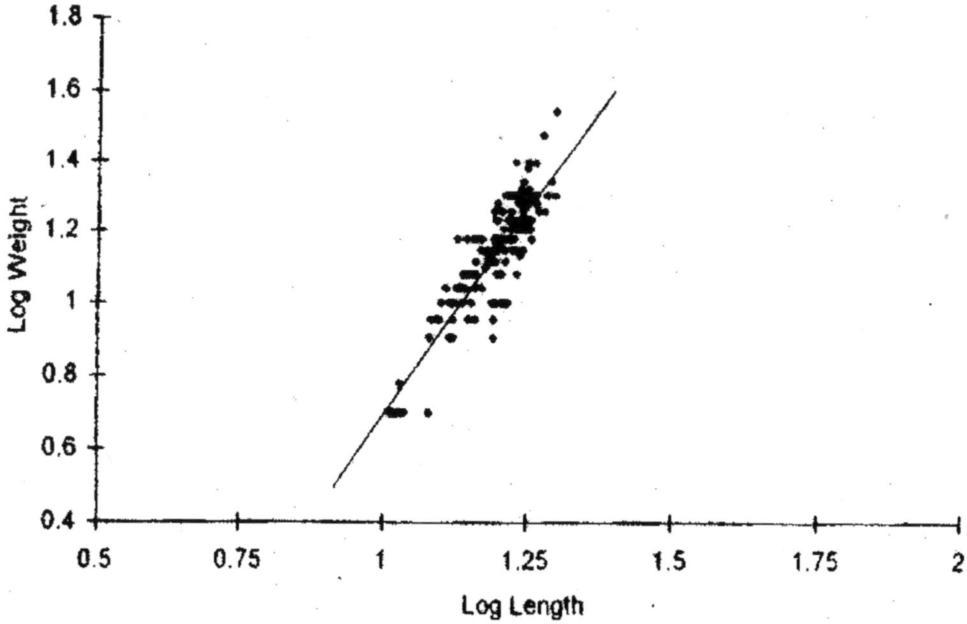

Figure 27.4: Log Length and Log Weight Relationship in Females of *P. lanceolatus*

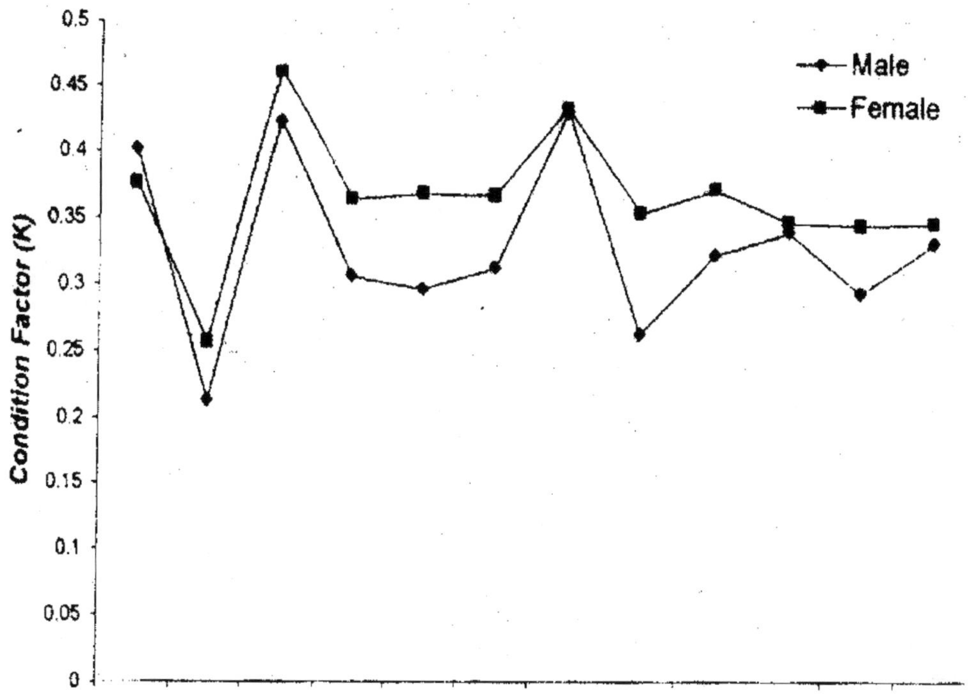

Figure 27.5: Condition Factor of Either Sex of *P. lanceolatus*

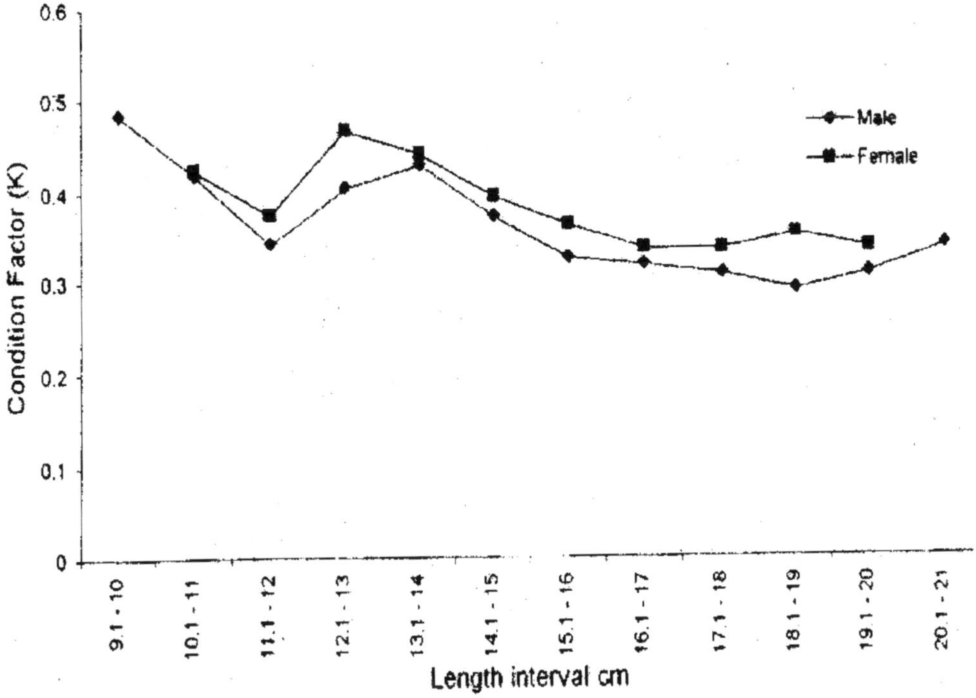

Figure 27.6: Relation Between Length and Condition Factor of Either Sex of *P. lanceolatus*

cube law, but depict negative allometric growth. That means increase in weight of either sex of *P. lanceolatus* was relatively less than that in length.

Similar departures from the cube relation were recorded by Torres (1991) in certain marine fishes of South Africa and King (1996) in several coastal water gobiids of Nigeria. Khaironizam and Norma-Rashid (2002) also studied negative allometry in *P. elongatus* in coastal waters of Malaysia.

Garcia *et al.* (1998) pointed out "biological interpretation of the numerical values of the parameters 'a' and 'b' is not always straight forward, except that when growth is isometric, 'a' can be interpreted as a conditional factor. In addition the 'b' values of males and females denote that they differ in growth rates from one another, females exhibiting relatively better growth than males. However in both cases the association of length and weight appears to be more or less equal.

The first flexion in the condition factor of *P. lanceolatus* appeared during January in the length class of 11.1–12.0 cm. (Figure 27.5) indicating that the fishes of either sex attains sexual maturity for the first time during this month at the size mentioned. The second flexion occurring in March with extension till May and the third turning up in July with continuation till November coupled with a gradual decline from the smallest to the largest length groups suggest that the fishes of all the size groups spawn during these two seasons.

Acknowledgement

The authors are grateful to Andhra University for providing facilities. One of the authors is grateful to U.G.C. for the award of Teacher Fellowship during the tenure.

References

Bailey, N. T.J. 1959. *Statistical Methods in Biology*. The English Universities Press Ltd., London, p. 1–200.

Beverton, R.J.H. and S.J. Holt. 1959. On the dynamics of exploited fish populations, *Fish. Invest. Lond. Ser.*, 2(19): 533.

Garcia, C.B., J.O. Duarte, N. Sandoval, D.Von Schiller, G. Melo and P. Navajas, 1998. Length-weight relationships of demersal fishes from the Gulf of Salamanca, Columbia. *Naga,* ICLARMQ, 21(3): 30–32.

King, R.P. 1996. Length-weight relationships of Nigerian Coastal water fishes. *Naga*, ICLARMQ, 19(3): 49–52.

Khaironizam, M.Z. and Y. Norma-Rashid, 2002. Length-weight relationship of mudskippers (Gobiidae: Oxydercinae) in the coastal areas of Selangor, Malaysia. *Naga,* 25(3&4): 20–22.

Kubbicki, M., G. Moutham, P. Thallot and L. Wantiez. 1993. Length-weight relationships of fish from the Lagoon of New Caledonia. *Naga*, ICLARMQ, 16(2–3): 26–30.

La Cren, E.D. 1951. The length-weight relationship and seasonal cycle in gonad weight and condition of Perch (*Perca fluvialitis*). *J. Anim. Ecol.*, 20(1): 201–209.

Tesh, F.W. 1971. Age and growth. In: *Methods for Assessment of Fish Production in Freshwaters*, (Ed.) WE. Ricker. IBP Hand Book No. 3, Blackwell, London, p. 93–123.

Torres, F.J. 1991. Tabular data on marine fishes from South Africa, Part 1: Length-weight relationships, Fish byte, 9(1): 50–53.

Chapter 28

Molecular Identification of Two Potential Probiotic Bacteria *Bacillus circulans* PB 7 and *Bacillus* sp. PP 9 and Three Fish Pathogenic *Aeromonas hydrophila* AH1, AH2 and AH3 in the Indian Major Carps *Catla catla* (Ham.), *Labeo rohita* (Ham.) and *Cirrhinus mrigala* (Ham.)

*Partha Bandyopadhyay**# *and Bidhan C. Patra***

**Latika Sea Food Private Limited, Jalalkhanbard, Contai – 721 401, Purba Medinipur, West Bengal, India*
***Aquaculture Research Unit, Department of Zoology, Vidyasagar University, Midnapore – 721102, West Bengal, India*

ABSTRACT

Two potential probiotic bacteria *Bacillus circulans* PB 7 and *Bacillus* sp. PP 9 isolated from *Catla catla* and *Cirrhinus mrigala* and during this course of time three fish pathogenic strains AH1, AH2 and AH3 of *Aeromonas hydrophila* also isolated from the Indian major carps *viz.*, catla (*Catla*

Corresponding Author: 60, Ramdulal Road, Kotalpur, Gurap (Balidaha), Hooghly – 712303, West Bengal, India, E-mail: parban_gurhoo@rediffmail.com.

catla), rohu (*Labeo rohita*) and mrigal (*Cirrhinus mrigala*), were collected from different parts of Midnapore district, West Bengal, India were identified with the help of different molecular approaches like RAPD–PCR, SDS–PAGE and 16S rRNA sequencing. The RAPD–PCR profile of *Bacillus circulans* PB 7 and *Bacillus* sp. PP 9 obtained by using primer OPC–02 9 showed the different banding patterns, whereas the replicate within the strain showed similar banding pattern, which was also indicated that two species were different. The RAPD–PCR profile AH1, AH2 and AH3 of *Aeromonas hydrophila* showed similar banding patterns, which may provide the evidence of taxonomic position of *Aeromonas hydrophila*. 12 per cent SDS–PAGE of strain *Bacillus circulans* PB 7 showed bands ranging from 16.6 to 73.7 KDa, whereas distinct bands as 24.6, 30.5, 36.2, 42.1, 52.8 and 56.7 KDa were observed from *Bacillus* sp. PP 9. The molecular weights of protein fingerprints of the three strains AH1, AH2 and AH3 of *Aeromonas hydrophila* was same and the molecular weights of them were 21.4, 32.4, 55.6 and 68.2 KDa. From 16S rRNA sequencing it was found that the strain *Bacillus* sp. PP 9 was closely similar to *Paenibacillus apiarius* and *Paenibacillus lautus*. The strain *Bacillus circulans* PB 7 had low G + C content, whereas *Bacillus* sp. PP 9 and *Aeromonas hydrophila* AH1 had high G + C content. In this study of identification of probiotic potential strains *Bacillus circulans* PB 7, *Bacillus* sp. PP 9 as well as the fish pathogenic strain *Aeromonas hydrophila* AH1 showed a clear documentation and identification of the isolates.

Keywords: *Probiotic bacteria, Aeromonas hydrophila, RAPD–PCR, SDS–PAGE, 16S rRNA sequencing, Indian major carps.*

Introduction

In the 40 years since the advent of molecular biology, developments in DNA technology have revolutionized biological research. The study of probiotic research is no exception and great advances should be made in this field. The time has now come to reap the benefits of work that has provided the means to improve the proper identification of the potential probiotic bacteria (Spanggaard *et al.*, 2000). Molecular techniques are revolutionizing genetic analysis of microbes and DNA based polymerization are being used for marker associated selection strategies, parentage listing, species identification/ characterization and population genetics studies (Bandyopadhyay *et al.*, 2003a; Bandyopadhyay *et al.*, 2003b, Das, 2003). The environmental probe based DNA as Random Amplified Polymorphic DNA (RAPD) has been used for genetic analysis and microorganism clarification (Martinez *et al.*, 1994 and Zanetti *et al.*, 1999). Recently protein-fingerprinting tools (SDS–PAGE of cell free extract of cell wall proteins) also used for clarification of microorganisms (Das, 2003). Like genomics, proteomics is the study of protein characteristics and function to obtain a global, integrated view of normal and abnormal cellular process, protein–protein interaction and regulatory networks of the protein level (Ausubel *et al.*, 1987). The protein fingerprinting through SDS–PAGE is a useful tool for identification of species and strains and also development of protein marker based studies (Bandyopadhyay and Patra, 2004).

To date, detection and identification of microbes in aquaculture has concentrated on the ribosomal RNA (rRNA), gene array. These genes are ideal target for diagnosis for several reasons; they occur in tandem array along the chromosome and this multi-copy arrangement provides multiple targets for primer or probe binding (Walters and McCutchan, 1990). There are also more variable regions of sequence with the genes and spacers. These have been found to vary between groups and species, enabling design of methodology that will separate organisms at the appropriate taxonomic level (Gasser and Hoste, 1995). It is so useful that enables identification of species or strains (Xiao and Desser, 2000).

Identification of fish microflora has typically relied on phenotypical and biochemical key characteristics. This is extremely time consuming and it is tempting to use a limited number of key characteristics. However, it is not known as to what extant, such tentative characterization reflects the real identity of the strains. Therefore the purpose of this study was to identify the two potential probiotic bacteria *Bacillus circulans* PB 7 and *Bacillus* sp. PP 9 isolated from *Catla catla* and *Cirrhinus mrigala* and the three fish pathogenic strains AHl, AH2 and AH3 of *Aeromonas hydrophila* isolated from the Indian major carps *viz.*, catla (*Catla catla*), rohu (*Labeo rohita*) and mrigal (*Cirrhinus mrigala*) from different parts of Midnapore district, West Bengal, India in molecular approaches like RAPD–PCR, SDS–PAGE and 16S rRNA analysis.

Materials and Methods

The experiment was conducted at Aquaculture Research Unit and Microbiology Research Unit of Vidyasagar University, Midnapore, India; Indian Institute of Technology, Kharagpur India; Central Institute of Fisheries Education, Mumbai, India and Central Institute of Freshwater Aquaculture, Bhubaneswar, India.

Random Amplified Polymorphic DNA–PCR Analysis

Extraction of DNA

Bacteria were streaked on nutrient agar plate. Pure colonies were subcultured and individual colonies were transferred to separate agar slants. The bacterial strains were grown at 37°C for 24 h. The cells were washed with sterile phosphate-buffered saline (PBS: 0.01 M NaH_2PO_4, $2H_2O$, 0.01 M $Na_2HPO_4.2H_2O$, 0.15 M NaCl. pH 7.2) and vortexed. The genomic DNA was prepared by standard DNA extraction methods (Ausubel *et al.*, 1987). The DNA was later resuspended in TE (10 mM Tris HC1, 100 mM EDTA, pH 7.8) buffer and stored at –20° C. The quantity of genomic DNA in each sample was measured at 260 nm using a spectrometer (Shimadzu–UV 365). The DNA concentration of each sample was adjusted to 40 ng/µl for PCR by diluting DNA in distilled, deionized water (Sudheesh *et al.*, 2002).

RAPD Conditions

PCR was performed using 10–50 ng of total DNA from each sample tested. The DNA was dissolved in 10 mM Tris–HCl, pH 9; 2.5 mM $MgCl_2$; 50 mM KCl; 0.1 per cent Triton X100; 150 µM each of dATP, dCTP, dGTP, dTTP; 0.1 µM oligonucleotide in a total volume of 50 µl, and 2 units of Taq DNA polymerase (Apligene) was added. PCR was carried out in a programmable thermal controller (Perkin–Elmer) for 45 cycles. Each cycle consisted of 1 min denaturation at 94°C, 1 min annealing at 35°C (5°C above the T_m of the used primer) and 2 min extension at 72°C. After the last cycle, there was a 5 min extension at 72°C.

RAPD Analysis

The random primers are selected to flank the DNA region to be amplified. Primers are oligonucleotides that are single stranded fragments of DNA, complementary to the 5′ end of the target DNA to be amplified. There must be nearly complete homology between the target DNA and the primer nucleotide sequence for primer annealing specifically at the sites of the DNA region to be amplified. Typically primers are 15–30 nucleotide long and it is necessary that the 5–6 bases at the 3′ end of the primers exhibit precise base paring with the target DNA. All primers for target amplification should have the same melting temperature (Temperature of annealing) and G + C content about 60–70 per cent. OPC–02–GTGAGGCGTC, a commercial primer from Operon Technologies (Alameda, CA, USA) was used with following 5′–3′ sequence.

Analysis of RAPD Product through Agarose Gel Electrophoresis

The PCR amplified samples were analysed on 1.8 per cent agarose gel eletrophoresis. Agarose gel (1.8 g) was taken in control flask and boiled in 100 ml of TBE buffer (Tris 50 nM, EDTA 100mM, Boric acid 20 ml, pH–8). Gel tray was prepared by pasting adhesive tape on both side and comb was placed. Then boil agarose was poured in the gel tray and allowed to solidify. After solidification the comb was removed gently and tray was placed in a gel tank. Then 1 × TBE buffer was poured in gel tank till cast gel submerges. Amplified DNA samples (6–8 µl), gel loading dye (2 µl) and distilled water (3–4 µl) were mixed together and loaded in wells. Known molecular marker (OPC–02) (250 bp) was also loaded along with the samples. The power pack was connected to the gel tray set at 100 V. The power pack was stopped when blue dye reached to the end of the gel and gel was transported to the glass tray for staining. Staining was done in glass tray using ethidium bromide for about 10–15 minutes. Then the gel was removed and observed under UV transilluminator. High molecular DNA remains nearer to well and smear of DNA, which indicates degradation of DNA. Picture was taken onto thermal paper using a video camera and video copy processor (Mitsubishi, Japan).

Protein Fingerprinting of Probiotic Bacteria by SDS–Polyacrylamide Gel Electrophoresis (SDS–PAGE)

Isolation of Protein

The probiotic bacterium isolated from the intestine of the Indian major carps were grown in 250 ml Erylenmyer flask containing 50 ml nutrient broth (Hi-media, India) for 16 at 30°C with shaking speed of 160 rpm. The culture broth were taken at the exponential phage and centrifuged at 5,000 rpm for 15 minutes. The cell pellets separated and were washed thrice with NSS. Cell pellets were dissolved in phosphate-buffered saline (PBS: 0.01 M NaH_2PO_4, $2H_2O$; 0.01 M Na_2HPO_4, $2H_2O$; 0.15 M NaCl; pH 7.2) and subjected to sonication for cell lysis. The cell lysets were centrifuged at 15,000 rpm for 30 minutes. Supernatants were used as crude source of protein.

Ammonium Sulphate Precipitation of Protein

Total crude protein was brought to 80 per cent saturation with solid enzyme grade ammonium sulphate [$(NH_4)_2SO_4$] and kept overnight at 4°C. The precipitate was collected by centrifugation (15,000 rpm for 30 min.) and dissolved in phosphate-buffered saline (PBS: 0.01 M NaH_2PO_4, $2H_2O$; 0.01 M Na_2HPO_4, $2H_2O$; 0.15 M NaCl; pH 7.2). The remaining salt in the protein was removed by the dialysis by using the same buffer for 2 days. The obtained protein was stored at 0°C for further study.

Preparation of Sample for SDS–PAGE

The protein sample was diluted at 1 : 1 volume by volume with 4 × SDS sample buffer [Tris–HCl (180 mM, pH 6.8) 7 ml; glycerol 2 ml; β–Mercaptoethanol 1 ml; SDS 400 mg; Bromophenol blue 2 mg) and heat 3–5 min at 100°C in a sealed screw-cap micro centrifuge tube. The protein molecular weight standard mixture was dissolved in 1 × X SDS sample buffer according to supplier's instruction.

SDS–PAGE

Electrophoresis was performed using a gel apparatus (GENI, Bangalore, India). Sodium dodecyl sulfate–polyacrylamide gel electrophoresis was carried out by the methods Schagger and Von Jagow

(1987) using 12 per cent separating and 5 per cent stacking polyacryalmide gels. Thoroughly cleaned and dried the glass plates and spacers, then assembled them properly. 2 per cent agar (melted in a boiling water bath) was then applied around the edges of the spacers to seal the chamber. Then prepared a sufficient volume of separating gel mixture and poured the gel solution in the chamber between the gel plates. Layer distilled water on top of the gel and leave to set for 40 minutes. After 40 minutes removed the water from the top of the separating gel and poured the stacking gel mixture, placed the comb in the stacking gel and allowed the gel to set for 1 hour. After polymerization comb was removed and poured the electrode buffer in both the chambers. Then the protein samples were loaded into the wells through micro syringe. Marking the position of wells on the glass plates with a glass marker. In the similar way loading the standard. The Protein Molecular Weight Marker of molecular weights 66.0 KDa, 45.0 KDa, 34.7 KDa, 24.0 KDa, 18.0 KDa, and 16.3 KDa were purchased from GENEI, Bangalore, India. Turn on the current to 10–15 mA for initial 10–15 minutes until the samples travel through the stackingel. Then continued the run at 30 mA until the bromophenol blue reaches the bottom of the gel (about 5 hours). After completion of the run, carefully remove the gel and washed in distilled water. The electrophoresis gel was stained with 0.125 per cent coomassie brilliant blue R-250 in 25 per cent ethanol and 10 per cent acitic acid. After destaining, the band of sample was compared with the control gel and the molecular weight markers.

Identification by 16S rRNA Sequence Data Analysis

Two probiotic bacterial strains *Bacillus circulans* PB 7 and *Bacillus* sp. PP 9 representing deferent physiological groups and different RAPD groups were picked for sequence analysis of the most variable regions of the 16S rRNA gene. Almost complete 16S rRNA gene fragments were amplifies *in vitro* as described earlier (Springer *et al.*, 1992). Following amplification, excess primer and deoxynucleotide triphosphates were removed using MicroSpin S-400 HR columns (Pharmacia Biotech, Sweden). Direct sequencing of the purified PCR products was performed using rDNA specific primers (Springer *et al.*, 1992). Cycle sequencing was performed using the Thermo Sequence fluorescent-labelled primer cycle sequencing kit (Amersham Life Science, Denmark) according to the instructions of the manufacturer and analyzed on an automated DNA sequencer (ALF Express, Pharmacia Biotech, Sweden).

Partial 16S rRNA sequences of the isolates were fitted into an alignment of about 8000 homologous full and partial primary structures using the respective automated tools of the ARB software package (Ludwig and Strunk, 1997). This partial sequences were used for maximum parsimony analysis and phylogenetic classification of the isolates.

Results and Discussion

The RAPD–PCR profile of *Bacillus circulans* PB 7 and *Bacillus* sp. PP 9 obtained by using primer OPC–02 were presented in Plates 28.1 and 28.2 respectively. Plate 28.1 indicated that lane 1–9 were the replicates of the strain *Bacillus circulans* PB 7 whereas in Plate 28.2, lane 1–4 were indicated the replicates of *Bacillus* sp. PP 9. In both plates M represented the molecular marker OPC–02. The replicates within the strain showed similar banding pattern, whereas the pattern is dissimilar between. This also indicated that two species were different. The RAPD–PCR profile of *Aeromonas hydrophila* AH1, AH2 and AH3 obtained by using primer OPC–O2 was presented in plate 3, where M indicated primer and lane 1, 2, 3 for the strain AH1, AH2 and AH3 respectively. All the strains showed similar banding patterns, which may provide the evidence of taxonomic position of *Aeromonas hydrophila*. The purpose of introducing RAPD was to perform a rapid, tentative grouping of the strain to reduce the number of strains for sequence analysis. All the strains generated a band profile in the first run and productivity of the individual RAPD fingerprints of primer OPC–02 were approximately 75 per cent. The RAPD–PCR assay differentiates strains based on differences in nucleotide sequences in the entire genome (Sudheesh *et al.*, 2002) and it has become increasingly important for identification of probiotic and pathogenic bacteria (Das, 2003). Molecular approaches that 'Interrogate' the whole genome appear to be the way forward to highlight what may be only minimal difference between strains (Sudheesh *et al.*, 2002). In the present study, the different strain showed different banding patterns, which were helpful for tentative grouping of the strain to reduce the number of strains for sequence analysis. RAPD has been used in development of molecular diagnostic techniques for bacteria (Martinez *et al.*, 1994). While there are limitations of the method if used in isolation for bacterial detection (Oakey *et al.*, 1998), but RAPD has been applied as the first step in identifying regions of the genome development that are useful as probes or targets for PCR (Argenton *et al.*, 1996 and Oakey *et al.*, 1999).

12 per cent SDS–PAGE of strain *Bacillus circulans* PB 7 was shown in Plate 28.4. Plate indicated lane B1–B4 were the replicates of the strain and M was the molecular weight marker. The protein bands were ranging from 16.6 to 73.7 KDa. All the bands were showing similar protein fingerprinting pattern, which is unique and can be useful for protein marker for identification of the molecular weight of the strain. Plate 5 represented the 12 per cent SDS–PAGE of the strain *Bacillus* sp. PP 9. In this plate M indicated as molecular weight marker, whereas Lane L1 and L2 were the replicates of the strain PP 9. All the bands were showing similar banding pattern and the distinct bands as 24.6, 30.5, 36.2, 42.1, 52.8 and 56.7 KDa were observed which may be of potential use as molecular marker for identification and determination of molecular weight of the various protein fractions of *Bacillus* sp. 12 per cent SDS–PAGE of *Aeromonas hydrophila* of three strains indicated AH1, AH2 and AH3 were shown in plate 6. The molecular weights of protein fingerprints of the three strains were same and the molecular weights of them were 21.4, 32.4, 55.6 and 68.2 KDa. This can be used for identification of *Aeromonas hydrophila* and also for determination of antigenic protein and further will be used for vaccination programme (Das, 2003).

Partial 16S rRNA sequence data analysis of *Bacillus circulans* PB 7, *Bacillus* sp. PP 9 and *Aeromonas hydrophila* AH1 were shown in Tables 28.1–28.3 respectively. High arrangements were found between traditional and molecular identification as established by traditional characterization earlier. The identification of the strains was also confirmed by the Nucleotide Division of National Centre for Biotechnology Information (NCBI), Bethesda, Maryland, USA. From 16S rRNA sequencing it was found that the strain *Bacillus* sp. PP 9 was closely similar to *Paenibacillus apiarius* and *Paenibacillus lautus* and it was also confirmed by Institute of Microbial Technology (IMTECH), Chandigarh, India. Further they also confirmed that the strain was *Bacillus* sp. The strain *Bacillus circulans* PB 7 had low G + C content, whereas *Bacillus* sp. PP 9 and *Aeromonas hydrophila* AH1 had high G + C content. The prokaryotic rRNA genes contain highly conserved sequences as in the eukaryots. The potential utility of conserved regions to identify or amplify the rRNA genes followed by exploitation of more variable regions of the genes spacers to detect or identify bacteria that may be difficult or even impossible to culture has long been recognized (Barry *et al.*, 1990 and Nilsson *et al.*, 1998). The same method was employed in the study of identification of probiotic potential strains *Bacillus circulans* PB 7, *Bacillus* sp. PP 9 isolated from *Catla catla* and *Cirrhinus mrigala,* as well as the fish pathogenic strain *Aeromonas hydrophila* AH1 isolated from Indian major carps (*Catla catla, Labeo rohita* and *Cirrhinus mrigala*). In this experiment a clear documentation and identification of the isolates were made. Such as assay could distinguish bacteria, which are potentially probiotic or pathogenic strains or types and such assays of higher resolution can examine of bacterial phenotype alone.

Table 28.1: Partial 16S rRNA Sequence Data Analysis of *Bacillus circulans* PB 7

1	tttatcggag	agtttgatcc	tggctcagga	cgaacgctgg	cggcgtgcct	aatacatgca
61	agtcgagcgg	acctgatgga	gtgcttgcac	tcctgatggt	tagcggcgga	cgggtgagta
121	acacgtaggc	aacctgccct	caagactggg	ataactaccg	gaaacggtag	ctaataccgg
181	ataatttatt	ttgcagcatt	gtgaaataat	gaaaggcgga	gcaatctgtc	acttgaggat
241	gggcctgcgg	cgcattagct	agttggtggg	gtaacggcyc	accaaggcga	cgatgcgtag
301	ccgacctgag	agggtgaacg	gccacactgg	gactgagaca	cggcccagac	tcctacggga
361	ggcagcagta	gggaatcttc	cgcaatgggc	gaaagcctga	cggagcaacg	ccgcgtgagt
421	gatgaaggtt	ttcggatcgt	aaagctctgt	tgccaaggaa	gaacgtcttc	tagagtaact
481	gctaaggaga	agtgacggta	cttgagaaga	aagccccggc	taactacgtg	ccagcagccg
541	cggtaatacg	tagggggcaa	gcgttgtccg	gaattattgg	gcgtaaagcg	cgcgcnngcg
601	gttctttaag	tctggtgtgn	aaacccgaag	ctcaacttcg		

Table 28.2. Partial 16S rRNA Sequence Data Analysis of *Bacillus* sp. PP 9

1	ctggcggcgt	gcctaataca	tgcaagtcga	gacgggacag	atgggagctt	gctccctgga
61	tgttagcggc	ggnacgggnt	gnagtaacac	gtgggtaacc	tgcctgtaag	actgggtata
121	actccgggaa	accggggcta	ataccggatg	gttgtttgaa	ccgcatggtt	cagacataaa
181	aggtggcttc	ggctaccact	tacagatgga	cccgcggcgc	antagctagt	tggtgaggta
241	acggctcacc	aaggcaacga	tgcgtagccg	acctgagagg	gtgatcggcc	acactgggac
301	tgagacacgg	cccatactcc	taccggaggc	agcattnggg	aatcttccgc	aatggacgaa
361	agtctgacgg	ancancgcct	cgtgagtgat	gaaggttttc	ggatcgtaaa	gctcttttgt
421	tagggaagaa	caagtgccgt	tcanataggg	cggcaccttg	acggtaccta	accagaaagc
481	cncngntaaa	tacttgccag	cngccgcggt	aatacctagg	tggcagccgt	tgtccggaat
541	tattgggcgt	aaagggctcg	caggcggttt	cttaagtctg	atgtgaaagc	ccccggctca
601	accggggagg	gtcattggaa	actggggaac	ttgagtgca		

Table 28.3: Partial 16S rRNA Sequence Data Analysis of *Aeromonas hydrophila* AH1

1	ggatccqaga	gtttgatcct	ggctcagatt	gaacgctggc	ggcaggccta	acacatgcaa
61	gtcgagcggc	aacgggaaag	tagcttgcta	cttttgccgg	cgagcggcgg	acgggtgagt
121	aatgcctggg	aaattgccca	gtcgaggggg	ataacagttg	gaaacgactg	ctaataccgc
181	atacgcccta	ccggggaaag	caggggacct	tcgggccttg	cgcgattgga	tatgcccagg
241	tgggattagc	tagttggtga	ggtaatggct	caccaaggcg	acgatcccta	gctggtctga
301	gaggatgatc	agccacactg	gaactgagac	acggtccaga	ctcctacggg	aggcagcagt
361	ggggaatatt	gcacaatggg	ggaaaccctg	atgcagccat	gccgcgtgtg	tgaagaaggc
421	cttcgggttg	taaagcactt	tcagcgagga	ggaaaggttg	atgcctaata	cgtatcaact
481	gtgacgttac	tcgcagaaga	agcaccggct	aactccgtgc	cagcagccgc	ggtaatacgg
541	agggtgcaag	cgttaatcgg	aattactggg	cgtaagcgca	cgcaggcggt	tggataagtt
601	agatgtgaaa	gccccgggct	caacctggga	attgcattta		

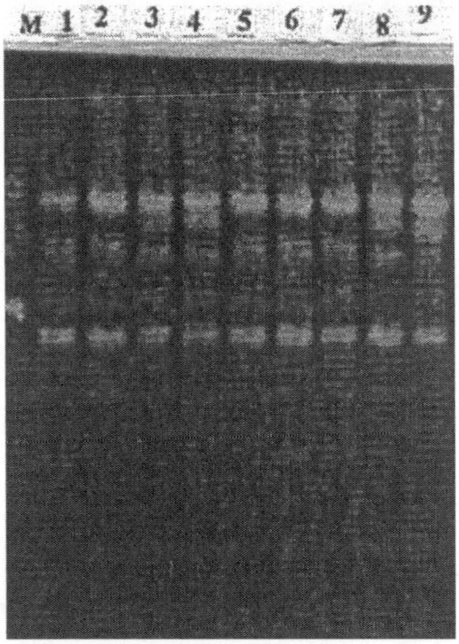

Plate 28.1: RAPD–PCR Profile of *Bacillus circulans* PB7 Generated After Amplification with Primer OPC 02

Plate 28.2: RAPD–PCR Profile of *Bacillus* sp. PP 9 Generated After Amplification with Primers OPC 02

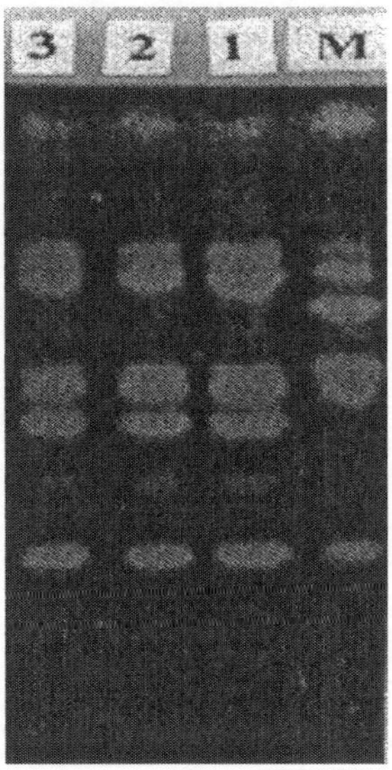

Plate 28.3: RAPD–PCR Profile of *Aeromonas hydrophila* AH1, AH2 and AH3 Generated After Amplification with Primers OPC 02

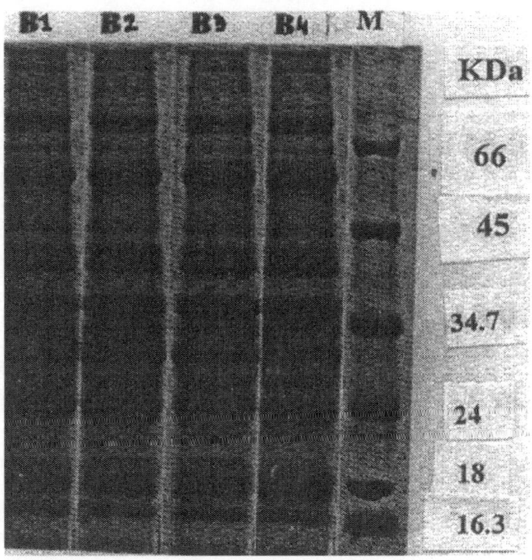

Plate 28.4: Protein Fingerprinting of *Bacillus circulans* PB7 Through SDS–PAGE Technique

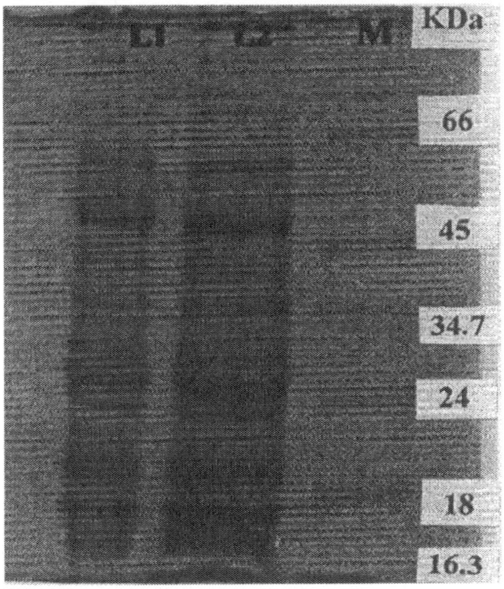

Plate 28.5: Protein Fingerprinting of *Bacillus* sp. PP 9 Through SDS–PAGE Technique

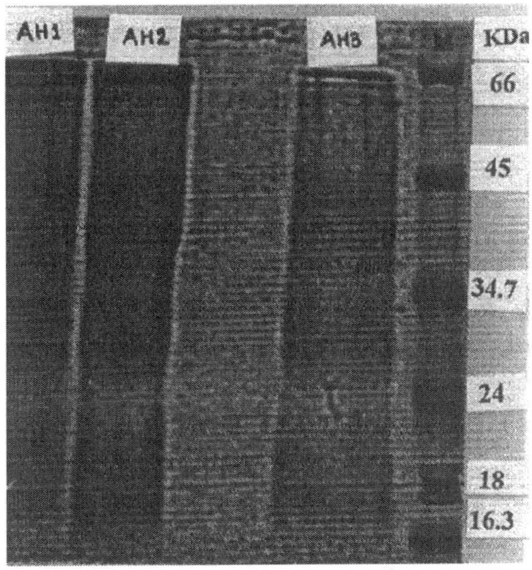

Plate 28.6: Protein Fingerprinting of *Aeromonas hydrophila* AH1, AH2 and AH3 Through SDS–PAGE Technique

Acknowledgement

The first is thankful to Indian Council of Agricultural Research for providing financial assistance during this research work.

References

Argenton, F., DeMas, S., Malocco, C., Dalla Valle, L. and L. Colombo. 1996. Use of Random DNA amplification to generate specific molecular probes for hybridization test and PCR based diagnosis of *Yearsinia ruckeri*. *Dis. Aquat. Org.*, 24: 121–127.

Ausubel, F.M., Brent, R., Kingston, R.E., Moore, D.D., Smith, J.A., Sideman, J.G. and K. Struhl. 1987. *Current Protocols in Molecular Biology*. Wiley, New York, USA.

Bandyopadhyay, P. and B.C. Patra. 2004. Probiotics for sustainable aquaculture and its utilization trial. In: *Course Manual of the Summer School on Development of Sustainable Aquaculture Technology for Fresh and Saline Waters*, Department of Zoology, CCS Haryana Agricultural University, India, p. 61–66.

Bandyopadhyay, P., Sarkar, B., Tewary A. and B.C. Patra. 2003a. Use of Random Amplified Polymorphic DNA (RAPD) for detection of some genetic changes in *Labeo rohita* (Ham.) of Midnapore, India. *J. Current Science*, 3(2): 285–288.

Bandyopadhyay, P., Sarkar, B., Tewary A. and B.C. Patra. 2003b. RAPD amplification is useful for the determination of polymorphism in the population of *Carassius auratus* (Lin). *J. Aquatic Biol.*, 18 (2): 199–203.

Barry, T., Powel, R. and F. Gannon. 1990. A general method to generate DNA probes for microorganisms. *Biotechnol.*, 8: 233–236.

Das, B.K. 2003. Protein fingerprinting of fish microbial pathogen. In: *Advances in Fish Disease Diagnostics*, (Ed.) Pani Prasad, K. CIFE, Mumbai, India, p. 177–181.

Gasser, R.B. and H. Host. 1995. Genetic markers for closely related parasitic nematodes. *Mol. Cell. Probes*, 9: 315–320.

Ludwig, W. and O. Strunk. 1997. ARB: A software environment for sequence data. http://www.mikro-biologie.tumuenchen.de/pub/ARB/documentation/arb.ps.

Martinez, I., Espelid, S., Johansen, A., Welsh, J. and M. McClelland. 1994. Fast identification of species and strains of *Vibrio* by amplification of polymorphic DNA. *J. Fish Dis.*, 17: 297–302.

Nilsson, W.B., Noonal, P.J. and M.S. Strom. 1998. Rapid identification of fish pathogens using RFLP analysis of bacterial 16S rhibosomal RNA. In: *Third International Symposium on Aquatic Animal Health*, 30th August–3rd September, Baltimore, USA, pp. 207.

Oakey, H.J., Ellis, J.T. and L.F. Gibson. 1998. Developmental of random DNA probes specific for *Aeromonas salmonicida*. *J. Appl. Microbiol.*, 84: 37–46.

Oakey, H.J., Gibson, L.F. and A.M. George. 1999. DNA probes specific for *Aeromonas hydrophila* (HG1). *J. Appl. Microbiol.*, 86: 187–193.

Spanggaard, B., Huber, I., Nielsen, J., Neilsen, T., Appel, K. F. and L. Gram. 2000. The microflora of rainbow trout intestine: a comparison of traditional and molecular identification. *Aquaculture*, 182: 1–15.

Springer, N., Ludwig, W., Drozanski, V., Amann, R. and K.H. Schleifer. 1992. The phylogenetic status of *Sarcobium lyticum*, and obligate intracellular bacterial parasite of small amoebae. *FEMS Microbiol. Letter*, 96: 199–202.

Sudheesh, P.S., Jie, K. and H. Xu. 2002. Random amplified polymorphic DNA-PCR of *Vibrio parahaemolyticus* and *V. alginolyticus* isolated from cultured shrimps. *Aquaculture*, 207: 11–17.

Waters, A.P. and T.F. McCutchan. 1990. Rhibosomal RNA: Nature's own polymerase-amplified target for diagnosis. *Parasitol. Today*, 6: 56–59.

Xiao, C. and S.S. Desser. 2000. Molecular characterization of mixozoan parasites from lake Sasajewun, Algonquin Park, Ontario, by riboprinting. *J. Eukariotic Microbiol.*, 47: 85 –89

Zanetti, S., Deriu, A., Dupre, I., Sanguinetti, M., Fadda, G. and L.A. Sechi. 1999. Differentiation of *Vibrio alginolyticus* strains isolated from Sardinian waters by ribotyping and a new rapid PCR fingerprinting method. *Appl. Environ. Microbiol.*, 65: 1871–1875.

Section II

AQUATIC ENVIRONMENT

Chapter 29

Study of the Influence of Aquaculture Development on Environment: A Remote Sensing Approach

P. Venkateswarlu, M.V. Rao**, Kiran*** and Ramamohan****

**Associate Professor, Centre for Environment, IST, JNTU, Hyderabad, India*
***Scientist, Oceanography Division, NRSA, Hyderabad – 37, India*
****Research Scholar, Centre for Environment, IST, JNTU, Hyderabad, India*

ABSTRACT

Aquaculture is the domestication and rearing of aquatic animals and plants. But, today the impact of aquaculture is seen on the socio-economic aspects and environment. Aquaculture plays a major role in the development of fishing industry and contributes in rural development, export revenue generation and replenishment of important aquatic species. Aquaculture is considered to be one of the production systems to increase protein consumption in Third World countries. The Aquaculture production in India and its role in the economy of the country as well as its impact on environment are discussed in this article. The Indian fisheries sector, which produced only 0.6 million tones of fish 50 years ago, now produces nearly 6.0 million tones through fish pond aquaculture development. Today, India is making strides with Blue Revolution, with a rapid increase of its fish production in small ponds and water bodies. This is a boon to small farmers, the nation' s nutrition and its Gross Domestic Product. The contribution of fisheries to GDP at the current price level is 1.3 per cent. The role of remote sensing in estimation of the aquaculture area in some selected parts along the east coast of India is presented. The economic impact of aquaculture is critically analysed considering certain statistics and the resultant help of high-resolution satellite remote sensing data. Tropical brackish water and saltwater aquaculture have contributed to the destruction of mangrove forests, wetlands and salt marshes, because they have been cleared for use as ponds. It is noted that, 60 per cent of tropical mangroves have been destroyed and turned into aquaculture farms. Apart from the destruction of land farms, water pollution from substances dissolved or suspended in pond effluents often is cited as a major impact. The best management practices and the guidelines issued by government of India for the development of fishpond aquaculture are discussed. It is found that there is a phenomenal development of pond aquaculture in India during the past decade and also significant hike in fish productivity. It is also evident that the pond culture activities in the country helped for the mitigation of rural poverty. On the other hand, the fresh water aquaculture has lead to the destruction of ecology, by way of polluting the drinking water source and destruction of land farms. The deliberate blockage of water inflow from various rivers and streams to the ponds and also the outflow from pond lead to disastrous consequences of floods all along the east coast during cyclones and very active South West monsoon seasons. The development of aquaculture can and does fit in well with the sound environmental ethics–the two go hand-in-hand. Precisive environmental quality is very much required in aquaculture compared to that of other industrial domains. Strict the wildlife. Clean oceans and strict controls on water quality ensures protected environment from the cultural and industrial ill interferences. The high-resolution data from satellites will be useful in mapping and monitoring of the study area from time to time and can provide direct information on land use patterns for proper management of the environment.

Introduction

"Aquaculture" means the propagation of fish seed or the raising of fish through husbandry during the whole or part of its life cycle. "Culture system" means any establishment, structure or facility employed in aquaculture, which includes bottom culture, cage culture, hanging-net culture, pen culture, pond culture, pole or stick culture, raceway culture, raft culture, rope culture and hatchery. Aquaculture, as with any other business, requires careful and objective planning and design to be successful. Edwards and Demaine (1997). The rural aquaculture may be defined as "the farming, usually by extensive and semi-intensive low-cost production technology, of aquatic organisms by small-scale farming at households or at communities".

Aquaculture is the domestication and rearing of aquatic animals and plants. It is related to agricultural development and fisheries management, both were initially the domains of biological scientists. The social scientific perspective into aquaculture projects was found a place while transferring technology and resource management. Like in any other project, economists were asked to provide information on feasibility study and economic impact statements in case of aquaculture projects. Economics is the most heavily weighted factor in any of the aquaculture ventures.

Aquaculture is recognized as an important component of the fisheries industry and figures prominently in all fisheries development plan. Since aquaculture production units are located largely in rural areas, it is merely assumed and accepted that aquaculture development is part of rural development. There are three areas in which aquaculture can make a significant contribution: rural development (Grover, Street and Starr, 1980; Liao, 1988, Mitchell *et al.*, 1986), generation of export revenue (Mitchell *et al.*, 1986, Rahman, Chowdury and Chowdury, 1985; South Carolina Sea Grant, 1989) and replenishment of wild stocks of commercially and recreationally important aquatic species (Amend, 1989; MaccKenzic, 1977; McCay, 1989).

The increasing population and declining fish stocks for capture, aquaculture has forced to look for innovative ways to secure an inexpensive source of protein for both the rural and urban poor. Of the production systems available today, aquaculture provides a mechanism to increase protein consumption in Third World countries. Even very poor farmer can supplement his or her family's protein through a backyard pond using a species that requires no feed and little care, provided adequate water, proper soils, and a source of post larvae are available. Landless rural poor and displaced fishermen can grow fish in cages or net enclosures in lakes, estuaries, and bays, because many aquaculture systems rely on either communally controlled or open-access waters (Bras, Olla, and Wickland, 1987; Nakamura and Trauth-Thomas, 1986; Santaella, 1989).

Aquaculture Production in India

India has vast potential for fishes both from inland and marine resources, since it has a long coastline of about 7500 km, 3 million hectares of reservoirs and 1.4 million hectares of brackish water. The Indian fisheries sector, which produced only 0.6 million tones of fish 50 years ago, now produces nearly 6.0 million tones. Of which, the inland sector registered 45.4 per cent of the total fish production. Inland production, including farming, is now catching up with production from the marine sector and is likely to overtake in the next millennium. Presently, freshwater aquaculture is growing at a healthy 6 per cent a year. In the inland fisheries, the freshwater aquaculture and brackish water aquaculture are considered to be the most important ones. The major states contributing are West Bengal (33 per cent), Andhra Pradesh (9.09 per cent), Bihar (8.71 per cent), Assam (6.92 per cent), Uttar Pradesh (6.49 per cent), Orissa (6.01 per cent), Tamil Nadu (4.82 per cent), Madhya Pradesh (4.07 per cent), Karnataka (3.89 per cent) and Maharashtra (3.4 per cent).

In India fisheries is a very important industry and aquaculture is well established, where the aquaculture production base is more coastal than inland. Inland aquaculture has emerged as a major fish producing system in India, with production currently (1998) around 1.7 million t/yr. Carp accounts for over 80 per cent of farmed fish. Major species cultured are rohu (*Labeo rohita*), catla (*Catla catla*), mrigal carp (*Cirrhinus mrigala*), grass carp (*Ctenopharyngodon idellus*), common carp (*Cyprinus carpio*), silver carp (*Hypothalmitchthys molitrix*), catfish (*Clarius batrachus*), singi (*Heteropneustes fossilis*), rainbow trout (*Onchorhynchus mykiss*), and giant river prawn (*Macrobrachium rosenbergii*).

Economics and Consumption of Aquaculture Products in India

Fisheries play an important role in the national economy, contributing 1.3 per cent to GDP at the current price level and providing full-time or part-time employment to 5.96 million people. It is also a major contributor to foreign exchange earning. During 1997-98, the estimated foreign exchange earning from 60 commodities of fish was about US$ 1200 million. The Indian market can absorb an estimated 4.5 million tones of freshwater fish. While it is true that India is known as a vegetarian culture, in fact 55 per cent of Indians are non-vegetarian. The Indian present annual per caput consumption of fish is 8 kilos per person, while the global average is 12 kilos. So there is lot of room for inland freshwater aquaculture to grow by availing all the freshwater bodies in the country.

The estimated area of backfish water available for aquaculture is 1.19 million ha. Traditional shrimp farming practices are popular in Kerala, West Bengal and Goa. The yields from this system vary from 300 to 1000 kg/ha/year. Intensive shrimp farming has become very common in recent years. Because of its high commercial value, giant tiger prawn (*Penaeus monodon*) is the dominant species in commercial production, although Indian white prawn (*Penaeus indicus;* around 5 per cent of total production) is also farmed in several places.

The main objectives for future fisheries development include enhancing fish production, generating employment, improving socio-economic conditions of fishers, promoting exports and providing food security by way of increasing *per caput* availability of fish. These objectives will be achieved through an integrated approach to marine and inland fisheries and aquaculture, taking into account the need for responsible and sustainable fisheries. India's marine fisheries production has reached a plateau and, at best, only marginal increase is predicted in the near future. Most major stocks are fully exploited and further increase has to come from exploitation of deep-sea resources. All future additional demand for fish will have to be met from aquaculture. A number of schemes have been instituted by state and central sectors to increase brackish-water aquaculture and fish production from tanks and ponds, lakes, reservoirs and rivers. The private sector has emerged as a major player in brackish-water aquaculture, particularly in shrimp farming.

Out of the total inland fishes, 0.226 million tones of fish produced in the State of Andhra Pradesh, consisting of 23 districts, from 96,000 hectares, which is nearly 14.2 per cent of the total inland fish production of India. Nearly 80 per cent of the total fish production of Andhra Pradesh state comes from only two districts, *viz.*, Krishna and West Godavari. As per the Andhra Pradesh state records 36 hectares is under fish pond cultivation in Krishna district.

In the 1960's India made headlines with its Green Revolution, by producing more than double of its wheat production from 1965 to 1972, using high-yielding varieties and improved technology. Today, India is pushing ahead with a Blue Revolution, in rapid increase of its fish production in small ponds and water bodies. This is a boon to small farmers, the nation's nutrition and its gross domestic product.

Impact of Aquaculture on Environment

Tropical brackish water and saltwater aquaculture have contributed to the destruction of mangrove forests, wetlands and salt marshes, because they have been cleared for use as ponds. The ecological importance of mangroves, wetlands and salt marshes is now well established. They are habitat for various animal species, they retard coastal erosion, and they are the spawning grounds for various finish, crustaceans, and mollusks. The role of these coastal resources in the subsistence of the rural poor, as resources of firewood, food, and building material has been identified. Fishermen also depend on mangroves, which provide nursery and breeding habitat for important to the fishery. The conversion of mangrove stands into shrimp and milkfish ponds renders a valuable economic resource unless these two social groups are causing ecological degradation. In this instance, the decline in species diversity is directly related to the decline in occupational diversity and the resultant social displacement. Here it is worth mentioning that 60 per cent of tropical mangroves have been destroyed and turned into aquaculture farms.

Apart from the destruction of land farms, environmentalists are criticizing aquaculture as the cause of various ecological perturbations. Water pollution from substances dissolved or suspended in pond effluents often is cited as a major impact. Aquaculturists are looking for ways to reduce effluent volume and to improve effluent quality, but because of a limited understanding of the principles of water quality within the aquaculture industry, efforts to reduce the water pollution potential of pond aquaculture are not moving ahead as fast as they should.

Commercial aquaculture depends upon fertilizer and especially feeds for elevating the production of fish and shrimp above that possible from natural nutrient supplies. Fertilizers contain nitrogen and phosphorus to stimulate phytoplankton productivity, which in turn augments the food web that culminates in aquacultural production. Even greater production of aquacultural species can be achieved when manufactured feed pellets are applied to supplement natural food sources. A large part of the absorbed feed is converted to carbon dioxide and water in respiration or excreted as ammonia and other metabolites. Only a portion of the absorbed feed nutrients is contained in the fish or shrimp at harvest.

Dense phytoplankton blooms in aquaculture ponds cause large daily fluctuations in dissolved oxygen. Concentrations are high during the day and low at night. Bottom water and the soil-water interface may become depleted of dissolved oxygen in intensive aquaculture ponds. It is common practice to mechanically aerate intensive aquaculture ponds to prevent low concentrations of dissolved oxygen in the water and at the sediment surface.

Aeration allows for greater production but as production increase the amount of ammonia excreted by the culture species also increase. Ammonia is a plant nutrient, but it is toxic to aquatic animals. At high production levels, it is necessary to exchange water and flush excessive ammonia from ponds.

Mechanical aeration creates strong water currents that erode pond bottoms and insides of levees. Eroded particles are suspended in the water, and suspended solids are flushed from ponds when water is exchanged. When ponds are drained for harvest, outflowing water and aquatic animal activity also suspends solids that are lost from ponds in effluents. Erosion of levees by rain and erosion of canals by flowing water is another source of suspended solids in pond effluents.

Chemical substances in addition to fertilizers are applied to aquaculture ponds are treatments for water quality and disease problems. These include liming materials, chlorine, zeolite, aluminum sulfate, calcium sulfate, copper sulfate, potassium permanganate, povidone iodine, alkyldimethyl-benzlammonium chloride (BKC), glutaraldehyde, formalin, enzyme preparations, and others.

Residues or degraded products of these substances can be contained in outflowing water. Some of these products such as liming materials and chlorine have been shown effective in improving culture conditions. However, most chemicals are not beneficial and should not be used. A pond ecosystem can assimilate pollutants and render them harmless to the environment. Nitrogen and phosphorus are absorbed by phytoplankton, but when the phytoplankton dies, these nutrients are recycled through microbial decomposition. Nevertheless, phytoplankton uptake of ammonia is a major factor preventing excessive ammonia concentrations in ponds.

Thus, if inputs of fertilizer, feed and chemicals are too much, water quality in ponds may deteriorate to levels that are harmful to the culture species. When this happens, aquaculture frequently use water exchange to flush polluted water from ponds and to replace it with higher quality water. Water flushed from ponds in response to water exchange may pollute natural, receiving water bodies.

Extensive aquaculture practices are being carried out which can be violent from the mushrooming growth of shrimp farms in an un-ruled manner, leaving in the wake of evident protest from environmentalists and social activists demanding the closure of these farms. Lack of knowledge on scientific management of aquaculture farming ends in out burst of viral diseases, incurring heavy losses is quite evident in recent past. Hence the holistic coastal aquaculture development can be sustainable and yield high production by judicious planning of the farms, ensuring supply of quality food, aeration of pond water, daily water exchange and high level of management efficiency.

Best Management Practices

The best approach to effluent management in pond aquaculture is to employ bet management practices or "BMPs". The BMPs would specify how certain aspects of farm operations would be done, and as long as a farm complies with the BMPs, it would be allowed to discharge water. Obviously, BMPs should be site specific, because different locations and operations will have different requirements for effective effluent management. Nevertheless, it is possible to list some general examples of BMPs that could be used:

1. Limits on stocking and feeding rates.
2. Reduction in nitrogen and phosphorus concentrations in feeds to lowest levels possible without impairing feed quality.
3. Prohibit the use of trash fish as feed
4. Conservative feeding practices to reduce the amount of wasted feed.
5. Limits on water use for water exchange.
6. Restrictions on the use of certain chemicals in ponds.
7. Specifications for pond and canal construction and aerator placement to minimize erosion and prevent excessive seepage.
8. Discharge of effluents, especially the final 25 per cent of pond volume, through sedimentation basins.
9. Prohibit discharge of brackishwater into freshwater bodies or onto agricultural land.
10. Require sediment disposal be done in an ecological responsible manner.

Once BMPs are developed for a farm, an inspector would need to visit occasionally to certify that they are being followed. This will be an added regulatory duly for the government, but it will be much simpler than establishing effluent standards and monitoring to verify compliance. A monitoring program of environmental conditions around large aquaculture farms or in areas with many small farms would be desirable to verify that BMPs are providing environmental protection.

The use of BMPs to reduce the environmental impacts of aquaculture also will have benefits to environmental conditions within ponds. Less wasted nutrients will reduce phytoplankton abundance, improve dissolved oxygen concentrations in the water and at the sediment surface, and lower ammonia concentrations. Use of better feeding methods will improve feed conversion ratios and lower production costs.

Guidelines to Aquaculture Projects in India

India too issued guidelines for aquaculture development. These guidelines are meant for prawn farmers, fisheries departments, pollution control boards and different administrative ministries of the center and state. The guidelines mandate is to see the Environmental Impact Assessment (EIA) for aquaculture units above 40 hectare size even at the planning stage. For farm often hectares or more, EIA statement would be required at clearance stage. The points would cover in EIA report are fish/ shrimp farming venture's impact on the water course, groundwater quality, drinking water source, agriculture activity and soil health. These aspects are to be monitored regularly by the State Pollution Control Boards that have to give a no objection certificate to any of those large aquaculture ventures at the time of lunch. Later the responsibility has been shifted to the Aquaculture Authority, which will work under the control of district administration. The guidelines also instructed the State Governments not to permit any aquaculture activity in natural mangrove areas or ecologically sensitive wetlands and swamps. Further suggested that the banning of groundwater for aquaculture, protection of the traditional rights and interests of the communities living in the aquaculture project vicinity and ensuring the sea front access to communities. Also imposed ban on the import of fish seeds, discouragement of wet feed use in the ponds and use of plant based stock regulators.

Creation of buffer zone between aquaculture projects and crop farms (minimum 50–100 meters), human settlement (at least 100 meters in case of village less than 500 population, 500 meters for villages of higher population and 2 km for major towns) are also specified. Spacing between two adjacent shrimp farms should be at least 20 meters is suggested. The environmental scientists suggest that there should be stringent rules similar to that of layout approval by the municipal authorities in case of urban housing development, while sanctioning approval of any aquaculture venture.

On the whole, it is say that the following measures need to be followed for better management of the aquaculture as well as environment.

1. Mapping and monitoring of damages to the ecologically sensitive land farms like mangroves, sand dunes, wetlands, marshy, areas etc., due to the development of aquaculture ponds.
2. Measurement and monitoring of pond effluent concentrations, impact of aquaculture ponds on water course, groundwater quality, drinking water source, agriculture activity and soil health, and
3. Monitoring of buffer zone distance between aquaculture projects and crop farms, human settlement and also spacing between two adjacent ponds.

In the present study, an attempt has been made to examine how remote sensing can help to address these issues.

Remote Sensing

It is clear from earlier studies that mapping and monitoring of different land farms are possible with the currently available high-resolution satellite sensor data. Some examples of degraded land farms due to the development of pond aquaculture using remotely sensed data are presented in this paper. It is very difficult to study the pond effluent concentrations and groundwater quality using present satellite sensors data. However, the changes in watercourse, impact of and aquaculture on drinking water source, agriculture farms and soil health are possible to map and monitor from time to time using satellite remote sensing data. Some such cases using remote sensing data were also illustrated.

It is known that there was a revolutionary change occurred in spatial resolution of satellite sensors. The ground resolution of satellite sensors improved from 76 meters (MSS data on board Landsat series) to 4 meters (Multi-Spectral sensor on IKNOS). Even one-meter resolution data available in panchromatic mode. These high-resolution data sets with some image enhancements (data merging linear enhancement etc.) are useful in monitoring the spacing between ponds, ponds and other land farms and also infrastructure facilities in the aquaculture projects. Advantages or high-resolution data in the present study have been demonstrated.

Wetlands

Wetlands are one of the most productive ecosystems with a great diversity. A wide variety of wetlands like marshes, swamps, lakes, impoundments, mangroves, tidal, marshes etc., exist in India. The importances of these wetlands have been considered increased to the recent past, because of their ecological significance, aquatic productivity and as a habitat of fish, birds and wildlife. Unfortunately, many wetlands have been declined for agriculture including aquaculture, industry and settlements in the country. The impact of aquaculture on some of the important wetlands is dealt with the help of satellite remote sensing data in this study.

Mangroves

Mangrove forests are self-ecosystems, which will regenerate the nutrients. The mangroves are habitat for various animals species, retard coastal erosion, and the spawning grounds for various aquatic animals such as crustaceans and mollusks.

Mangroves are assessed (based on RS data) in terms of extent, density of the community, condition, diversity (habitat evaluation, dominant community types, identifying potential habitats, heterogeniety in ecologically rich areas leading to identification of pure and homogeneous communities and heterogeneous or mixed communities).

Single largest block of mangrove swamps are found in Sunderbans (Part of India and Bangladesh). Second largest formations in the Indian subcontinent are at Bhitarkanika (Orissa), followed by Andaman and Nocobar group of islands (18 per cent of the total mangroves). The other prominent mangroves areas are found in Krishna (near Machalipatnam) Godavari (Korangi mangrove forest) deltaic regions (A.P.), Pichavaram, Vedaranyam, Chorao Island (Goa), Gulf of Kuchchh and Kori and in Andaman and Nicobar Islands. Based on the satellite data, an estimated area of 4460 sq. kms. currently occupied by mangroves in India, which is less by 2280 sq. kms. from 1987 estimates. This is mainly due to overexploitation and conversion to agriculture, aquaculture and industrial establishment. Mangroves of the Godavari-Krishna deltaic regions and Tamil Nadu are under pressure from aquaculture industry. The destruction of Corangi mangroves for the development of aquaculture ponds is clearly seen on the IRS-ID LISS-III+PAN merged imagery (Figure 29.1). West coast mangroves are mostly scrubby and degraded except Kori creek. Kori creek (Indus delta) mangroves showed significant improvement (1986–446.9 sq. km and 1992–938 sq.km). However, recent images show salt encrustation at many places on mangrove habitat.

Figure 29.1: IRS-ID LISS-III+PAN Merged Image Depicting the Mangrove Degradation Due to Aquaculture Near Kakinada

Lakes/Lagoons

The lagoon is an area of salt or brackish water separated from the adjacent sea by a low-lying sand/ shingle barrier. Many large expanses of relatively protected, shallow marine water also differently called as lakes. Following this definition, many bodies of water near to the coast and connected with it either at present or in the recent past, come under the category of lagoons/lakes. Along the Indian coast the eastern and southern parts have many coastal bodies of water that can be assigned to the category of lagoons/lakes. No estimates area available on the total extent of area. Only few of them are well known and have been investigated in varying detail. These are the Vembanad lake on southwest coast, the Pulicat lake, the Kolleru lake and the Chilka lake along east coast. The lagoons/ lakes have been used as fishponds since the ancient time. With the advent of aqua boom, the lagoons/lakes have been suffered from the development of pond aquaculture. If this trend continues lagoons/lakes may soon disappear. In addition to this, lagoons/lakes also suffer with the silt deposit due to heavy freshwater discharges mainly from rivers. The growth of freshwater weeds was also seen in some of the lagoons/lakes. It is only possible to monitor the encroachments and other pond aquaculture and wetland paddy cultivation in lake or at the fringes of the lake through satellite remote sensing data. Kolleru lake (16° 33′ 54.52″ N and 16°45′ 33.86″ N and 81° 05′07.52″ E and 81° 21′ 02.84″ E), one of the largest freshwater lake situated in Krishna and West Godavari districts of Andhra Pradesh state, India, how it is suffered from the man made activities like development of fish pond aquaculture and wetland paddy cultivation has been illustrated in; Figure 29.2 using IRS-ID/LISS-III data of 1999. Ninety per cent of the lake was occupied for such man made activities.

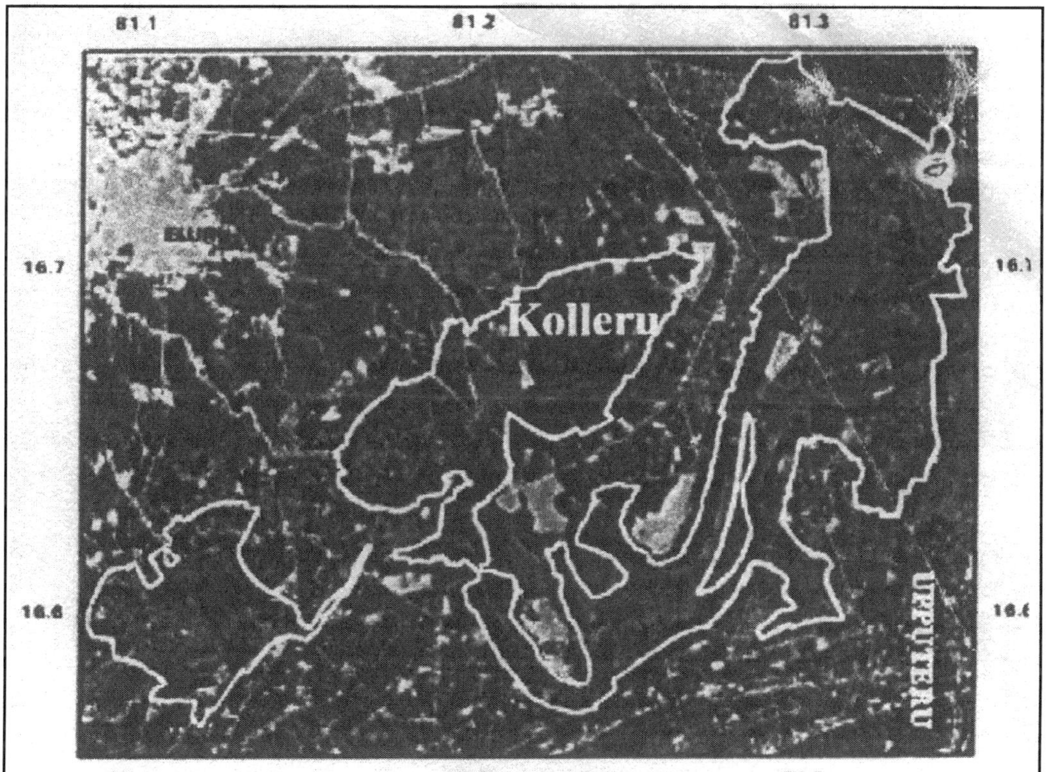

Figure 29.2: Satellite View of Kolleru Lake During March, 1999 Showing Fish Ponds in the Lake

Other Coastal Wetlands

Other coastal wetlands like estuary, creek, backwater, bay, tidal/mud flat and marsh are expected to have threat from the pond aquaculture development. As was discussed earlier, having seen the benefits from the pond aquaculture these land farms have been started converting to pond aquaculture. These land farms are present all along the east and west coasts of India. Some of the Islands in estuary and creeks have been used for the development of pond aquaculture. It is clearly seen from the satellite imagery that how a creek near Nuvalrevu, Andhra and Orissa border, has been occupied for the development of pond aquaculture (Figure 29.3). It was also seen the pond aquaculture development near Etimoga, the Island between two branches of Krishna river, on IRS-1C/LISS-III image (Figure 29.4). Similarly, the development of fish ponds are seen in the tidal/mud flats and marsh areas near Upputeru (outlet of Kolleru lake water into Bay of Bengal (Figure 29.5). Conversion of fertile paddy fields into fishponds is very much seen in the coastal mandals of Krishna and Godavari districts of Andhra Pradesh. One of such example is shown in Figure 29.6.

Beach Ridges

Parapet like ridges of gravel or cobbles built up by strong waves on long ward side of backshore. These are sub-parallel ridges of sands, shell or pebble, varying in amplitude from a few inches to many feet and varying also in distance between them. The depressions between them are known as swales. The number of ridges may vary from two or three to over a hundred. They are located behind the

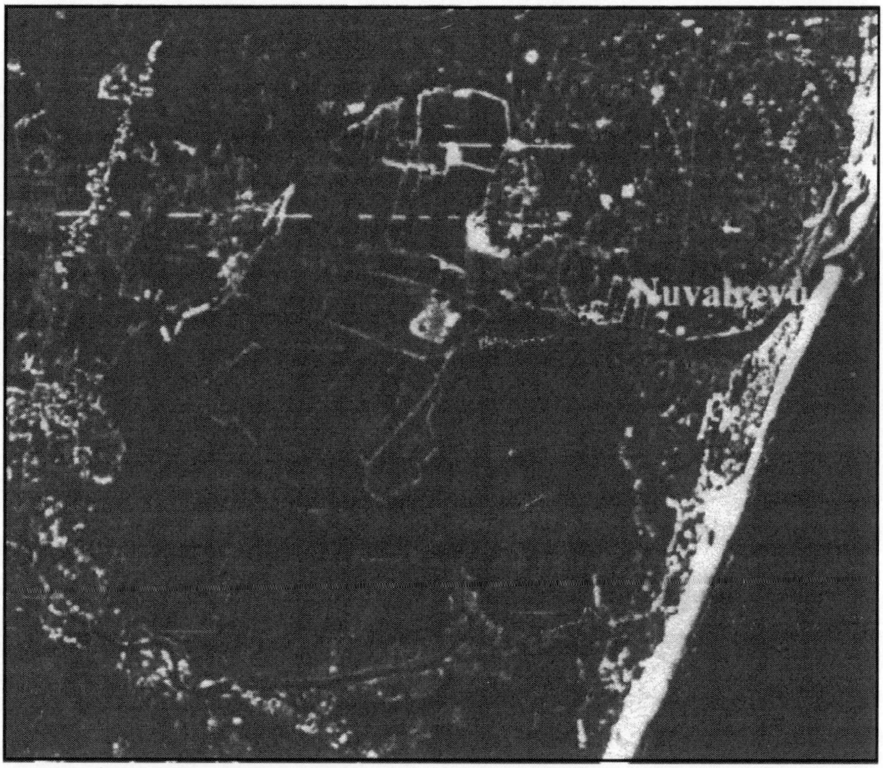

Figure 29.3: Satellite Image Showing the Development of Fish Ponds in the Creek Near Nuvalrevu, A.P. and Orissa Border

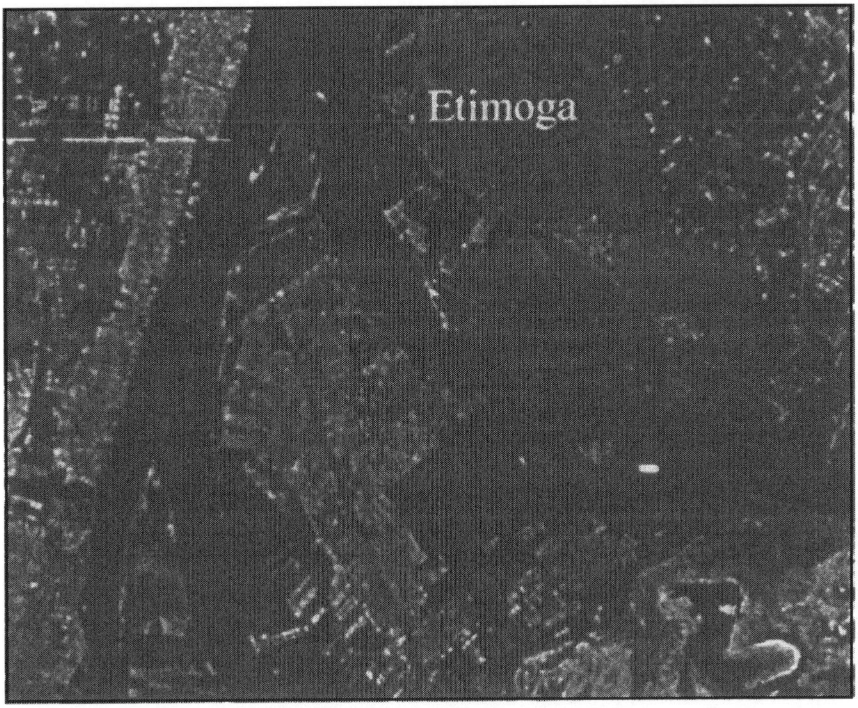

Figure 29.4: IRS-ID/LISS-III Image Showing the Development of Aquaculture Ponds in the Islands of Krishna River

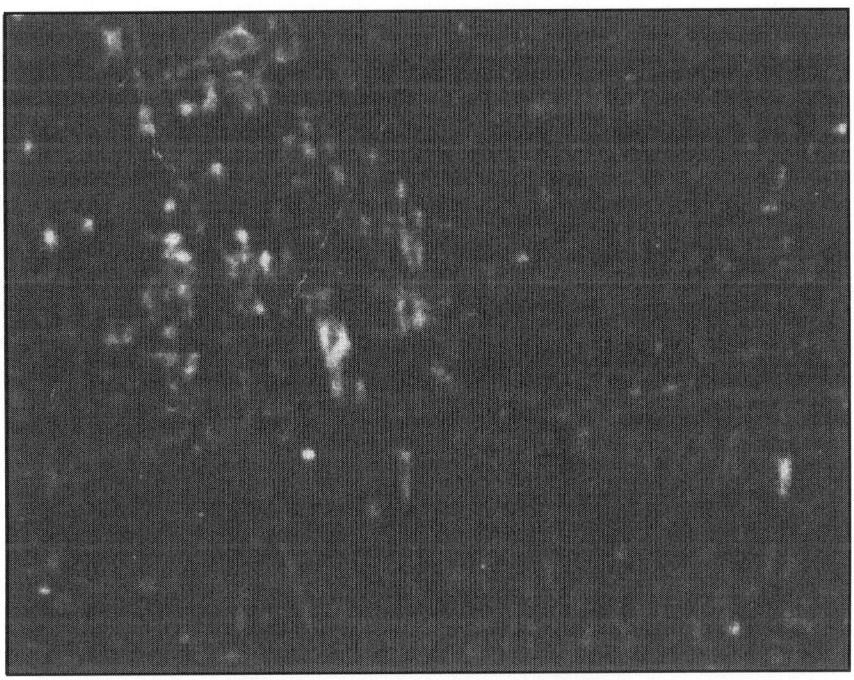

Figure 29.5: Development of Fish Ponds in the Marsh Areas of Kalidindi Mandal as seen from Satellite Image

Figure 29.6: Conversion of Agriculture Lands into Fish Ponds Near Ogirala, Krishna District as seen from Satellite Image

present shore and are to be distinguished from the ridge and runnel features occurring between tidemarks, whereas in deltaic environments, the ridges are perched on a muddy substratum. The beach ridges are seen all along the east coast of India. Most of these beach ridges are treated as waste lands. With the changed scenario, these land farms are converted for fishpond aquaculture. This activity is seen in some of the coastal mandals of Krishna district, A.P. (Figure 29.7).

Figure 29.7: Satellite Image Showing the Development of Fish Ponds on the Beach Ridges in Kalidindi Mandal

Advantages with the High Resolution Data

For any ground implementation, the information required is higher scale (say 1 : 7000 or 1 : 1000). To prepare such a maps, high-resolution data from satellites are required. Currently 1-meter resolution data from IKONOS in panchromatic mode is available. An exercise with 1 meter panchromatic data of Machalipatnam is carried out to demonstrate, how a detailed information can be generated using such high-resolution data (Figure 29.8). This type of high-resolution data is highly useful in monitoring buffer zone activities. It is also useful in monitoring of the water circulation in the creeks and how this circulation impact on aquaculture farming in the upstream area of Kandaleru creak near Krishnapatnam, Nellore district using high resolution satellite data (Figure 29.9).

Conclusion

There is a phenomenal development of pond aquaculture in India during the past decade and also significant increase in fish production. It is also evident that the pond culture activities in the country helped for the development of rural poor.

On the other hand, the freshwater aquaculture has contributed to the destruction of ecology, by way of polluting the drinking water source and destruction of land farms.

Figure 29.8: IKONOS 1 Meter Image Merged with IRS PAN and LISS Depicting the Aquaculture Ponds Near Machalipatnam

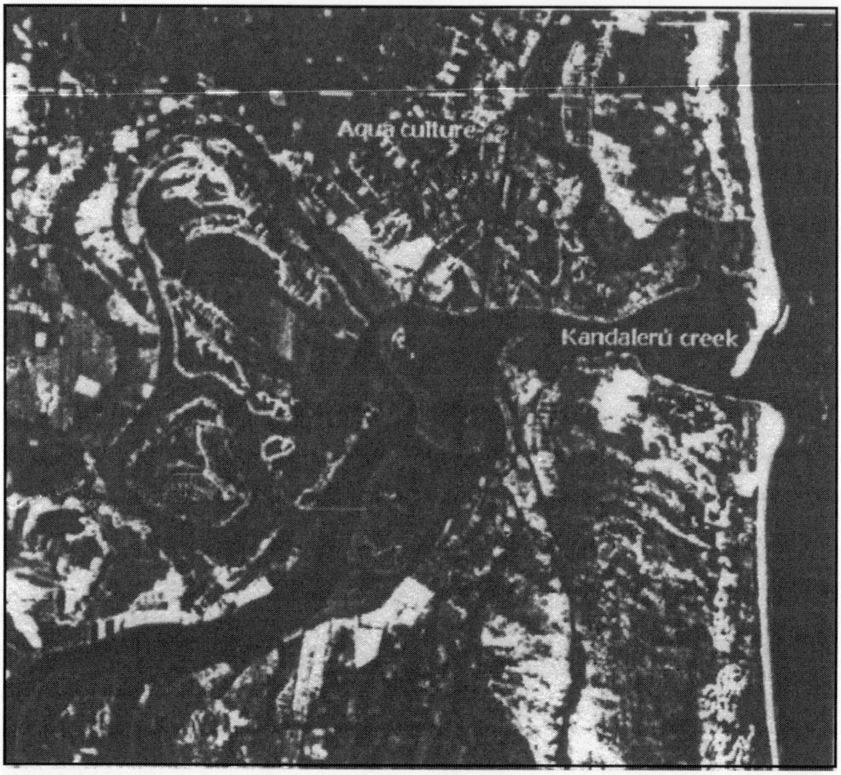

Figure 29.9: IRS-ID LISS-III and PAN Merged Image Illustrating the Chocking of Creek Circulation Due to Siltation

The deliberate blockage of water flow from various rivers and streams due to development of pond aquaculture lead to disastrous consequences of floods all along the east coast during cyclones and very active South West monsoon season.

The development of aquaculture can and does fit in well with the sound environmental ethics– the two go hand-in-hand. Aquaculture relies more than most industries on the maintenance of high environmental standards, without which the industry would fail.

Strict laws govern the sting and style of development, but industries can and do develop in harmony with the environment and the wildlife. Clean oceans and strict controls on water quality ensure that the environment is protected for the industry and people of today and for future generations.

The high-resolution data from satellites will be useful in mapping and monitoring of any study area from time to time and can provide direct information on land use patterns for proper management of the environment and also provide direct and strong proof on the encroachments, in the absence of *in situ* information.

References

Amend, Donald F. 1989. Alaska's regional aquaculture associations' co-management of salmon in southern southeast Alaska. In: *Cooperative Management of Local Fisheries,* (Ed.) E. Pinkerton. University of British Columbia Press, Vancouver, p. 125–134.

Brass, Jane, Bori Olla and Robert Wicklund. 1987. Social economic, and cultural considerations for saltwater cage culture of Florida red tilapia in northeastern Haiti. *Paper Presented at the Gulf and Caribbean Fisheries Institute Meeting*, Curacao, Netherlands Antilles, 9–13 November.

Edwards, P. and Demaine. 1997. *Rural Aquaculture: Overview and Framework for Country Reviews*. RAP Publication 1997/36. Regional Office for Asia and the Pacific, Food and Agriculture Organization of the United Nations, Bangkok, Thailand.

Grover, John H., D.R. Street, and P.D. Starr.1980. Review of Aquaculture Development Activities in Central and West Africa (Research and Development Series no. 28, Auburn University), Auburn, AL, Auburn University press.

Liao, I-Chiu. 1988. East meets West: An eastern perspective of aquaculture. *Journal of the World Aquaculture Society*, 19: 62–73.

MacKenzie, Clyde L. 1977. Development of an aquaculture program for rehabilitation of damaged oyster reefs in Mississippi. *Marine Fisheries Review*, 39: 1–13.

McCay, Bonnie, J. 1989. Co-management of a calm revitalization project: The New Jersey "spawner sanctuary" program. In: *Cooperative Management of Local Fisheries: New Directions for Improved Management and Community Development*, (Ed.) E. Pinkerton. University of British Columbia Press, Vancouver, p. 103–124.

Mitchell, J.R., H. Ackefors, J. Bardach, E. Egidius, A. Munro, S. Tal, H. Rosenthal, C. Sinderman, and M. Bilio. 1986. Statement Committee. In: *Realism in Aquaculture: Achievements, Constraints, Perspectives*, (Eds.) M. Bilio H. Rosenthal, and C. Sinderman. European Aquaculture Society, Bredene, Belgium p. 580–585.

Nakamura, Royden, and Judith Trauth-Thomas. 1986. The Lummi India (USA) aquaculture training program (1969–1972): Marine biology and aquaculture technology in socio-economically depressed, environment. *Proceedings Symposium Coastal Aquaculture*, 4: 1030–1038.

Rahman, Reza S., Haque Chowdury, and Mahrun N. Chowdury. 1985. Evolution of shrimp culture in Sathhira and its impact on landless, small and marginal farmers/fishermen: A rapid appraisal. Consultation on Social Feasibility of Coastal Aquaculture, BOBP/MIS/2, Swedish National Board of Fisheries and the Bay of Bengal program.

Report. 1996. Kolleru falls prey to shrimp farmers, encroachers. *The Hindu*, Daily Newspaper, 5 May, 1996, Hyderabad.

Report, 1996. Political sharks gobbling up Kolleru. *Indian Express*, Daily Newspaper, 30 August, 1996, Hyderabad.

Santaella, Eladio. 1989. Implementation of national aquaculture programmes. In: *Aquaculture: A Review of Recent Experience*. OECD, Paris, p. 187–192.

South Carolina Sea Grant Consortium. 1989, January. The strategic plan for aquaculture development in South Carolina, Vol. 2.

Chapter 30

Use of Biogas Slurry as a Fish Pond Manure

Sarbjit Singh Sooch**#, Asha Dhawan** and Davinder Pal Singh*

***Assistant Professor, Department of Civil Engineering, College of Agricultural Engineering, PAU, Ludhiana*
***Zoologist (Fisheries), Department of Zoology and Fisheries, PAU, Ludhiana*
****B. Tech. Student (Ex.), PAU, Ludhiana*

ABSTRACT

Effects of animal manures (Poultry droppings, biogas slurry and cow dung) on water quality parameters plankton production and survival and growth of fish were studied at PAU, Ludhiana. The water quality parameter were remained in optimum range. Plankton production (Phytoplankton and Zooplankton) was observed maximum in ponds manured with biogas slurry. However the growth of *Cyprinus carpio* in terms of percentage increase in length and weight gain was observed maximum in the pond manure with poultry droppings, followed by biogas slurry and cow dung.

***Keywords**: Organic manures, Water quality, Fish production and Fish growth.*

Introduction

India has made impressive and rapid strides in the development of fisheries, particularly in the area of aquaculture. Increasing aquaculture production is no longer the main target but its sustainability

---

E-mail: sarbjit_sooch@sify.com.

is more important. The sustainable aquaculture depends upon the provision of eco-friendly, economically and socially viable culture system.

The supplementary diet contribute to 40–60 per cent of the total production cost in intensive fish culture system. Increasing cost of feed stuffs (grains, fish meal, oil, cakes etc.) used in the preparation of supplementary diet has stressed the scientists to look for some cheaper substitutes. Many wastes and by-product of agriculture, animal husbandry and agro-industries have food value and can be used as pond manures and fish feed ingredients.

Moreover fish is an excellent converter of poor quality ingredients into high level animal protein *i.e.* the nutrients for natural food of fish (Plankton) which had a significant and positive influence on fish growth. Some fish species also feed directly upon the undigested fractions of these wastes (Schroedar, 1980) which may be low in nutritional value, but microorganisms adhere to them are of high protein value. Further these wastes can also be incorporated into the production of cheap low protein pelleted feed (Devaraj and Keshavappa, 1985).

Review of Literature

Many organic manures like cow dung, poultry manure, duck manure, pig dug and sewage etc., have been tried. Much work has been done on the productivity potential of cow dung, which is now universally used as the organic manure (Banerjee *et al.*, 1969) in pisiculture practice.

Schoonbee *et al.* (1979) observed marked fluctuations in dissolved oxygen level in manured ponds (cattle manure) but anerobic conditions did not prevail at any time. Burn and Stickney (1980) reported increase in ammonia with increase in level of poultry waste in the ponds. Sandhu (1982) found that phosphate contents of water increased by manuring (farm yard manure and poultry droppings of 5–17 t/ha/yr).

Yadav and Garg (1992) observed that addition of poultry droppings @ 30–428 kg dry matter/ha and cow dung @ 24 t/ha/yr respectively did not affect the water quality and hydrobiological factors remained within the optimum level.

The animal manure increase the biological productivity of the pond through various pathways, which is judged through the qualitative and quantitative estimation of plankton production (Schroeder, 1980).

The type of manure also influence the plankton production. Biogas slurry has been found to be one of the potential manurial input in enhancing the growth of plankton than cow dung (Rai, 1981, Kalayani and Shetty, 1987, Sehgal *et al.*, 1991).

Animal manures have direct or indirect effect on the survival and growth of fish. Above 90 per cent survival of *C. carpio* was reported in ponds fertilized with poultry manure (Kaur and Kaur, 1987) whereas 100 per cent survival of *C. carpio* was reported from ponds receiving cowdung and biogas slurry (Kaur *et al.*, 1987). Dhawan and Toor (1989) recorded maximum growth of *C. mrigala* in ponds treated with poultry droppings followed by cow dung (both @ 20 g/ha/yr) than unmanured pond. Growth rate of 71.15, 48.33, 38.15 and 29.67 g of *C. carpio* was reported from ponds receiving poultry droppings (18 t/ha/yr), cow dung (10 t/ha/yr), biogas slurry (19 t/ha/yr) and control (without manure) respectively (Kaur, *et al.*, 1989).

Materials and Methods

Experiments were conducted from 19th April 2002 to 25th June, 2002 at the Fisheries Research Complex of Punjab Agricultural University, Ludhiana to study the effect of biogas slurry on the growth of *Cyprinus carpio* (common carp). Physico-chemical parameters of water, plankton production were also analysed to assess whether biogas slurry had any effect on them.

Six cemented tanks (1–6) each with an area of 0.002 hectare (5.85 × 2.77 m) and a depth of 1.22 m were selected and manuring was done with cow dung, poultry waste and biogas slurry. Cow dung, poultry waste and biogas slurry were added at desired levels *i.e.*, @ 18 t/ha/yr in a tank (700 g/ha/week). The tanks were stocked with fingerlings of *Cyprinus carpio* (common carp). The different water quality parameters like water temperature, secchi disc transparency, hydrogen ion concentration (pH), dissolved oxygen (DO), alkalinity and free carbon dioxide were analysed. Pond productivity was assessed in terms of plankton production which include both phytoplankton and zooplankton by taking. Plankton samples simultaneously at the time of sampling water for its physico-chemical analysis.

Results and Discussion

The effects of different manures on water quality parameters, pond productivity and fish growth are shown in Tables 30.1–30.3 respectively.

Table 30.1: Effect of Different Manures on Water Quality Parameters

Parameters	*Days*	*Manures*			*Optimum Range of Water Quality Parameters*
		Biogas Slurry	*Cow Dung*	*Poultry Droppings*	
Water temperature (°C)	0	28.75	28.75	27.75	
	20	31.50	31.00	31.50	15–37.5
	40	27.50	27.50	26.50	
	60	33.00	32.00	31.50	
Secchi Disc	0	23.25	19.87	24.62	—
Transparency (cm)	20	20.95	18.60	21.75	
	40	17.87	17.05	19.60	
	60	17.45	16.60	19.10	
Hydrogen Ion	0	8.87	8.42	8.19	
Concentration (pH)	20	8.95	8.91	8.63	6.5–9.6
	40	9.07	8.22	8.17	
	60	9.41	8.18	8.59	
Dissolved Oxygen (mg/l)	0	7.40	6.60	5.50	
	20	15.30	12.00	7.90	7.8–9.0
	40	12.40	5.30	5.60	
	60	13.40	10.40	12.90	
Alkalinity (mg/l)	0	275	247	365	
	20	249	231	223	137–230
	40	130	152	158	
	60	166	162	206	
Free Carbon dioxide (mg/l)	0	0	0	0	
	20	0	0	0	15.00
	40	0	8	16	
	60	0	8	0	

Table 30.2: Effect of Different Manures on Pond Productivity

Pond Productivity	*Days*	*Manures*		
		Biogas Slurry	*Cow Dung*	*Poultry Droppings*
Phytoplankton (ml/l)	0	34	5	7
	20	10	7	18
	40	27	20	28
	60	22	5	13
Mean		23.25	9.25	16.5
Zooplankton (ml/l)	0	30	10	15
	20	44	22	18
	40	48.52	47	67
	60	35	42	50
Mean		39.38	30.25	37.5

Table 30.3: Effect of Different Manures on Fish Growth

Days	*Fish Growth*	*Manures*		
		Biogas Slurry	*Cow Dung*	*Poultry Droppings*
Initial (0 day)	Length (cm)	12.50	14.08	12.30
Final (60 days)		13.76	14.58	14.60
Length gain (%)		10.08	3.55	18.70
Initial (0 day)	Weight (gm)	35	39	38
Final (60 days)		42	42	50
Weight gain (%)		20	7.69	31.57

In the present study, phytoplankton was observed maximum in pond manured with biogas slurry (23.25 ml/l). Similarly zooplankton was observed to be maximum in the pond manured with biogas slurry (39.38 ml/l) followed by poultry droppings (37.5 ml/l) and cow dung (30.25 ml/l). Biogas slurry had been observed to be one of the potential manurial input enhancing the growth of plankton (Phytoplankton and Zooplankton) than cow dung (Rai, 1981, Kalayani and Shetty, 1987). This high value of plankton in pond manured with biogas slurry was because of high nutrients contents of organic manures which enhance the plankton production.

Fish growth (in terms of percentage increase in length and weight gain) was observed to be maximum in the pond manured with poultry droppings (length gain = 18.7 per cent, weight gain = 31.57 per cent) followed by biogas slurry (length gain=10.08 per cent), weight gain=20 per cent) and cow dung (length gain = 3.55 per cent, weight gain=17.6 per cent). Maximum growth of fish was observed in case of poultry droppings, this has also been observed by Dhawan and Toor (1989).

Conclusion

Water quality parameters *viz.*, water temperature, pH, dissolved oxygen, alkalinity, secchi disc transparency and free carbondioxide were analysed, which remained in optimum range. The most critical parameter *i.e.*, dissolved oxygen in all the treatments was in optimum range and maximum dissolved oxygen values were in case of biogas slurry. Plankton production (Phytoplankton and Zooplankton) was observed maximum in ponds manured with biogas slurry followed by poultry droppings and cow dung. The growth of *Cyprinus carpio* in terms of percentage increase in length and weight gain was observed maximum in case of pond manured with poultry droppings (length gain = 18.7 per cent, weight gain = 31.57 per cent) followed by biogas slurry (length gain = 10.08 per cent, weight gain = 20 per cent) and cow dung (length gain = 3.55 per cent, weigh gain = 17.6 per cent).

The organic wastes are available in large quantities all over the world can be used for supplementary diet for fish production. Moreover recycling of organic wastes for fish culture served the dual purpose of cleaning the environment and reaping economic benefits.

References

Banerjee, R.K., G.S. Singit and P. Ray. 1969. Some observations on the use of poultry manure as fertilizer in rearing major carp fry. *Indian J. Fish.*, 16: 29–34.

Burn, R.P. and Stickney, R.P. 1980. Growth of *Tilapia auria* in ponds receiving poultry wastes. *Aquaculture*, 20: 117–121.

Devaraj, K.V. and Keshavappa, G.Y. 1985. Preliminary studies on poultry-cum-fish culture. *Mysore J. Agric. Sci.*, 19(3): 189–193.

Dhawan, A. and Toor, H.S. 1989. Impact of organic manure and supplementary diet on plankton production and fish growth and feeding of an Indian major carp, *Cirrhina mrigala* (Ham.) in fish ponds. *Biol. Waste*, 29: 289–298.

Kalayani, R. and Shetty, H.P.C. 1987. Use of biogas slurry as a fertilizer for production of carps. *The First Indian Fisheries Forum*, December 4–8.

Kaur, K. and Kaur, S. 1987. Efficacy of poultry droppings in common carp *Cyprinus carpio* (Linn) culture: Effect on survival and growth. *Zoologica Orientalis*, 4(1&2): 13–18.

Kaur, K., Kaur, K. and Dhawan, A. 1989. Effects on organic wastes on the growth and ovarian maturation of *Cyprinus carpio. Proc. Nat. Sem. Freshwat. Aqua.*, p. 81–83.

Kaur, K., Sehgal, G.K. and Sehgal, H.S. 1987. Efficiency of biogas slurry in carp. *Cyprinus carpio* var. *communis* (Linn.) culture. Effects on survival and growth. *Wastes*, 22(2): 139–146.

Rai, S.P. 1981. Role of biogas plant in integrated aquaculture for rural development. In: *Summer Institute on Farming System Integrating Agriculture Livestock and Fish Culture*, 6th July to 4th August, CIFRI, Barrackpore, West Bengal.

Sandhu, J.S. 1982. Role of organic wastes in composite fish culture. *Ph.D. Thesis*, P.A.U., Ludhiana.

Schoonbee, H.J., Nakani, V.S. and Prinsloo, J. 1979. The use of cattle manure and supplementary feeding in growth studies of the Chinese Silver Carp in Transkri, South Africa. *S. Afr. J Sci.*, 75(2): 489–495.

Schroeder, G.L. 1980. Fish farming in manure loaded ponds. ICLARM–SEARCH *Conf. Agri. Aquaculture Fmg. Systems*, Manila, Phillipines, p. 73–86.

Sehgal, H.S., Kaur, K. and Sehgal, G.S. 1992. Zooplankton response to biogas slurry in carp ponds. *Bioresource Technology*, 41: 111–116.

Yadav, N.K. and Garg, S.K. 1992. Relative efficacy of different doses of organic fertilizer and supplement feed utilization under intensive fish farming. *Bioresource Technology*, 42: 61–65.

Chapter 31

Water Quality and Aquatic Biodiversity of Merakhong River, Manipur

Haobijam Dhamendra and Laishram Kosygin
Lake Monitoring and Research Laboratory, Loktak Development Authority, Ningthoukhong – 795126, Manipur, India

ABSTRACT

Water quality, flora and fauna of Merakhong river were studied during September 2000–June 2002. Well mark seasonal variations have been recorded in respect of physico-chemical characteristics. The river water is considered polluted as it has high concentration of free CO_2 (11.9 mg/l), nitrite-nitrogen (0.035 mg/l), inorganic phosphorus (0.110 mg/l) and aquatic flora and fauna, which can thrive well in polluted environments.

Key words: Merakhong river, Water quality, Biodiversity, Pollution.

Introduction

Merakhong river arises from Kangchup hills and finally fall into the Loktak Lake near Ishok, Bishenpur District, Manipur. Some of the important villages through which the river passes are Salam, Sagoltongba, Yurenjam, Maitram, Mekola, Leirenjam, Kodompokpi, Gobindagram, Thiyam Langpok, Utlou, Laitonjam etc. It belongs to the Chindwin-Irrawaddy drainage system of Myanmar. The river water is being used for various purposes for domestic and agriculture purposes by the local community. It discharges the untreated domestic wastewater as well as agricultural runoff from the surrounding agricultural fields into the Loktak Lake of Manipur. Study on the water quality and biodiversity of the river is of importance, since it drains into Loktak Lake, which is considered as lifeline of the state and there is no report on the ecology of this river. In the present study an attempt has been made to study the water quality, flora and fauna of the river.

Materials and Methods

Surface water samples were collected from the downstream of Merakhong river near Nambol Laitonjam on quarterly basis (September/post-monsoon, December/winter, March/pre-monsoon and June/monsoon) during 2000 and 2002. Samples were collected between 9 and 11 h from 10 cm depth and were transported to the laboratory immediately. Physico-chemical parameters were analyzed following APHA (1989) and Trivedi and Goyal (1986).

Fishes were identified following Jayaram (1981), Talwar and Jhingran (1991) and for snails following Rao (1989). Aquatic macrophytes and planktons were identified following Adoni *et al.* (1985) and APHA (1989) and by consulting other relevant literatures.

Results and Discussion

Physico-chemical Characteristics of Water

The results of physico-chemical characteristics of the Merakhong river are summarized in Table 31.1. Surface water temperature fluctuates from 17.5°C to 27.6°C with an average value of 23.3°C. It showed a positive correlation with free carbon dioxide (Table 31.2). The pH value is within the tolerance limit of Indian Standard (1982). The lower value of pH was observed during monsoon and post-monsoon periods. It showed negative correlation with free carbon dioxide.

Table 31.1: Data on Physico-chemical Characteristics of Merakhong Rivers During September, 2000–June, 2002

Sl.No.	*Parameters*	*Range*		*Mean*	*SD*
		Min.	*Max.*		
1.	Water temperature (°C)	17.5	27.6	23.33	3.633
2.	pH	5.9	7.9	6.95	0.709
3.	Dissolved oxygen	4.6	8.2	6.36	1.28
4.	Free CO_2	4.6	18.5	11.95	4.72
5.	Total dissolved solids	108.00	181.00	144.87	25.91
6.	Chloride	8.50	20.00	14.76	3.63
7.	Calcium	4.80	10.20	7.67	1.67
8.	Magnesium	4.20	8.40	6.33	1.47
9.	Sodium	7.00	23.00	17.25	5.03
10.	Potassium	2.00	8.00	4.37	1.84
11.	Nitrite-nitrogen	0.018	0.056	0.0352	0.0139
12.	Nitrate-nitrogen	0.314	0.798	0.59	0.19
13.	Total nitrogen	1.987	3.827	2.74	0.604
14.	Inorganic phosphorus	0.071	0.127	0.11	0.039
15.	Organic phosphorus	0.137	0.301	0.24	0.067
16.	Total phosphorus	0.231	0.497	0.35	0.082

Note: All the values are in mg/l, except temperature and pH.

Table 31.2: Pearson's Correlation Coefficient for Physico-chemical Characteristics of the Merakhong River (n = 8)

Parameters	*Temp.*	*pH*	*DO*	*Free CO_2*	*TDS*	*Cl*	*Ca*	*Mg*	*Na*	*K*	*Nitrite-N*	*Nitrate-N*	*Total N*	*Inorganic P*	*Organic P*	*Total P*
Temp.	–	– 0.774	– 0.88	0.857	0.416	– 0.469	0.263	0.477	– 0.028	– 0.092	0.291	– 0.489	– 0.412	– 0.071	– 0.42	– 0.378
pH		–	0.949	– 0.942	0.067	0.815	0.279	– 0.137	0.592	0.496	0.259	0.872	0.719	0.1001	0.721	0.638
DO			–	– 0.973	0.011	0.746	0.172	– 0.19	0.463	0.349	0.119	0.772	0.559	0.212	0.628	0.615
Free CO_2				–	0.015	– 0.658	– 0.226	0.154	– 0.446	– 0.243	0.012	– 0.789	– 0.501	– 0.265	– 0.586	– 0.607
TDS					–	0.308	0.681	0.816	0.635	0.473	0.752	0.429	0.154	0.496	0.374	0.543
Cl						–	0.464	0.027	0.76	0.821	0.589	0.795	0.772	– 0.017	0.814	0.658
Ca							–	0.585	0.743	0.3303	0.589	0.461	0.071	0.127	0.33	0.331
Mg								–	0.218	0.278	0.385	0.271	– 0.107	0.358	0.356	0.462
Na									–	0.633	0.776	0.714	0.595	0.255	0.505	0.538
K										–	0.704	0.681	0.836	– 0.032	0.849	0.68
Nitrite-N											–	0.407	0.458	– 0.029	0.404	0.317
Nitrate-N												–	0.775	0.364	0.878	0.892
Total N													–	0.032	0.776	0.561
Inorganic P														–	0.13	0.584
Organic P															–	0.88
Total P																–

The dissolved oxygen (DO) and free CO_2 concentrations showed marked seasonal fluctuations (Figure 31.1). Highest value of DO was observed during December and lowest during June. It was negatively correlated with free carbon dioxide. The free CO_2 concentration was usually high during the present study with an average value of 11.95 mg/l. Its concentration was highest during September and lowest during December. The river water is not suitable for fish culture as average value of free CO_2 exceeds the tolerance limit (Indian Standard, 1982).

The total dissolved solid varies from 108.0 to 181.0 mg/l with an average value of 144.87 mg/l. The concentration of potassium was highest during the pre-monsoon period and declines during the monsoon and post-monsoon periods (Figure 31.2). The calcium content was usually low and may be categorized as calcium poor water. Concentrations of sodium, chloride, and magnesium occur within the tolerance limit of Indian Standard (1982).

The higher concentration of nutrients, particularly nitrogen and phosphorus, are naturally to be expected in the polluted waters (Munawar, 1970). The nitrite-nitrogen concentration ranges from 0.018–0.056 mg/l with an average value of 0.035 mg/l. Presence of nitrite-nitrogen in the present

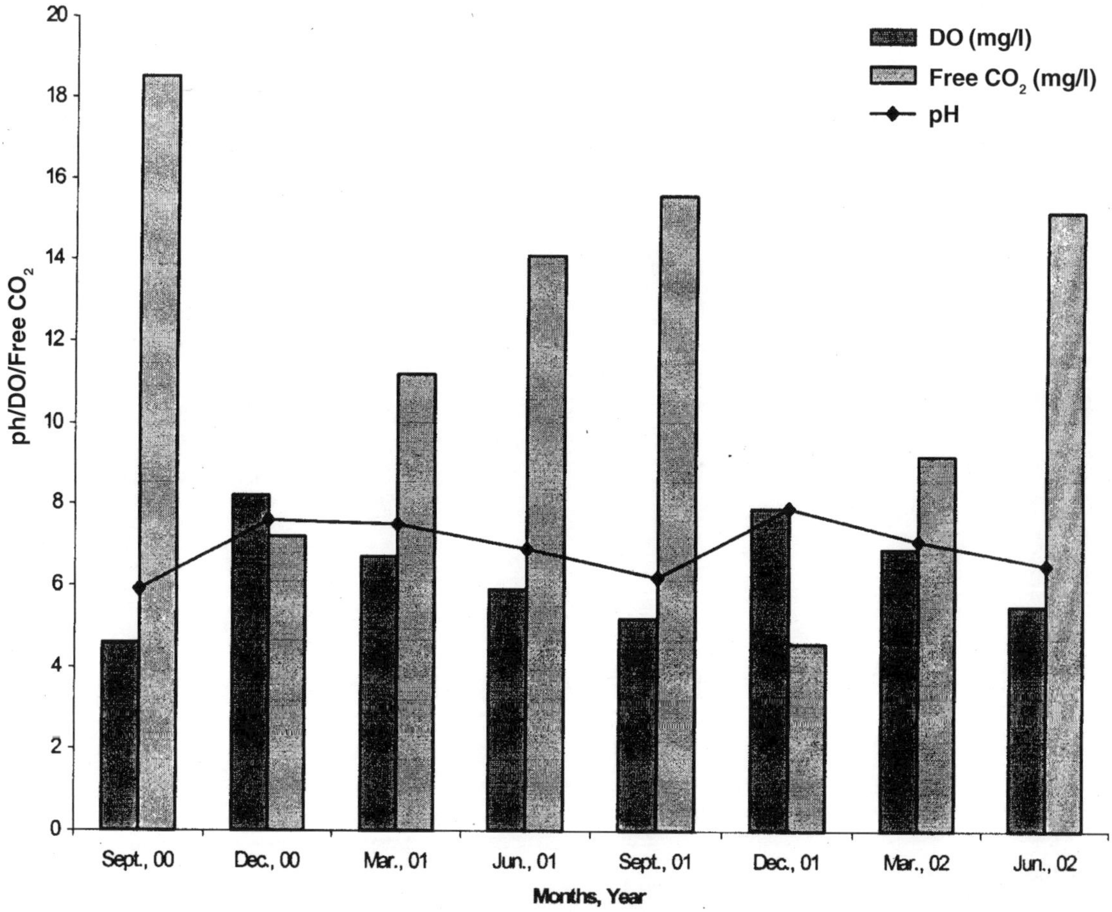

Figure 31.1: Seasonal Variations in pH, Dissolved Oxygen (DO) and Free CO_2 in Merakhong River During September, 2000–June 2002

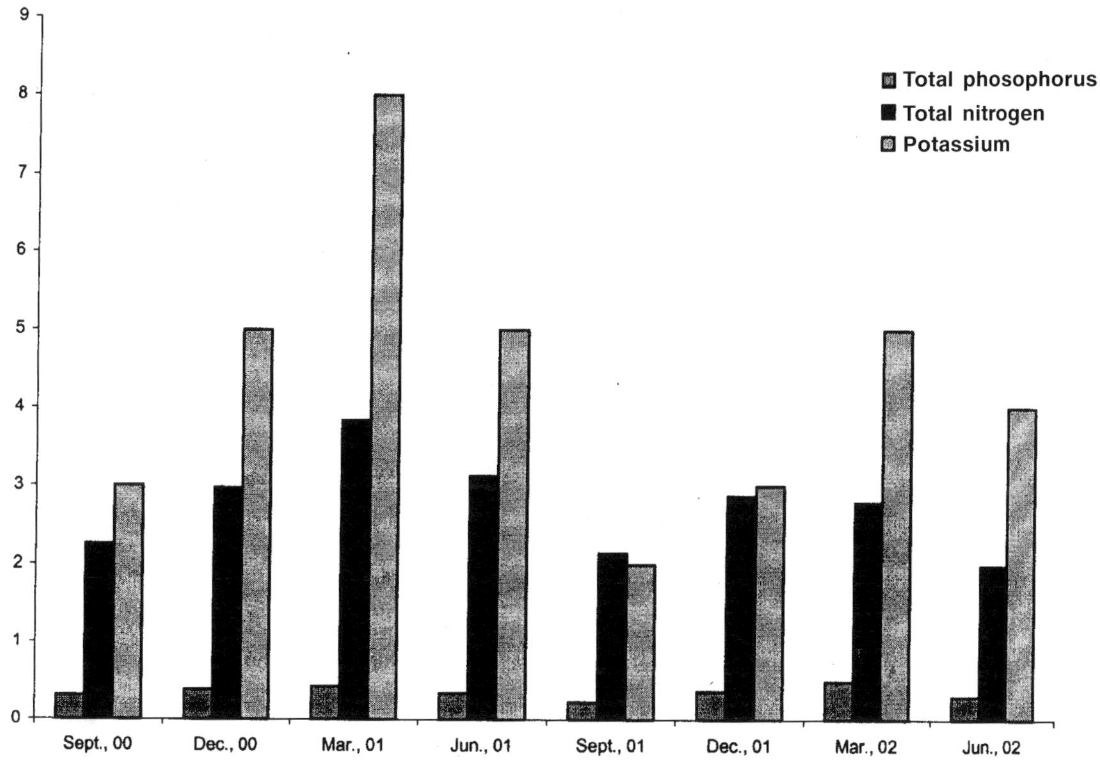

Figure 31.2: Seasonal Variation in Total Nitrogen, Total Phosphorus and Potassium in Merakhong River During September, 2000–June 2002

study indicates the organic pollution in the Merakhong river. The nitrate-nitrogen concentration was highest during the winter and lowest during the post-monsoons. The phosphorus content during all the seasons exceeded the critical level of algal bloom (0.03 mg/l) suggested by the Sawyer *et al.* (1945). It was comparatively higher during the pre-monsoon period (Figure 31.3). The average concentration of inorganic phosphorus in the present investigation was found to be higher than the standard permissible limit (0.050 mg/l) recommended by USEPA (1976) for any tributary discharging to a lake. The concentrations of nutrients were higher in winter (December) and pre-monsoon (March), decreasing in the monsoon season and lowest in post-monsoon (September). This may be due to the fact that during winter and pre-monsoon period there was less flow of water and nutrients were accumulated in the river. Some of the important sources of nutrients in the rivers are runoffs from the surrounding agricultural fields and the catchment hills, domestic and municipal waste from the human settlements and deposition of animal waste.

Aquatic Biodiversity

List of fishes, snails, macrophytes and planktons collected and identified during the present study are shown in Table 31.3. The collection included 20 species of fishes (belonging to 16 genera and 10 families), 3 species of snails (belonging to 3 genera and 3 families) and 10 species of macrophytes (belonging to 9 genera and 8 families). Most of the fish species collected in the present study were widely distributed forms, which can thrive well in the polluted waters.

Table 31.3: List of Fishes, Snails, Macrophytes and Planktons of Merakhong River

Fishes	*Snails*	*Macrophytes*	*Planktons*
Family: Cyprinidae			
1. *Catla catla* (Hamilton)	**Family: Viviparidae**	**Family: Poaceae**	**Phytoplanktons**
2. *Cirrhinus mrigala*	1. *Angulyagra oxytropis*	1. *Echinochloa stagnina* Retz.	1. *Endorina* sp.
3. *Ctenopharyngodon idellus* (Val.)	(Benson)	2. *Leersia hexandra* Sw.	2. *Euglena* sp.
4. *Labeo gonius* (Hamilton)			3. *Diatoma* sp.
5. *Labeo rohita* (Hamilton)	**Family: Pilidae**	**Family: Onagraceae**	4. *Navicula* sp.
6. *Puntius sophore* (Ham)	2. *Pila* sp.	3. *Ludwigia claveliana* Gomez	5. *Nitzschia* sp.
7. *Ampblypharyngodon mola* (Ham)		de la Manza	6. *Phacus* sp.
8. *Esomus danricus* (Ham)	**Family: Unionidae**		7. *Spirogyra* sp.
	3. *Lamellidens* sp.	**Family: Convolvulaceae**	
		4. *Ipomea aquatica* Forsk	
Family: Cobitidae			**Zooplanktons**
9. *Lepidocephalus guntea* (Ham)		**Family: Amaranthaceae**	1. *Cyclops* sp.
10. *Lepidocephalus irrorata* (Hora)		5. *Alternanthera pheloxeroides*	2. *Daphinia* sp.
		(Mart) Grised	3. *Parameicium*
Family: Bagridae			4. *Vorticella* sp.
11. *Mystus bleekeri* (Day)		**Family: Pontederiaceae**	
		6. *Eichhornia crassipes* (Mart)	
Family: Clariidae		Solms.	
12. *Clarias batrachus* (Linnaeus)			
		Family: Araceae	
Family: Heteropneustidae		7. *Colocasia esculenta* (Linn.)	
13. *Heteropneustes fossilis* (Bloch)		Schott.	
Family: Symbranchidae		**Family: Hydrocharitaceae**	
14. *Monopterus albus* (Zuiew)		8. *Hydrilla verticillata* (L.F.) Royle	
		9. *Hydrilla* sp.	
Family: Gobiidae			
15. *Glossogobius giuris* (Ham)		**Family: Lemnaceae**	
		10. *Lemna* sp.	
Family: Anabantidae			
16. *Anabas testudinouc* (Bloch)			
Family: Belontiidae			
17. *Colisa fasciatus* (Schn)			
Family: Channidae			
18. *Channa orientalis* (Schn.)			
19. *Channa marulius* (Hamilton)			
20. *Channa punctatus* (Bloch)			

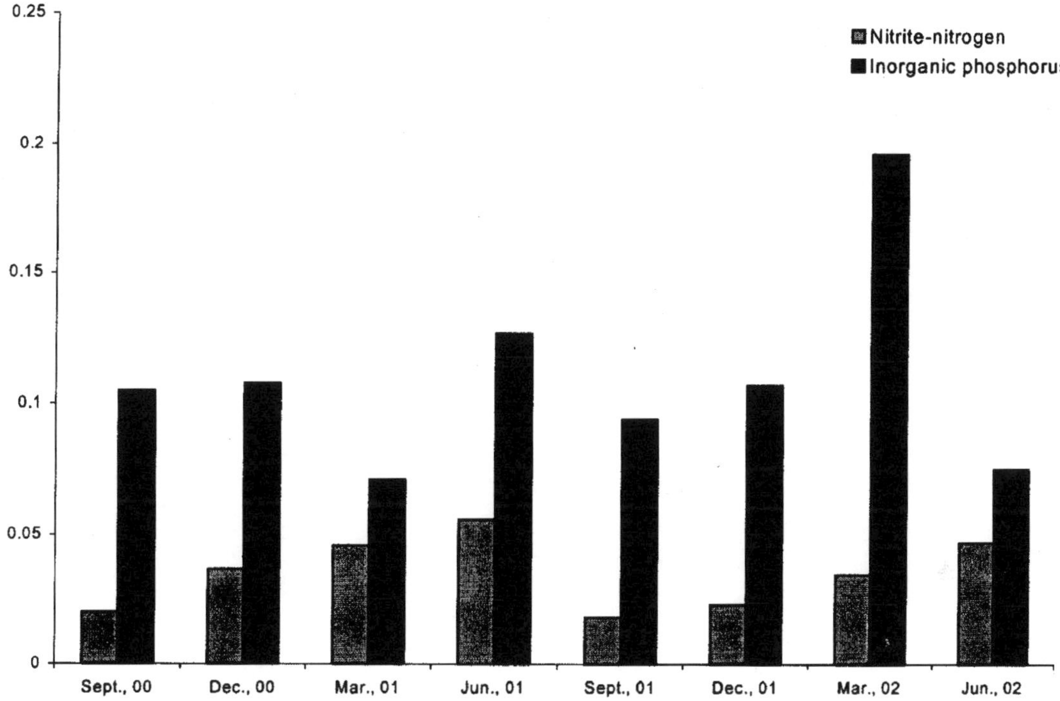

Figure 31.3: Seasonal Variations in Nitrite-Nitrogen and Inorganic-Phosphorus in Merakhong River During September, 2000–June 2002

In the present study 7 genera of phytoplanktons and 4 genera of zooplanktons were identified. All the genera of algae observed were included in the sixty most pollution tolerant genera listed by Palmer (1969). It suggests the occurrence of organic pollution in the river water.

From the foregoing account, it is clear that down stream of the Merakhong river is polluted as it has high concentration of free CO_2, nitrite-nitrogen and phosphorus. It is further confirmed by its floral and faunal composition of the river, which can thrive well in polluted environmental conditions. The impacts of accelerating human activities have been actually felt throughout the river ecosystem resulting in undesirable changes and being accompanied by environmental degradation. Therefore, proper conservation measures should be taken up to save the river from further deterioration.

Acknowledgement

We are grateful to Shri Th. Manihar, Project Director, Loktak Development Authority, Manipur for providing laboratory facilities. We are also thankful to India Canada Environment Facility (ICEF) for extending financial assistance to carry out the work.

References

Adoni, A. D., Joshi, G, Ghosh K, Chaurasia, S. K., Vaishya, A.K. Yadav, M. and Verma, H. G. 1985. *Work Book on Limnology*. Pratibha Publishers, Sagar, pp. 216.

APHA. 1989. *Standard Methods for the Examination of Waters and Wastewaters*, 17th Edition. American Publication Health Association, Inc., New York.

Indian Standard Institution. 1982. Indian standard-tolerance limit for inland surface waters subject to pollution (second revision), IS: 2296. Indian Standard Institution publication, pp. 18.

Jayaram, K.C. 1981. *Freshwater Fishes of India, Pakistan, Bangladesh, Burma, and Sri Lanka Handbook.* Zoological Survey of India, Kolkata, pp. 475.

Munawar, M. 1970. Limnological studies on freshwater ponds of Hyderabad, India. II. The biocoenose. Distribution of unicellular and colonial phytoplankton in polluted and unpolluted environments. *Hydrobiologia,* 36: 105–128.

Palmer, C.M. 1969. Keys to water quality indicative organisms (South Eastern United States). Dept. of Interior, Federal Water Pollution Control Administration, Ohio, U.S.A.

Sawyer, C. N., Lackey, J. B., and Lenz, R. T. 1945. Investigation of the odor nuisance occurring in the Madison lakes particularly Monona, Waubesa and Kegonsa, from July 1943–July 1944. Rept. Governor's Committee, Modison, Wis., p. 92 –100.

Subba Rao. N.V. 1989. *Handbook: Freshwater Molluscs of India*. Zoological Survey of India, Kolkata, pp. 289.

Talwar, P.K. and Jhingran. A.G. 1991. *Inland Fishes of India and Adjacent Countries*, I. Oxford and IBH Publ. Co. Pvt. Ltd., New Delhi, Mumbai, Kolkata, pp. 541.

Trivedy, R.K. and Goyal, P. K. 1986. *Chemical and Biological Methods for Water Pollution Studies.* Environmental Publication, Karad, pp. 248.

U.S.E.P.A. 1976. *Quality Criteria for Water*. EPA–440/9–76–023, Washington, D.C.

Chapter 32

Heavy Metal Accumulation in Seaweeds in Polluted and Unpolluted Estuaries

K. Sasikumar* and A.N. Subramaniyan**

**Department of Botany, Annamalai University, Annamalai Nagar - 608 002, Tamil Nadu, India*
***Center of Marine Environmental Studies, Ehime University, Japan*

ABSTRACT

Three seaweeds in the Vellar and Uppanar estuaries when analysed for heavy metals, showed that Cu, Zn, Cd and Pb varied in them with their habitats. However, *Chaetomorpha aerea* and *C. linum* did not show much variations between the two habitats. *Entromorpha intestinalis* exhibiting low levels of Cu, Zn, Cd and Pb, could be used as a pollution indicator. Among the four metals, concentration of Zn was high in all the seaweeds followed by Pb, Cd and Cu.

Keywords: *SIPCOT industries, Heavy metals, Seaweeds, Bioaccumulation.*

Introduction

In India, seaweeds are used as raw materials for the production of agar, alginate and seaweed liquid fertilizer. Unfortunately, several toxic pollutants, unknown to the biota early, have been introduced in large quantities into the aquatic environment. Metals are natural components of seaweeds and as such are harmless to marine life, but they can build up to high concentration as a result of man's activities, as in mine tailing or industrial effluents and may then cause a risk to human consuming seafood. Group of Expert of Scientific Aspects of Marine Pollution (GESAMP, 1974) has compiled brief reviews on a number of elements especially cadmium, lead, copper and zinc which have been considered as potential pollutants. Heavy metals may accumulate, unnoticed, to toxic levels. Thus problems associated with heavy metal contamination were first highlighted in the industrially advanced countries because of their large industrial discharges and especially by incidents of mercury and cadmium pollution in Sweden and Japan (Kurland *et al.*, 1960; Nitta, 1972; Goldberg, 1976).

Some heavy metals such as Cu, Zn, Cd and Pb are essential for the growth and well being of living organisms including man. However, they are likely to show toxic effects when organisms are exposed to levels higher than normally required. Other elements such as Pb and Cd are not essential for metabolic activities and exhibit toxic properties. Heavy metals form the major group of toxic pollutants among the different pollutants, as these metals tamper the harmony of the ecosystem. Most of the organisms have the ability to bioaccumulate metals from the surrounding water into their bodies.

The concentration of metals in different species living in the same environment may vary as some species contain high, whereas others contain small quantities. These variations are due to the differences in assimilation and excretion capacities, or both, as the concentration of a metal in an animal is being determined by the balance between these two processes. Studies on the concentration of heavy metals in the polluted environment were reported by Badsha and Goldspink (1982), Jaffer and Ashraf (1988), Ashraf *et al.* (1991) indicating the heavy metals to concentrate in the different compartments of aquatic ecosystem. The enriched concentration of heavy metals in aquatic ecosystem due to acidification is reported earlier by Almmer and Dickson (1974) and Campbell and Stokes (1985). The physico-chemical form of metals are affected by acidification and these changes influence the bioavailability and toxicity to many organisms (William and Campbell, 1981).

Hence, in the present investigation, an attempt has been made to study the physico-chemical and heavy metal accumulation such as Cu, Zn, Cd and Pb of seaweeds occurring in the Vellar estuary which is unpolluted and the Uppanar estuary polluted from coconut husk retting grounds and effluent discharges from SIPCOT industrial complex (Small Industries Production Council of Tamil Nadu).

Materials and Methods

The Vellar estuary (Station 1; lat. 11° 20′N, long. 79° 46′E) is a small estuary, blooming with fishing and aquaculture activities (Figure 32.1). Uppanar estuary (Station 2; lat. 11° 59′N, long. 79°50 E) is also a small estuary but polluted with industrial effluents of SIPCOT industrial estate, Cuddalore. For studying the distribution, algae were collected seasonally throughout the year from October 2001 to September 2002. Meteorological data such as monthly rainfall and humidity were obtained from the Meteorological Department, Parangipettai and Cuddalore. Physical parameters such as atmospheric temperature, surface water temperature and pH were determined in the field itself. Chemical parameters such as salinity, dissolved oxygen and nutrients were studied using standard methods (Grasshoff *et al.*, 1983) in the laboratory. All these samples were dried at 60°C for 48 h and made into a fine powder. Heavy metal concentration in the samples was determined through Atomic Absorption Spectroscopy (AAS). Solvent extraction procedure was adopted for preconcentration of heavy metals from water as a preliminary step to their determination by AAS. Results are expressed in mg/l. Tissue digestion of the dried, powdered samples of different organs was carried out by ashing the samples in a muffle furnace at 550°C for two hours. The ash is further digested in conc. HNO_3, perchloric acid and sulphuric acid and dissolved in 1N HNO_3. A series of standard metal solutions (Cu, Zn, Cd and Pb) were prepared in the optimum concentration range with metal free double distilled water. The results are expressed in µg/g dry wt.

Table 32.1: Meteorological Data from October 2001 to September 2002

Parameters	*Station*	*Monsoon*	*Post-monsoon*	*Summer*	*Premonsoon*
Rainfall (mm)	I	333.5	12.8	5.8	8.8
	II	297.7	10.8	4.6	7.2
Relative humidity	I	64	90	85	88
08.30 h (%)	II	58	84	77	81
Relative humidity	I	80	76	77	69
17.30 h (%)	II	67	74	73	58

Table 32.2: Data Collected on Physico-chemical Parameters from October 2001 to September 2002

Parameters	*Station*	*Monsoon*	*Post-monsoon*	*Summer*	*Pre-monsoon*
Atmospheric	Station I	26	28	34	30
Temperature (°C)	Station II	25.5	26.4	32.6	31.2
Surface water	Station I	22	23	30	28
Temperature (°C)	Station II	21	22	29	27
Dissolved Oxygen	Station I	6.6	5.8	4.2	3.6
(mg/l)	Station II	4.8	5.9	3.7	3.9
pH	Station I	7.9	7.8	8.4	8.1
	Station II	8.0	7.9	8.25	8.18
Salinity (%)	Station I	24.8	26.1	33.4	31.6
	Station II	20.5	22.6	30.8	28.4
Phosphate (mg/l)	Station I	0.182	0.006	0.009	0.174
	Station II	1.04	0.054	0.062	1.01
Silicate (mg/l)	Station I	6.56	5.48	1.54	1.42
	Station II	7.74	6.62	1.44	1.56
Nitrate (mg/l)	Station I	0.420	0.2600	0.340	0.270
	Station II	0.451	0.290	0.438	0.248
Nitrite (mg/l)	Station I	1.057	1.042	0.378	0.392
	Station II	1.081	1.048	0.079	0.082

The heavy metal accumulations of the dominant seaweeds *Enteromorpha intestinalis* and *Chaetomorpha linum* from the Vellar estuary were compared with the dominant seaweeds *Enteromorpha intestinalis* and *Chaetomorpha aerea* from the Uppanar estuary throughout the year.

Results

Meteorological factors such as rainfall and humidity as well as other physico-chemical factors such as atmospheric and surface water temperature, dissolved oxygen, pH, salinity and nutrients etc., are presented in Tables 32.1 and 32.2. The heavy metal constituents such as Copper, Zinc, Cadmium and Lead were analysed and the seasonal data are presented in Figures 32.1–32.4, respectively.

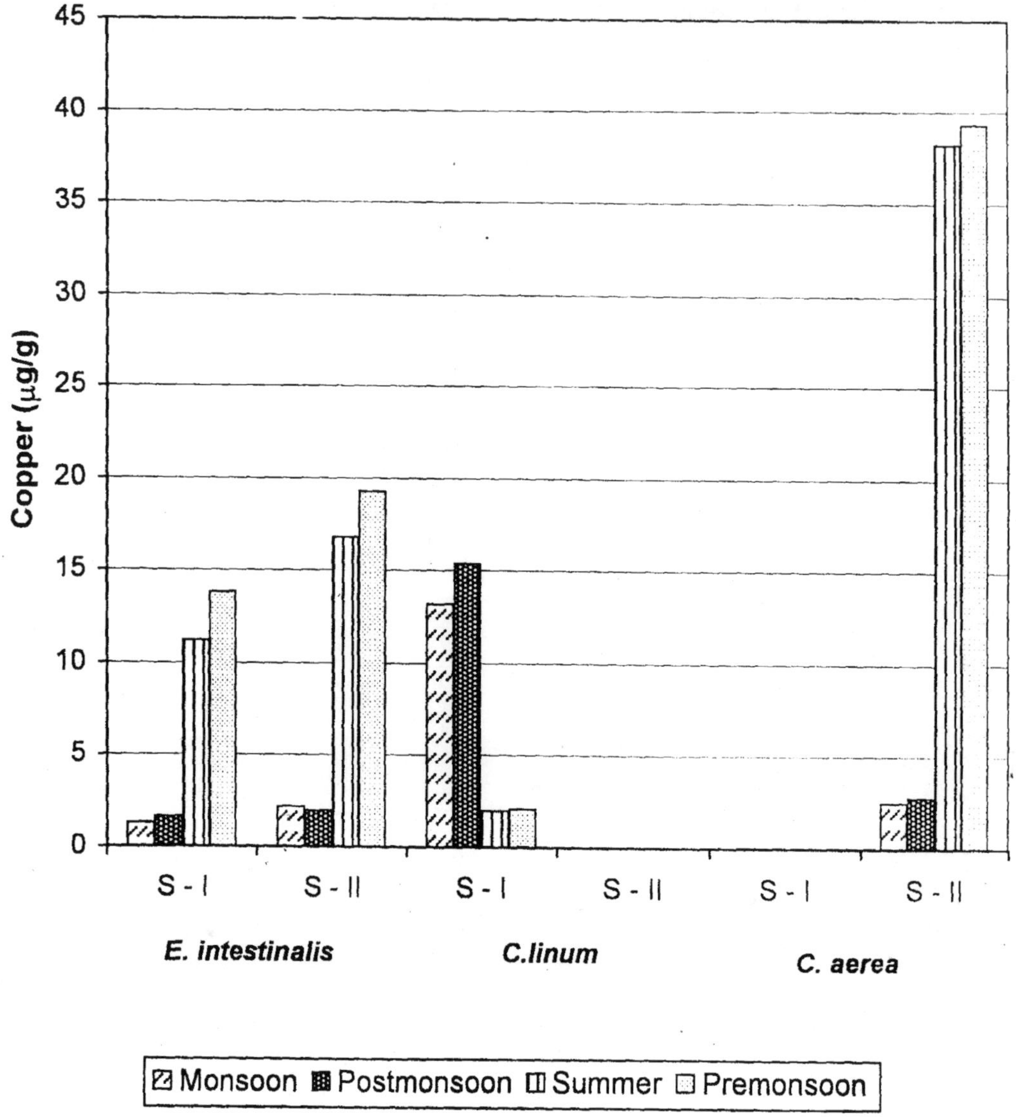

Figure 32.1: Seasonal Variations of Copper Content (µg/g)

Copper

At Station 1, Cu concentration in *E. intestinalis* varied from 1.65 to 19.8 µg/g. Minimum (1.65 µg/g) and maximum concentrations (19.8 µg/g) were recorded during the post- and pre-monsoon seasons, respectively. Cu concentration in *Chaetomorpha linum* varied from 2.30 to 16.2 µg/g. Maximum concentration (2.30 µg/g) was recorded during the premonsoon season and maximum concentration (16.2 µg/g) was recorded during the postmonsoon season (Figure 32.1). *E. intestinalis* accumulated more Cu than *Chaetomorpha linum.*

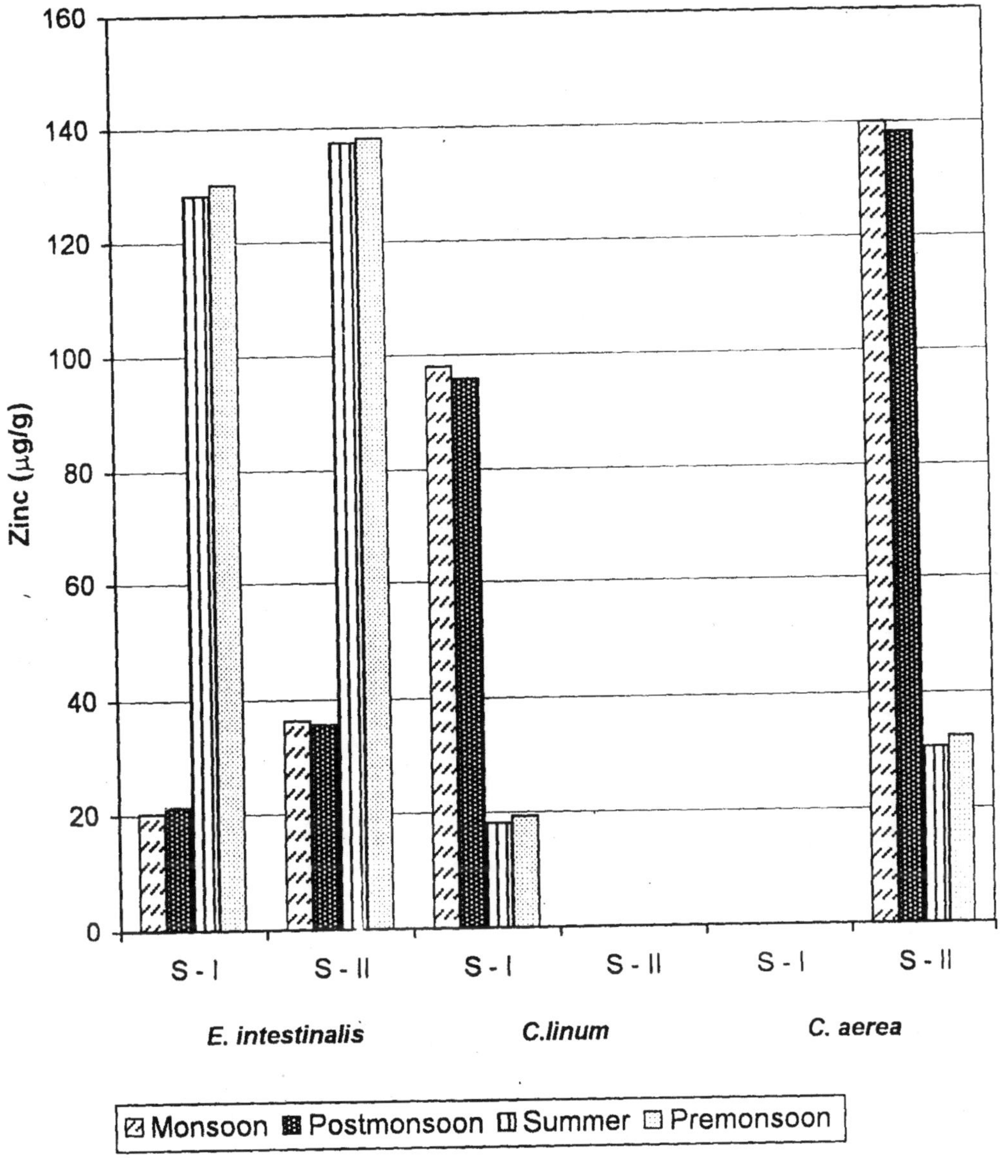

Figure 32.2: Seasonal Variations of Zinc Content (µg/g)

At Station 2, Cu concentration in *E. intestinalis* varied from 2.40 to 16.3 µg/g. Minimum concentration (2.40 µg/g) was recorded during the postmonsoon season. Maximum concentration (16.3 µg/g) was recorded during the premonsoon season. Cu concentration in *Chaetomorpha aerea* varied from 2.50 to 39.30 µg/g. Minimum concentration (2.50 µg/g) was recorded during the monsoon season. Maximum concentration (39.30 µg/g) was recorded during the postmonsoon season (Figure 32.1). C. *aerea* accumulated more amount of Cu than *E. intestinalis*.

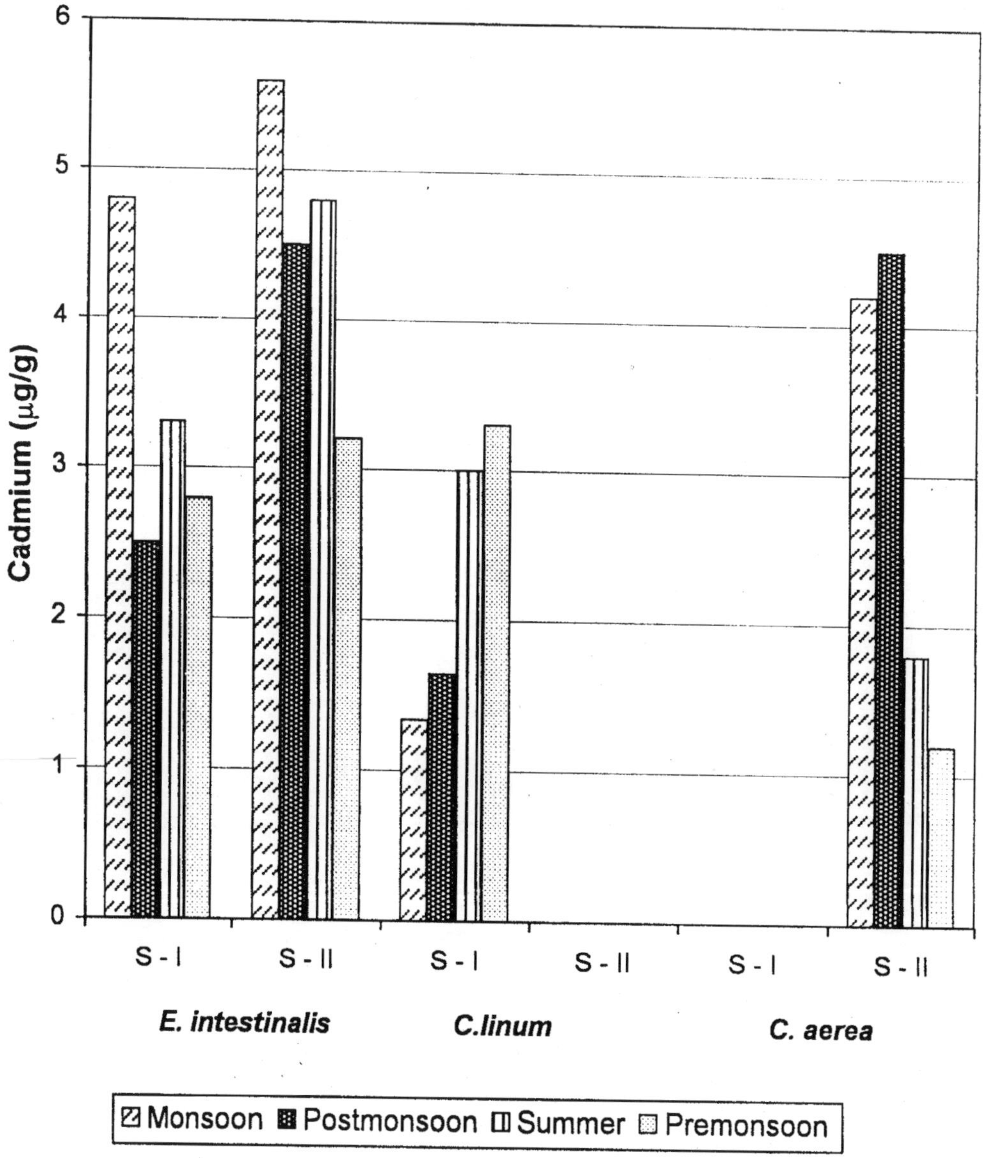

Figure 32.3: Seasonal Variations of Cadmium Content (μg/g)

Zinc

At Station 1, Zn concentration in *E. intestinalis* varied from 21.40 to 130 μg/g. Minimum concentration 21.4 μg/g was recorded during the postmonsoon season. Maximum concentration (130 μg/g) was noted during the premonsoon season. Zn concentration in *Cheatomorpha linum* varied from 19.3 to 98 μg/g. Minimum concentration (19.3 μg/g) was recorded during the premonsoon season. Maximum concentration (98 μg/g) was recorded during the monsoon season (Figure 32.2). *E. intestinalis* accumulated more amount of Zn than *C. linum*.

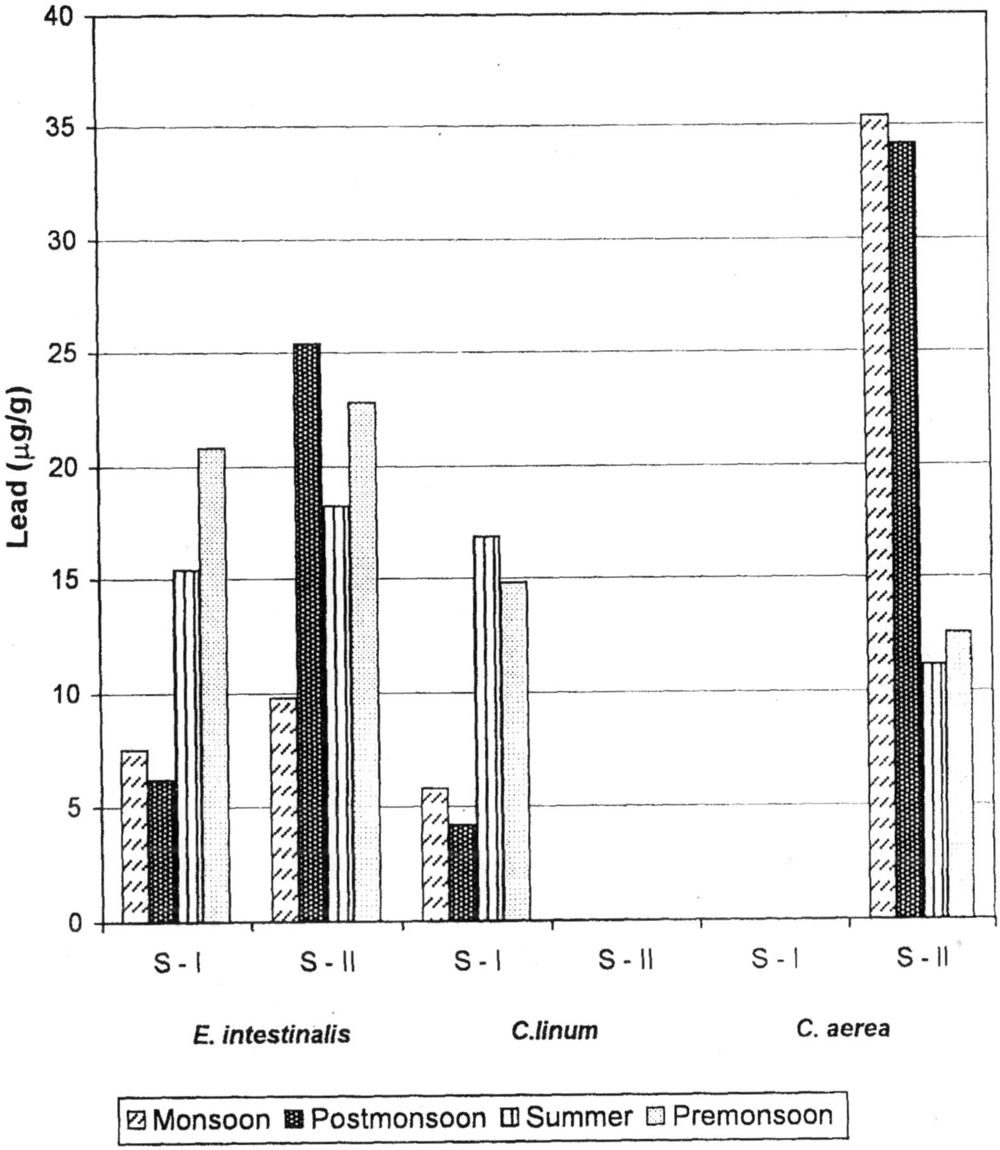

Figure 32.4: Seasonal Variations of Lead Content (µg/g)

At Station 2, Zn concentration in *E. intestinalis* varied from 35.6 to 138 µg/g. Minimum concentration was recorded during the postmonsoon season, and the maximum during the premonsoon season. Zn concentration in *C. aerea* varied from 30.6 to 140 µg/g. Minimum concentration (30.6 µg/g) was recorded during the summer season. *C. aerea* accumulated more amount of Zn than *E. intestinalis* (Figure 32.2.).

Cadmium

At Station 1, Cd concentration in *E. intestinalis* varied from 2.50 to 4.8 µg/g. Minimum concentration was recorded during the postmonsoon season. Maximum concentration was recorded during the monsoon season. Cd concentration in C. *linum* varied from 1.65 to 3.3 µg/g. Both minimum and maximum were recorded during the premonsoon season (Figure 32.3).

At Station 2, Cd concentration in *E. intestinalis* varied from 2.5 to 5.6 µg/g. Minimum concentration (2.5 µg/g) was recorded during the postmonsoon season. Maximum concentration (5.6 µg/g) was noted during the monsoon season. Cd concentration in C. *aerea* varied from 1.8 to 4.5 µg/g. Minimum concentration (1.8 µg/g) was recorded during summer and maximum concentration (4.5 µg/g) during the postmonsoon season (Figure 32.3).

Lead

At Station 1, Pb concentration in *E. intestinalis* varied from 7.5 to 20.8 µg/g. Minimum concentration (7.5 µg/g) was recorded during the monsoon season. Maximum concentration (20.8 µg/g) was noted during the premonsoon season. Pb concentration in *C. linum* varied from 5.8 to 16.8 µg/g. Minimum concentration (5.8 µg/g) was recorded during the monsoon season and maximum concentration (16.8 µg/g) during the summer season (Figure 32.4). Pb concentration was higher in *E. intestinalis* than in *C. linum.*

At Station 2, Pb concentration in *E. intestinalis* varied from 9.8 to 25.4 µg/g. Minimum concentration (9.8 µg/g) was recorded during the monsoon season and maximum concentration (25.4 µg/g) during the premonsoon season. Pb concentration in *C. aerea* varied from 11.20 to 35.40 µg/g. Minimum concentration (11.20 µg/g) was recorded during the summer season and maximum concentration (35.40 µg/g) during the monsoon season (Figure 32.4). Pb concentration was higher in *C. aerea* than *E. intestinalis.*

Discussion

The concentration of heavy metals in the seaweed sample reveals are higher at Station 2 in the polluted area (Figures 32.1–32.4) compared to Station 1 the non-polluted area. Sources of heavy metals to Station 2 are from the discharge of industrial and domestic wastes in addition to various harbour activities. Moreover, differences in metal accumulation of seaweeds might be related not only to different metals levels in water but also to different ecological conditions such as rainfall, temperature salinity, dissolved nutrients and different geological structures (Munda and Hudnik, 1991).

The decreasing order of preferential accumulation of metals in *E. intestinalis* is Zn > Pb > Cu > Cd. Similar trend has been observed by Munda and Hudnik (1991) at Rovinj in *Enteromorpha* sp., Sivalingam (1978) at Penang in *Entermorpha flexuosa* and by Rajendran *et al.* (1993) at Kasimedu in *E. intestinalis*. A different trend has been reported by Guvan *et al.* (1992), who have showed the order of preferential accumulation of metals in *Hepnea valentiae* as Zn > Cu > Pb > Cd. Similar trend has been observed by Kesava Rao and Indusekhar (1986) who have recorded the order of preferential accumulation of metals in *Padina boergeseni* is Zn > Pb > Cu > Cd and a similar trend has been observed by Sheila Devi Ramachandran *et al.* (1994) at Malaysia in *Padina* spp.

Seasonal variation in the distribution of heavy metals in the seaweeds shows that of the 2 species, *E. intestinalis* (Stations 1 and 2) accumulates higher concentrations of Cu, Zn, and Pb during the premonsoon season and Cd during the monsoon season. *C. linum* (Station 1) exhibits recorded high concentration of Cu, during the postmonsoon season, Cd and Pb during the summer season and Zn during the monsoon season. This trend is similar to that reported by Ganesan (1992). *C. aerea* (Station 2) shows high concentration of Cu, and Cd during the postmonsoon season and Zn and Pb during the monsoon season. The heavy metal levels are statistically different between seasons, but not between the stations. Among the four heavy metals taken for the study, zinc is substantially at elevated levels in all the seasons and in water followed by lead and cadmium. It is observed that the concentration of these heavy metals increases in premonsoon and decreases in monsoon season. The industrial effluents released into the stream at Station 2 are perhaps acidic in nature, ultimately reduce in the monsoon. Under these conditions, the solubility of metals increases making the stream heavily concentrated with metals. Stumm and Morgan (1981), William and Campbell (1981). Campbell and Stokes (1985) Wren and Stephenson, (1991) also have reported that concentration of lead and cadmium in water depends upon the acidity of surrounding medium.

Acknowledgement

Our sincere thanks are due to K. Kathiresan, CAS in Marine Biology, Annamalai University, Parangipettai and Prof. L. Kannan, Research Director for their suggestion and constant encouragement in the present work. Our thanks are also due to the authorities of Annamalai University for the facilities provided for the present investigation.

References

Almmer, B., Dickson, W., Ekstrom, C., Hornstrom, E. and U. Miler. 1974. Effect of acidification on Swedish lakes. *Ambio.*, 3: 30–36.

Ashraf, M., Tariq, J. and M. Jaffer. 1991. Contents of trace metals in fish, sediment and water from three freshwater reservoirs in the Indus river. *Pakistan Fisheries Research*, 12(4).

Badsha, K.S. and C.R. Goldspink. 1982. Preliminary observations on the heavy metal content of four species of freshwater fish in NW England. *J. Fish Biol.*, 21 : 251–267.

Campbell, P.G. and P.M. Stokes. 1985. Acidification and toxicity of metals to aquatic biota. *Can. J. Fish Aqat Sci.*, 42: 2034–2049.

Ganesan, M.. 1992. Ecobiology of seaweeds of the Gulf of Manner with special reference to hydrography and heavy metals. *Ph.D. Thesis*, Annamalai University, India, pp. 162.

Gesamp. IMO/FAO UNESCO/WMO/WHO/IAEA/UN/UNEP. 1974. Joint Group of Experts on the Scientific Aspects of Marine Pollution, Review of harmful substances, Supplement to the Report of Sixth session, Geneva, 22–28 March 1974, Geneva, World Health Organization.

Goldberg, E.D.. 1976. The Health of the Oceans. Paris, *UNESCO*, pp. 172.

Grasshoff, K., Ehrhadt, M. and K. Kremling. 1983. *Method of Sea Water Analysis*, Verlag demic gymb, Weinheim, pp. 419.

Guvan, K.C., Topcuoglu, S., Kut, D., Esen, N., Erenturk, N., Saygi, N. and E. Cevher. 1992. Govener, Band Ozturk, B. Metal uptake by Black sea algae. *Bot. Mar.*, 35: 337–340.

Jaffer, M. and Ashraf, M.. 1998. Selected trace metal concentrations in different tissues of fish from coastal waters of Pakistan (Arabian Sea. *Indian J. Mar. Sci.*, 17: 231–234.

Kesava Rao, C.H. and V.K. Indusekhar. 1987. Chromium, lead and cadmium contents of certain seaweeds from Saurashtra coast. *Phykos*. 26: 1–7.

Kurland, L.T., Faro, S.W. and H. Siedler. 1960. Minamata disease: The out-break of a neurological disorder in Minamata, Japan and its relation to ingestion of seafood containing mercury compounds. *World Neurol.,* 1: 370–395.

Munda, I.M. and V. Hudnik. 1991. Trace metal content in some seaweeds from the northern Adriatic. *Bot. Mar.,* 34: 241–249.

Nitta, T. 1972. Marine pollution in Japan. In: *Marine Pollution and Sea Life* (Ed.) M. Ruivo. West Byfleet Survey, Fishing News Books, p. 77–81

Rajendran, K., Sampathkumar, P., Govindasamy, C., Ganesan, M., Kannan, R. and L. Kannan. 1993. Levels of trace metals (Mn, Fe, Cu and Zn) in some Indian seaweeds. *Baseline,* 26: 283–285.

Sheila Devi Ramachandra., Phang siew Moi. and Soo Loong Tong. 1994. Heavy metals content of some Malaysian seaweeds. *Proceedings of the First Asia-Pacific Conference on Algal Biotechnology,* Phang (Eds.), University of Malaya, p. 339–343.

Sivalingam, P.M.. 1978. Biodeposited trace metals and mineral content studies of some tropical marine algae. *Bot. Mar.,* 21: 327–330.

Stumm, W. and J.J. Morgan. 1981. *Aquatic Chemistry: An Introduction Emphasising Chemical Equilibria in Natural Waters.* Wiley-Interscience, New York, pp. 583.

William, O.N. and P.G.C. Campbell. 1981. The effects of acidification on the geochemistry of Al, Cd, Pb and Hg in freshwater environments: A literature, review. *Environ. Pollut.,* 71: 91–130.

Wren, C.D. and G.L. Stephenson. 1991. The effect of acidification on the accumulation and toxicity of metals to freshwater invertebrates. *Environ. Pollut.,* 71: 2–4.

Chapter 33

Assessment of Copper Concentrations in Two Freshwater Reservoirs of Nanded, Maharashtra State

*G. Gyananath**#, S.V. Shewdikar*, T.A. Kadam*, S.K.G.K Charyulu** and R.S. Rao****

**School of Life Sciences, **School of Earth Sciences, Swami Ramanand Teerth Marathwada University, Vishnupuri, Nanded – 431 606, Maharashtra, India*

****Department of Chemical Engineering, CBIT, Hyderabad*

ABSTRACT

Heavy metals of vital importance are widely distributed in the environment. Several studies have been carried out in recent years to understand the distribution of heavy metals in different water bodies. Among the metals, Copper (Cu) is of environmental interest and an assessment of Cu in two freshwater reservoirs of Nanded, using Flame Atomic Absorption Spectroscopy. The results indicated that Cu was detected in both Amduri and Shikarghat reservoirs of raw and treated water samples. The changes in copper concentrations in natural waters and its possible effects on water quality are discussed. This study suggests that the source of copper in two reservoirs may be dissimilar.

Keywords: Copper, Trace amounts, Environment, Seasons, Freshwater reservoirs.

Introduction

Copper is one of the significant metals to mankind and is being used since prehistoric times.

Email: gyananath@yahoo.com.

Copper reaches the aquatic environment through wet or dry deposition, mining activities, surface runoff, industrial, domestic and agricultural waste disposal. The various studies shows that Cu is ubiquitously distributed in the environment and its occurrence in significant amounts in the water samples of reservoirs may suggest the availability of free Cu ions in the water. Since higher levels may show adverse biological and toxic effects, the water samples need constant monitoring of Cu concentration to protect the aquatic environment.

The variations in copper concentrations in water are different in different aquatic environments and are discussed by many scientists earlier with the possible health disorders due to copper. Copper is toxic to aquatic biota and the toxicity of the metal varies with factors such as pH, hardness, presence of other metals and Copper species present in water (Taylor and Demayo, 1981). Chakoumakos, (1977) attributed alkalinity and pH are to be the major factors controlling Copper speciation. McKnight *et al.* (1983) suggested that the aqueous chemistry of copper is dominated by its tendency to form complexes principally with legends. Patel *et al.* (1985) reported dissolved Cu in Bombay harbor waters which receives large amounts of industrial and municipal wastes. Ajmal and Uddin (1986) reported elevated levels of Cu in the treated hand pump water samples compared to the running water from the City of Aligarh (India) attributing it to non-compliance with the standards. DeLeon *et al.* (1986) and Whitehead *et al.* (1988) reported Cu in river environments, Mississippi river and Mediterranean river. Arens (1999) elaborated various factors to be considered concerning the corrosion of Cu^{++} tubes in drinking water.

The occurrence of Cu has caused a series of severe systematic diseases in Germany (Cu induced Liver Cirrhosis), besides Cirrhosis another type of disease has occurred which like wise appeared to be induced by Cu in tap water (Eife *et al.*, 1999). Buchanan *et al.* (1999) made an assessment of copper levels in drinking water to determine the association of copper with Gastrointestinal (GI) illness. Pizarro *et al.* (1999) reported that there was no association between aggregate Cu in drinking water within the range of 0 to 50 µg/l, but a level of 30 µg/l of ionized Cu was associated with nausea, vomiting or abdominal pain.

The present study is an attempt to understand the distribution of a heavy metal, copper based on seasonal variations in two reservoirs since seasonal variations of Cu in aquatic environment may prove useful to understand long-term implications of Cu in the reservoirs. Since the reservoirs are primarily used for irrigation and drinking water, study of Copper contamination of source water may provide useful database for understanding of environmental health effects in humans.

Methodology

Water samples from different stations of two fresh reservoirs, Shikarghat and Amduri reservoirs, of Nanded district, Maharashtra were collected from June 1998 to May 1999 in the screw cap polyethylene bottles for heavy metal analysis. The bottles were cleaned with the tap water rinsed and soaked for 24 hrs in 1 per cent Nitric acid (HNO_3) and again rinsed with high purity (AIMIL; Water Purification System). The bottles were then dried and cooled to room temperature, capped and labeled. Nitric acid (1 ml acid/100 ml water sample) was added to the bottle and each bottle was filled to brim with the water sample to avoid any air space. The water samples, before and after treatment, obtained and analysed from both Shikarghat and Amduri reservoirs for each month for the entire year. The samples collected were transported to the laboratory and the analysis was carried out using Perkin-Elmer model 'Analyst 100 and 300' Atomic Absorption Spectrophotometer (AAS). Air-acetylene flame was used exclusively and instrument was operated according to the optimal experimental conditions.

Results and Discussion

Variations in Copper Levels of Amduri and Shikarghat Reservoirs

Monthly changes in the Copper concentrations of raw water samples of both Amduri (6 stations) and Shikarghat (1 station) reservoirs along with the variations of one treated sample from each reservoir are shown in Figure 33.1. Figures 33.1 (A) and (B) shows the monthly copper variations of both treated and raw waters of Amduri and Shikarghat reservoirs. The average copper levels at all stations were

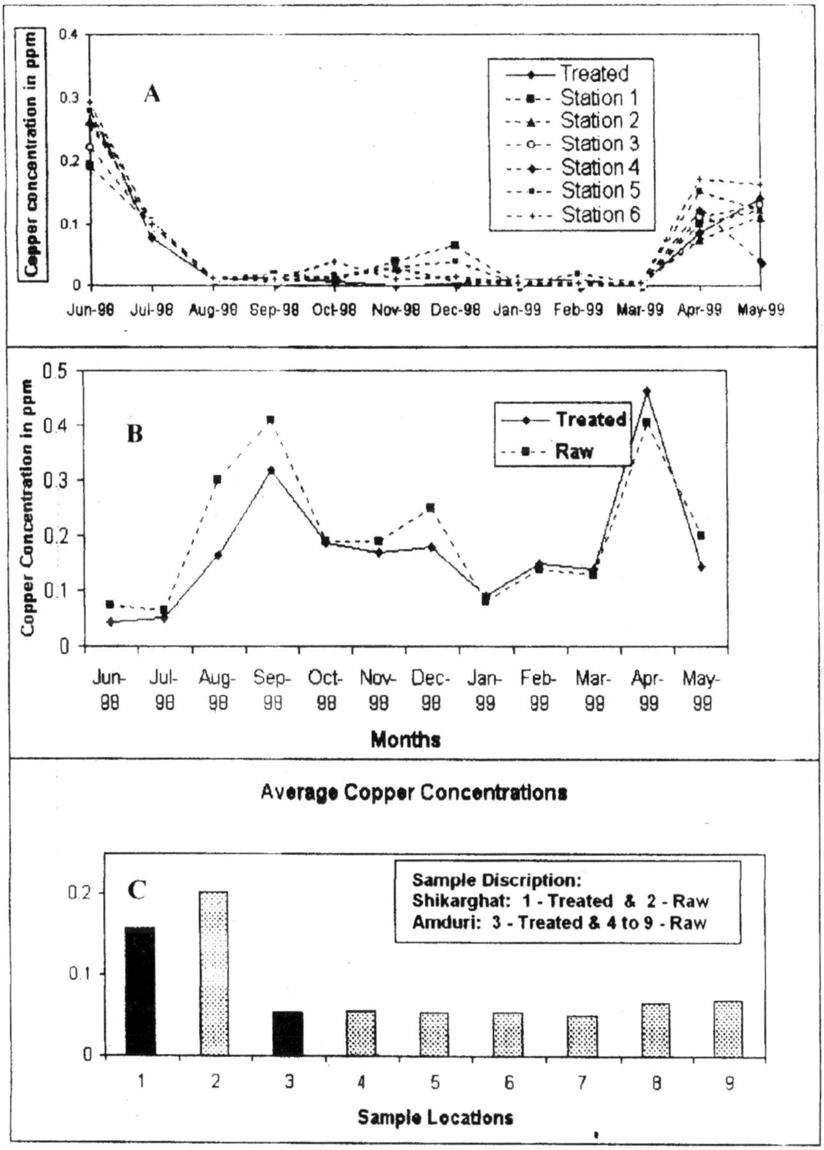

Figure 33.1: Copper Concentrations (ppm) of Treated and Raw Water Samples (A) Monthly Variations of Amduri Reservoir; (B) Monthly Variations of Shikarghat Reservoir and (C) Average Values in Both Amduri And Shikarghat Reservoirs

separately shown as bar graph in Figure 33.1 (C). The copper values at all the stations were compared with the Bureau of Indian Standards (BIS).

Copper was above the detectable limits at all the stations of Amduri and the range of the concentrations is between 0.295 to 0.001 ppm (Figure 33.1A). The copper levels at Amduri are above the permissible limits during the months of June–July 1998 and April–May 1999 *i.e.* almost in summer. For the remaining months though the copper values are below the permissible limits, all are above detection limits and the situation is alarming. The variations are similar at all stations including the treated waters (range 0.282–0.001 ppm), suggesting that the copper levels are ineffective to treatment. It is also observed that the concentrations are progressively increasing from the beginning of the summer and reaching a maximum towards the end of the summer.

Figure 33.1 (B) shows the copper levels at Shikarghat and the copper levels are much above the detectable range for all the months. The range of the concentrations is varying between 0.462 to 0.042 ppm. Unlike in Amduri the copper levels at Shikarghat are above the permissible limits during the entire period of study. From the Figure 33.1(B) it is observed that the copper values are maximum during August–September 1998 and April 1999. The Copper concentrations were more in rainy season and comparatively less in mid summer. The trends and variations of both treated and raw water samples of Shikarghat are appears to be similar.

Figure 33.1 (C) shows the bar diagram of average values of copper levels for both Amduri and Shikarghat reservoirs, indicating that the average values of raw and treated waters are similar in both the reservoirs.

As compared to the Amduri reservoir, there are wide fluctuations in copper levels in Shikarghat reservoir throughout the year. It is obvious from the results that there are no definite uniform seasonal variations at the two reservoirs. The copper concentrations of treated water samples of Shikarghat reservoir show distinct changes during rainy season and in Amduri the changes are prominent in summer. It is also evident that a distinct marked difference in the magnitude of copper concentrations at both the reservoirs. This suggests that either the source or the process of contamination of copper in both the reservoirs is different. In light of these results it is interesting to note that the Amduri reservoir is catering the needs of a village only and the Shikarghat reservoir is used by floating pilgrims of Holy Gurudwara shrine as well as the villagers. The main source of copper is padogenic and the different catering needs may result the differing copper levels in these reservoirs.

References

Arens, P. 1999. Factors to be considered concerning the corrosion of copper tubes. *Env. J. Med. Res.*, 28: 4(6).

Ajmal, M. and Uddin, R. 1986a. Studies on the heavy metals in the groundwater of the City of Aligarh (U.P.), India.

Environment Monitoring and Assessment, 6: 195–205.

Buchanan, S.D., Disekar, R.A. 3rd, Sinks, T., Olson, D.R., Daniel, J. and Floodman, T. 1999. Copper in drinking water, Nebraska. *Int. J. Occup. Environ. Health.*, 5(4): 256–261.

DeLeon, I.R., Byrne, C.J., Peuler, E.A., Antoniee, S.R., Seaeffer, J. and Murphy, R.C. 1986. Trace organic and heavy metal pollutants in the Mississippi river. *Chemosphere*, 15: 795–805.

Eife, R., Weiss, M., Muller-Hocker, M., Lang, T., Barros, V., Sigmund, B., Thanner, F., Welling, P., Lang, H., Wolf, W., Rodeck, B., Kittel, J., Schramol, P. and Reiter, K. 1999. Chronic poisoning of Copper in tap water: II: Copper intoxication with predominantly systemic symptoms. *Env. J. Med. Res.*, 4(6): 224–228.

McKinght, D.M., Feder, G.L., Thurman, E.M., Wershaw, R.L. and Westall, J.C. 1983. Complexation of Copper by aquatic humic substances from different environments. *Science and Total Environment*, 28: 65–76.

Patel, B., Bangera, V.S., Patel, S. and Balani, M.C. 1985. Heavy metals in the content of drinking water, Bombay Harbor area. *Pollution Bull.*, 16: 22–28.

Pizarro, F., Olivares, M., Speisky, H., Lonnerdal, B. and Uauy, R. 1999. *Safety of World Health Organization Provincial Guideline Value for Copper*.

Samanidou, V. and Fytianos, K. 1987. Partitioning of heavy metals into selective chemical fractions in sediments and rivers in Northern Greece. *Science of the Total Environment*, 67: 279–285.

Taylor, M.C. and Demayo, A. 1981. Copper. In: *Guidelines for Surface Water Quality*, Vol. I: Inorganic Chemical substances, inland waters. Directorate, Environment, Canada.

Whitehead, N.E. and Huynh-Nhoc Arson, S.R. 1988. Trace metals in two Northern Mediterranean waters. *Air and Soil Pollution*, 42: 7–18.

Chapter 34

Ecological Imbalance by Reservoirs

V. Srihari and C.R. Suribabu

Senior Lecturers, School of Civil Engineering, Shanmugha Arts, Science, Technology and Research Academy (SASTRA) Deemed University, Thanjavur – 613 402

ABSTRACT

Reservoir construction has acquired a global scenario during the last three decades. This can be realized by a wide use of hydropower resources, intensive development of irrigation and municipal water supply, recreation and flood control. To meet out the increasing demands for various purposes will necessitates the construction and operation of reservoir on a global basis. Providing barrier across the freely flowing river affects various elements of the environment, hydrological regime, aquatic and terrestrial ecosystems, sectors of the national economy, social-economic conditions. Due to submergence of large watershed area in upstream of solid barrier results the higher nutrient concentration and sedimentation. As a result, many reservoirs experience the consequences of accelerated rates of eutrophication, including excessive algae growth, reduced transparency, low dissolved oxygen concentrations and loss of volume due to sediment accumulation. Planning and development of natural resources without endangering the environment is an important issue. For effective utilization of water resources in an area, the integrated watershed development and reservoir planning and management approach is to be considered for maintaining the ecological balance. This paper highlights the various impacts of reservoirs on the environment and control measures to be adopted.

Keywords: *Dams, Reservoirs, Reservoir induced sesmicity, Eutrophication.*

Introduction

Water is most precious gift of the nature, which has to be conserved for increasing future demands. The reservoir construction has acquired tremendous importance by the wide usage of water. The manner in which man utilizes the water can be classified into two types, *i.e.*, consumptive usage and non-consumptive usage. In first case the amount of water is taken out of streams or from the ground or from reservoirs for the purpose of domestic consumption, industrial purposes, irrigation. In second case, retaining at a place for the purpose of hydropower generation uses water, flood control, navigation, recreational, aquaculture development purposes. Human being mainly depends on ground water and surface water for his needs. Rivers are considered to be one of the most important water resources. Most of the river water flows towards ocean without proper use. So, construction of dams across river becomes essential to satisfy the above needs. Construction of reservoirs has not only benefits but also have some adverse effects on the environment. The various adverse impacts of the reservoirs can be visualized by the following means:

1. Land submergence: Loss of flora and fauna, Loss of wild life, Loss of monuments and Water logging.
2. Earth Quake possibility or Reservoir Induced Sesmicity
3. Silt Deposition
4. Quality of water: Eutrophication, Pollutant movement and Health aspects

Land Submergence

Construction of dams and Reservoirs occupies huge space of land. In most of the cases forest land, agricultural land, villages, structures and monuments may undergo submergence. Planting new forests in an equivalent area to be submerged can compensate the loss of forest area. Once the forest land undergoes submergence surely there is a chance of loss of wild life. To save the wild life, sanctuaries are to be developed. Proper conservation measures of these sanctuaries are necessary for preservation and protection of flora and fauna. Due to submergence, many of the people become homeless. It is necessary to provide liberal and compassionate rehabilitation and resettlement package to the affected people and keep the submergence area of the reservoir to be minimum. Sometime in order to minimize the submergence zone the dam height can be reduced, but the necessity of storage puts a limit on the dam height, which cannot be reduced beyond certain minimum value under consideration of optimum planning techniques and requirements. Sometimes important monuments, local temples, pilgrimage centers may undergo submergence. Care should be taken to save the monuments and relocation of any movable structures.

One of the most adverse impacts of a reservoir on its irrigated area is water logging, which reduces the crop yield. It happens because of insufficient drainage and rise of water table near the ground surface. Water logging leads to salinity. Due to this, surface irrigation will be affected. To prevent water logging combined usage of surface and groundwater proves to be more effective. The irrigation water supply should be designed combinedly by providing 60 per cent of water from surface source and 40 per cent of water from the underground for each crop. Pumping can be regulated carefully so that the water table should not rise close to the ground surface. Table 34.1 gives the information on number of reservoirs having capacity more than 100 Mm^3 in the world.

Reservoir Induced Sesmicity

The storage of huge mass of water on up stream side and no load in down stream side may induces unbalanced forces. Due to this tremors may produce through the surrounding area, which may finally results in earthquakes. The incident of Koyna (6.7) earthquake in Maharashtra in 1967 was one such example in India. Before construction of reservoir, the potential of induced seismicity of reservoir should be studied and reservoir has to be designed to withstand earthquake shocks of greater magnitude. Worlds highest dam Nurek in USSR, which is 300m high, already withstood a seismic shock of intensity 7 on Richter scale. In India so far two dams Koyna and Bhatsa experienced some earthquakes. The Tehri dam is said to lie in a seismic gap between the great earthquakes of Kangra in 1905 and Bihar in 1934. The risk posed by the earthquakes can be mitigated by careful control of the water level in the reservoir. It is also suggested that the contour bunding, bench terracing and gully plugging etc, can be done to improve the stability of reservoir slopes. It is better to construct rock fill dam in earthquake prone zones.

Table 34.1: Number of Reservoirs having More than 100 Mm³

Continent	*Upto 1900*	*1901–1950*	*After 1951*	*Total*
Europe	9	104	407	520
Asia	5	46	529	580
Africa	1	15	89	105
North America	25	342	518	885
Latin America	I	22	182	210
Australia	0	10	65	75

Constructions of massive rock fill dam slump or be deformed but will not break or be overtopped. In this case, it is essential to provide considerable free board than normally designed value.

Silt Deposition

The rate of sediment deposition reduces the life of the reservoir. The silt deposition rate depends on soil erosion in the catchment area. The silt can be carried to the reservoir from the catchment area through runoff. The silt load in the runoff depends on nature of the soil, topography of the area, vegetation cover and rainfall intensity. As silting increases a certain percentage of total storage gradually reduces. The reservoir planning must include the consideration of probable rate of silting. So that life of the reservoir may be determined. Construction of other small tanks, reservoir in the basin will reduce the silting to some extent. Secondly providing dead storage for accumulation of silt or silt scouring structures, so that the silt deposits should not interfere the live storage of the reservoir.

Eutrophication

Eutrophication is one of the most severe water pollution problem, is due to enrichment of waters by plant nutrients. The excessive nutrient resulting in the dominance of blue green algae, which form a dense floating material on the surface, which reduce the light penetration deep into the waters. The aquatic weeds like water hyacinth, water lettuce etc., and can interfere water bodies and impairing the quality of water, all of which produce a toxic water which is dangerous to fish and potable water supply system. The eutrification basically a natural process which gets accelerated by increased nutrient supply by various sources like natural runoff, submergence of vegetation, discharge of domestic sewage, industrial waste, agriculture waste etc. The decay of deposited organic matter at the bottom of the reservoir is responsible for decrease of dissolved oxygen and prevails anaerobic conditions, which is harmful to several aquatic species and produces bad odour. The physico-chemical changes in the quality of water bring about a change in flora and fauna. Undesirable species may replace desirable ones. The accumulation of nutrients in a reservoir is depending on various factors like rainfall,

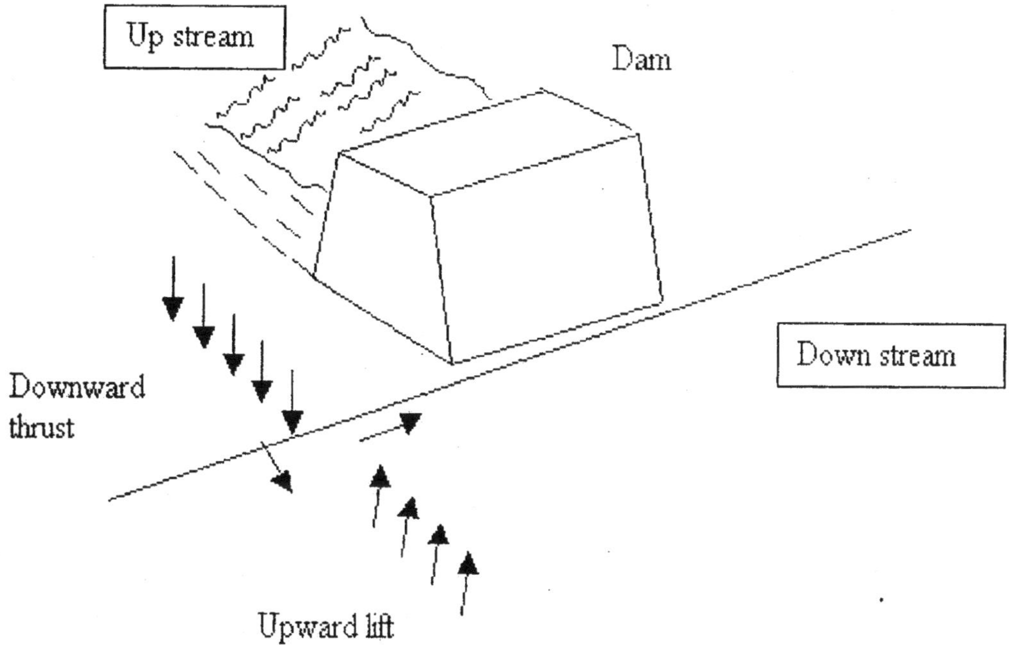

Figure 34.1: Reservoir Induced Sesmicity

atmospheric deposition, urban and rural run-off, domestic and industrial waste etc. Limiting the various parameters such as nutrients, algae species, productivity etc., can control Eutrophication. The nutrients enter the reservoir mainly through point sources such as domestic and industrial waste. So the diversion of these wastes directly to downstream side may reduce nutrient concentration. By providing proper treatment to the sources of the waste may also reduce nutrient concentration. A large portion of nutrients can be removed by dredging the sediments of the reservoir. By covering the sediments with suitable inert material like clay of fly so that mixing of nutrients from the deposit can be avoided.

Pollutant Movement

Pollutants may enter into the reservoir by direct discharge of domestic and industrial wastewater into the reservoir, by runoff and seepage, by reaction between water and sediments deposit. The run-off may pick up soluble, suspended and set table matter from the soil and transport them to the reservoir. The set table matter is responsible for silting. The behavior and movement of pollutants in reservoir depends on various factors such as turbulence or rapid mixing may circulate pollutants throughout the water body of the reservoir. Temperature and depth are the responsible factors for the dispersion of pollutants. If depth is shallow, pollutants can mix up with entire water, whereas in deep waters mixing may be up to certain depth. The nature of the pollutant whether degradable or non-degradable is another parameter that affects the quality of water. Every water body has a self-purification capacity regarding to biodegradable pollutants depending upon dissolved oxygen availability. The rate of inflow, outflow, seepage and evaporation are important factors that influence the accumulation of pollutants in water. The sediments absorb sometimes pollutant. Reservoir pollution control can be achieved by regular monitoring of pollutant movement in the reservoir.

Health Aspects

The construction of large reservoir to store huge quantity of water is considered to be a potential factor for increasing various diseases. The stagnated water in the reservoir is the main source of mosquitoes breeding. During floods, reservoirs may be filled with highly turbid, contaminated water. To maintain ecological balance the minimum water quality standards are given in the Table 34.2. The presence of pathogenic bacteria, viruses are responsible for the most of the water borne diseases. To control diseases, the state government should direct the public health department to monitor the incidence of water borne diseases in their respective areas and make the ready for preventive measures. The continuous operation of the reservoir itself controls the growth of malaria larvae. During monsoon rains the larvae stay around the periphery get drowned and destroyed. When water is drawn for power generation and irrigation the larvae are washed out and destroyed.

Table 34.2: Environmental Quality Standards for the Conservation of the Living Environment in Reservoirs

Purpose of Water Use	*Standard Values*						
Water Supply	*pH*	*BOD mg/l*	*SS mg/l*	*DO mg/l*	*No. of Coliform Groups MPN/100ml*	*Total Nitrogen mg/l*	*Total Phosphorus mg/l*
(a) Up to primary treatment	6.5 to 8.5	< 1	< 25	< 7.5	< 50	< 0.2	< 0.02
(b) Up to secondary treatment	6.5 to 8.5	< 2	< 25	< 7.5	< 1000	< 0.2	< 0.02
(c) Highly sophisticated treatment	6.5 to 8.5	< 3	< 25	< 5	< 5000	< 0.2	< 0.02
Industrial Uses	6.5 to 8.5	< 10	< 100	< 2	–	< 1	< 0.1
Fishery	6.5 to 8.5	< 5	< 50	< 5	–	< 0.6	< 0.05
Agriculture	6.5 to 8.5	< 8	< 100	< 2	–	–	–
Recreation and conservation of environment	6.5 to 8.5	< 10	< 50	< 5	–	< 1	< 0.1

Minimum Flow Concept

Generally, the nature experiences drought and wet period periodically. Detailed hydrological analysis of inflow and rainfall in the catchment area will help to understand such an extremities. Whenever reservoir experiences such a low inflow over a year, reservoir will be operated only sufficient quantity of storage is built up. Due to this, no release will be made in the downstream side for portion of the year and riverbed will be kept dry. Cauvery river system of Tamil Nadu state is one of an example. The part of Cauvery River, which falls in Tamil Nadu boundary, has been facing low inflow rate since 2002. It is necessary to maintain minimum flow in order to maintain ecosystem of basin, growth of flora and fauna and maintaining the groundwater level around the bank of the river. It is a normal practice in any river basin farmer to extract groundwater excessively for irrigation during non-monsoon period or premonsoon period. This ultimately leading to the depletion in groundwater level. This kind of problem is encountered in most of the intensive cultivable area like Cauvery delta district. Palar basin of Tamil Nadu is expected to become like a river in desert in next 25 years as per the prediction of flow past few years. During rainy season, the Palar is fed by the ten tributaries in the course of its flow to sea. But, following continuous failure of monsoon in last three years not only the Palar but its tributaries also have dried up. The farmers and beneficiary of this basin are shown their opposition in construction of dams across tributaries of the Palar river in Tamil Nadu. The beneficiaries of Palar basin are interested in maintaining minimum flow in the river course, which can supplement the daily requirements of water for domestic use and minor irrigation. To augment the drinking water supply for Chennai city in the year of 1970's, the rural area called Minjur near city was identified as potential source of groundwater area. In the absence of surface water sources in the late years, almost the half of the entire requirement of Chennai city is being met through abstraction of groundwater from six well fields situated in the Araniar-Koratalaiyar (AK) river basin. Due to extensive pumping of groundwater, the groundwater level gone down by giving room to seawater intrusion. To bring into its original potential level of Minjur aquifer needs intensive rainwater harvest and stopping further pumping. These well fields need to be converted as recharge well by diverting the water from AK river basin which used flow reasonable level during north-east monsoon period.

Conclusion

Construction of large reservoir is becoming impossible in the most part of the world due to increased land value and intensification of the economic development of river valleys. The further increase of height of a dam influence the other ecological and socio-economic factors.

Any reservoir project should be undertaken after complete study of the economic significance and their impact on environment with special attention to preserve the ecological balance and the quality of water. The reservoir operation policy should be updated based on regional requirement with due consideration of political interference. Comprehensive mathematical model should be developed to reflect the consequences of reservoir on environment with respect to water quality, down streamside ecological balance, socio-economic problem rehabilitation and resettlement, climatic fluctuations and any unforeseen environmental impacts.

References

Avakan, A.B. 1982. Modern problems of construction, multi-purpose use and investigation of reservoirs. *Vodnye Resursy*, No. 2.

Beeton, A. 1969. Changes in environment and biota of the great lakes. In: *Eutrophication, Causes, Consequences and Correctives*, OECD, USA, p. 150–170.

Emery, R.N and Clopfer, D.C. 1976. The Distribution of Transuranic elements in a freshwater pond ecosystem. In: *Environmental Toxicity of Aquatic Radionuclide: Models and Mechanisms*. Ann Arbor Science, Ann Arbor, Michigan, pp. 269.

Fillos, J. and Swanson, M.R. 1975. The release of nutrients from river and lake sediments. *Journal of Water Pollution Cont. Fed.*, 47: 1032–1041.

Larry, W. Canter. 1985. *Environment Impact of Water Resources Projects*. Lewis Publications, INC, Michigan.

Mehta, M.C.1999. *The Role of Judiciary and Citizens in Environment Protection and Sustainable Development*. 19(5), No. 2, p. 1–9.

Mogali, J. Nandan and Robert H. Kennedy. 2001. Limnology and water quality-management of central India reservoirs. ST/ESCAP/SER.C/209, p. 74–81.

Patel, G.C. 1991. Sardar Sarovar Project; what it is and what it is not. In: *Proceeding of International Conference on What Future for Large Rivers*. Orleans France.

Yasuhiro, Morioka. 1999. Water pollution control in Japan. *Water Resources Journal*, United Nations, ST/ESCAP/SER.C/200, p. 83–95.

Chapter 35

Air Borne Fungal Spores at Sharavathi Reservoir, Karnataka, India

K. Harish Kumar, B.R. Kiran, R. Purushotham, E.T. Puttaiah and S. Manjappa

Department of Environmental Science, Kuvempu University, Shankaraghatta – 577 451, Karnataka, India

ABSTRACT

The occurrence of fungal spores and hyphal fragments in the atmosphere of Sharavathi reservoir (Site 1) and its downstream stretch (Site 2) was, monitored for five months from August–December, 2004. Open plate method was employed for monitoring the mycoflora. During this study 11 types of fungal spores having diverse group were recorded *viz.*, *Penicillium* sp., *Aspergillus* sp., *Cladosporium* sp., *Fusarium* sp., *Glomercularia* sp., *Helimanthosporium* sp., *Alternaria* sp., *Curvularia* sp., *Rhizopus* sp., *Cephalosporium* sp. and *Trichoderma* sp. While, *Penicillium* and *Aspergillus* species were common in both the sites. The highest fungal flora was found in August and lowest in December. The factors responsible for the occurrence of fungal spores are discussed in detail.

Keywords: *Fungal spore, Reservoir.*

Introduction

The main sources of viable airborne mycoflora are soil, water and vegetation. These are transported to the atmosphere and are always associated with fine, dusty, particulate matter, survival of microflora depends on the morphological features, micro meteorological, climatological conditions and on the interaction between these parameters. From the atmosphere, microbes can reach the soil and waterbodies contaminate food or water supplies and or invades the people who may become healthy carriers or each develops a disease caused by them.

Fungi produce spores that becomes airborne, some also produce mycotoxins or volatile organic compounds. Some of the most common fungi that are found in the indoor and can cause health problems are *Penicillium* sp. *Aspergillus* sp. and *Alternaria* sp. The enumeration of fungi forms a major component of the bioparticles in the air. Nair *et al.* (1986) included the information obtained on the airborne fungal spores of different, regions of India. Sreeramulu (1967) and Chanda and Mandal (1978) reviewed the available literature and showed the diversity and richness of airborne fungal spores in Indian environment. In the present study the mycoflora in the air of Sharavathi reservoir site and its downstream stretch was monitored for a five months and the observations on the relative abundance and seasonal occurrence of different fungal spores type are present.

Materials and Methods

Air monitoring was carried out during August to December 2004 with open plate containing potato dextrose agar media exposing the plates for 5 minutes. Triplicate samples were collected at two sites (Site 1–Sharavathi reservoir itself, Site 2–down stream stretch of the reservoir) every month. The study area is located at latitude 14 ° 41' 24" N, longitude 74° 50' 54" E. The plates are incubates at 28° C till the appearance of growth. The fungal colonies were counted and identified by different literature (Gilman, 1949; Tandon, 1968; Ellis, 1976 and Subramanian, 1971).

Results and Discussion

The prevailing air temperature at Sharavathi reservoir varied between 28 and 30°C there was rainfall in most of the months (June to September).

The fairly high temperature prevailing throughout the year coupled with rainfall naturally promoted a good growth of vegetation and accumulation of organic matter supporting luxuriant growth of fungi. Table 35.1 shows the occurrence of fungal spores at Site 1 and 2 respectively. A total of 11 species of spores were identified *viz.*, *Penicillium* sp., *Aspergillus* sp, *Alternanthara* sp. *Curvularia* sp, *Rhizopus* sp, *Helimanthosporium* sp., *Cladosporium* sp, *Fusarium* sp, *Glomerularia*, *Cephalosporium* sp. and *Trichoderma* sp. In this study, the dominant type was *Aspergilli,* which mostly consisted of the genera *Aspergillus* and *Penicillium* sp. In India, few studies which attempted to delineate the two genera in the aeromycoflora showed that *Aspergillus* is more prevalent than *Penicillium* (Agarwal *et al.*, 1969 and Rati and Ramalingam, 1976). It is therefore presumed that a similar situation may occur at Sharavathi reservoir stretch. This spore is also known to be more frequent in urban areas than in rural areas (Subba Reddi, 1970 and Bajaj, 1978). *Alternaria,* third in order of abundance has been reported as number one in the United States of America (Morrow *et al.*, 1964). In India its frequency is rather high in northern parts than in south (Nair *et al.*, 1986). A simultaneous survey using both cultural and visual means may throw light on this differential frequencies of *Alternaria* sp. Spores were encountered on all the days of the months trapping was done (Figure 35.1). The month of August witnessed the maximum incidence accounting to 35.08 and 38.84 per cent for Sites 1 and 2 respectively. At Site 1, the month of September recorded 26.32, per cent followed by October (19.30 per cent), November (12.28 per cent) and December (7.02 per cent). Similarity, in Site 2 the total count for spores were 24.79 per cent, 17.36 per cent, 11.57 per cent and 7.44 per cent for the month of September, October–November and December respectively. The incidence of hyphal fragments (665–900/sq. cm) was considerably high. Most of the fragments were brown, septate and the hyaline ones rarely appeared. On several occasions they were trapped in association with the conidia of *Cladosporium*, *Cephalosporium*, *Curvularia* and

Alternaria. They were more numerous during August–October since many of the air borne hyphal fragments remain viable and germinate in the manner of fungal spores (Pady and Gregory, 1963), they may be considered as an important constituent of the fungus air spora and as part of the dispersal mechanism of the parent fungus. Gregory (1973) reported that environmental factors are the most important which affect the total number of fungal population in the atmosphere. Hence, seasonal variation in the fungal concentration of Sharavathi reservoir and its downstream stretch area depends upon the climatic factors.

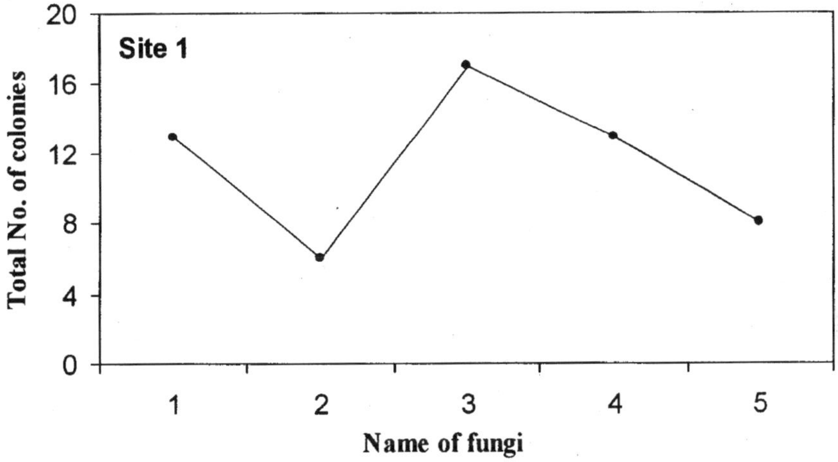

1. *Rhizopus* sp., 2: *Cephalosporium* sp., 3: *Penicillium* sp., 4: *Aspergillus* sp., 5: *Trichoderma* sp.

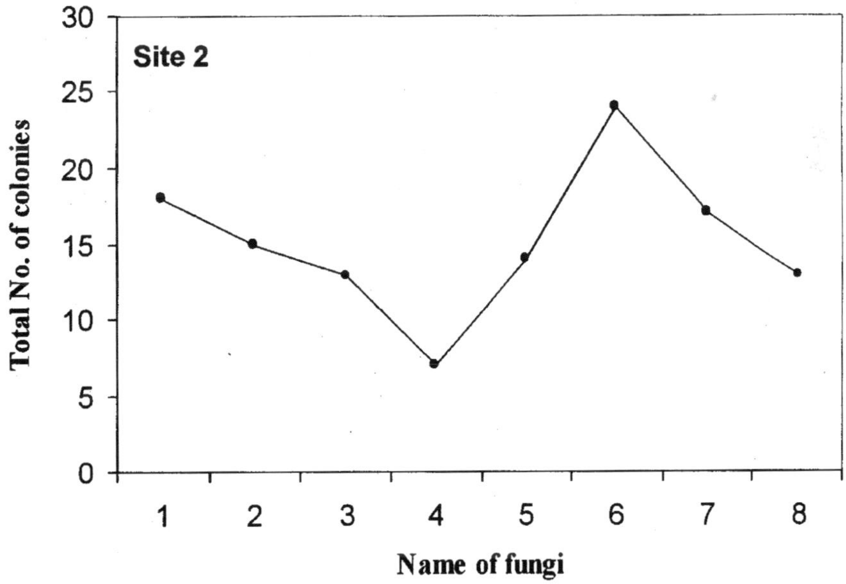

1. *Aspergillus* sp., 2: *Cladosporium* sp., 3: *Fusarium* sp., 4: *Glomerularia* sp., 5: *Helimanthosporium* sp., 6. *Penicillium* sp., 7: *Alternaria* sp., 8: *Curvularia* sp.

Figure 35.1: Total Number of Colonies of Different Fungal Species

Table 35.1: Distribution of Different Types of Fungal Spores at Sharavathi Reservoir (Site 1) and its Stretch (Site 2)

Name of the Species	August	September	October	November	December	Total (%)
			Site 1			
Rhizopus sp.	5	3	2	2	1	22.81
Cephalosporium sp.	3	2	1	0	0	10.52
Penicillium sp.	5	4	3	3	2	29.82
Aspergillus sp.	4	3	3	2	1	22.81
Trichoderma sp.	3	3	2	0	0	14.04
Percentage	35.08	26.32	19.30	12.28	7.02	
			Site 2			
Aspergillus sp.	6	4	3	3	2	14.88
Cladosporium sp.	7	4	2	1	1	12.40
Fusarium sp.	7	3	2	0	1	10.74
Glomerularia sp.	3	2	1	1	0	5.79
Helimanthosporium sp.	4	3	3	2	2	11.57
Penicillium sp.	9	6	4	3	2	19.83
Alternaria sp.	6	4	4	2	1	14.05
Curvularia sp.	5	4	2	2	0	10.74
Percentage	38.84	24.79	17.36	11.57	7.44	

References

Agarwal, M.K., Shivu, D.N. and Mukherji, K.G. 1969. Studies on the allergenic fungal spores of Delhi, India, metropolitan area. *J. Allergy*, 44: 192–203.

Bajaj, A. 1978. Study of viable spores in air at two different sites of Nagpur. *J. Palynol.*, 14: 136–149.

Chanda, S. and Mandal, S. 1978. Aerobiology in India with reference to respiratory allergy. *International Aerobiology News Letter* (Special Issue), 8: 7–13.

Ellis, M.B. 1976. *Dematiaceous Hyphomycetes*. Commonwealth Mycological Institute, Kew, England.

Gelman, J.C. 1959. *A Manual of Soil Fungi*. Oxford and IBH Publ. New Delhi, pp. 608.

Jamaluddin and Tandon, M.P. 1976. Some new market disease of vegetables and fruits. Tandon M.P. *Indian Phytopathology*, 29: 74–75.

Morrow, M.B., Meyer, G.H. and Prince, H.E. 1964. Prince: A summary of airborne mold surveys. *Ann. Allergy*, 22: 575–587.

Nair, P.K.K., Joshi, A.P. and Gangal, S.V. (Eds.). 1986. *Air Borne Pollen Spores and Other Plant Materials of India: A Survey*. CSIR Centre for Biochemicals, Delhi.

Pady, S.M. and Greogry, P.H. 1963. Numbers and viability of airborne hyphal fragments in England. *Trans. Br. Mycol. Soc.*, 46: 609–613.

Rati, E. and Ramalingam, A. 1976. Airborne *Aspergilli* at Mysore. *Aspects of Allergy and Applied Immunology*, 9: 139–149.

Sreeramulu, T. 1967. Aerobiology in India. *J. Scient. Ind. Res.*, 26: 474–480.

Subba Reddi, C. 1970. A comparative survey of atmospheric pollen and fungus spores at two places twenty miles apart. *Acta. Allergol.*, 25: 189–215.

Subramanian, C.B. 1971. *Hypomycetes.* ICMR, New Delhi.

Chapter 36

Study of Water Quality Parameters in Selected Areas of Karnataka State

Oinam Jayalakshmi Devi* and S.L. Belagali*, Regina Sharmila Dass** and G.R. Janardhana**

**Department of Studies in Environmental Science*
***Plant Molecular Biology and Mycology Laboratory, DOS in Botany*
University of Mysore, Manasagangotri, Mysore – 570 006, Karnataka, India

ABSTRACT

The different physico-chemical and microbiological parameters of water samples from some selected areas of Karnataka were investigated. A total of 47 water samples were collected from different districts over five months. Most of the physico-chemical parameter are within the permissible limits of BIS and WHO standards for drinking water. However, some samples show pH lower than the normal limit. The fluoride content in Hassan District was very high. One of the samples from Chitradurga contained very high amount of Nitrite than the desirable limit. Water samples were subjected to microbiological analysis also. Some samples were found to be highly contaminated with coliforms while some other samples were suitable for human consumption. So, it was concluded that a few water samples were unsuitable for drinking and irrigation purpose.

Keywords: *Water quality, Fluoride, Alkalinity, Nitrite, pH, DO, COD, TDS.*

Introduction

There are many different materials that may pollute surface water or groundwater. There are so many parameters to assess the quality of the water bodies. R.E. Khadsan *et al.*, carried out drinking water quality analysis of some borewell waters of Chikhilli town, Maharastra (2004). There are of course exceptions to the rule and water quality may improve downstream, behind dams and weirs, at points where tributaries groundwater enter the main stream and in wetland.

Rivers frequently act as conduits for pollutants by collecting and carrying waste water from catchments and ultimately discharging it into the ocean. Several researchers have studied the physico-chemical characteristics of Jamuna River (Bhargava *et al.*, 1985; Chakraborty *et al.*, 1959).

Water quality is closely linked to the surrounding environment and land use. Other than in its vapour form, water is never pure and is affected by community uses such as agriculture, urban and industrial use, and recreation. The modification of natural stream flows by dams and weirs can also affect water quality (S.R. Ambika *et al.*, 2004).

Water quality is commonly defined by its physical, chemical, biological and aesthetic (appearance and smell) characteristics.

There are many different materials that may pollute surface water or groundwater. There are so many parameters to assess the quality of the water bodies. The present article discusses the results of water quality parameters in Karnataka state.

Materials and Methods

Water samples of bore well, open well, pond, tap and channel water were collected from different districts of Karnataka (47 samples were collected from different districts over five months). 1 liter of water was collected and analyzed in the laboratory. pH determination was carried out by electrometric method using a digital pH meter. Calcium hardness by EDTA titrimetric method. Nitrate, Nitrite, Fluorides were analysed by UV spectrophotometer. While chloride were analyzed by titrimetric method. The concentrations of Sodium and Potassium were estimated by flame photometer. Heavy metals were analysed by Atomic Absorption Spectrophotometer.

In the microbiological studies, coliforms were analysed by Millipore-membrane filter technique. Samples were collected from differents district of Karnataka in the sterile plastic containers and kept for two days in the refrigerator and analysed. After filtering the water samples using Millipore-membrane filter apparatus the filter discs were transferred onto sterile EMB media and incubated for 24–48 hours. These parameters were determined by employing Standard Methods (APHA, 1975; Trivedi and Goel, 1986).

Results and Discussion

The results obtained for the analysis of water samples of different districts of Karnataka over five months are shown in Tables 36.1 and 36.2. Drinking water standards are prescribed by the Indian Council of Medical Research (ICMR), World Health Organization (WHO) and Bureau of Indian Standards (BIS).

pH

One of the water samples had pH less than the permissible limit. But it is not harmful. And other remaining samples were within the permissible limit.

Total Dissolved Solids

TDS of water sample studied ranged between 0 to 720mg/l, whereas desirable limit TDS is 500 mg/l due to high concentration of TDS in water, the water of some studied area in Chamarajanagar District is not suitable for textile, paper and food industries where water boilers are used.

Table 36.1: Physico-chemical and Microbiological Analysis of Water Samples

Location	pH	TDS	DO	COD	Alkalinity	Acidity	Ca Hardness	Fe^{2+}	Cl^-	SO_4^{2-}	Nitrate	Nitrite	Fluoride	Na^+	K^+	Chromium	Zinc	No. of Colonies
Bangalore District																		
Tap water	6.5	ND	3.7	ND	144.5	10	50.4	0.11	41.2	0.18	1.95	3.95	1.2	E	2.1	ND	ND	ND
Pond water	7.8	400	3.4	ND	356.5	15	101	ND	41.2	0.4	1.85	4	1.45	E	6.25	ND	ND	10
Belgaum District																		
Hidkal dam	8.5	ND	6.6	ND	62	15	47.6	ND	8.5	ND	0.01	ND	1.51	1.2	0.4	ND	0.42	10*
Borewell water	7.3	200	5.4	ND	418.5	16	90	ND	89.4	0.15	2.15	0.4	1.56	10.1	0.1	ND	0.26	ND
Openwell water	7.3	320	3.7	ND	344.2	16	84	ND	117	0.37	2.9	1.45	1.6	12.1	ND	ND	ND	5*
Bellary District																		
Tap water	7.7	80	3.5	ND	16	ND	75.6	ND	61.8	0.4	0.95	1.15	1.4	E	1.15	ND	ND	ND
Pond water	7.9		5.6	ND	243	ND	47.6	ND	89.4	0.55	2.1	1.45	1.05	E	0.6	ND	ND	ND
Borewell water	7.2	200	3.4	ND	13.5	29	129	ND	240	0.35	3.5	1.25	1.51	E	6.75	ND	ND	3
Chikkamagalur District																		
Tap water	6.8	440	3.5	ND	46	29	16.8	ND	27.5	0.17	1	1.21	1.52	1.9	0.4	ND	0.2	ND
Tap water	6.4	480	3.7	8.72	92	20	16.8	ND	55	ND	1.85	1.65	1.1	25.2	2.7	ND	0.21	ND
Chamrajanagar District																		
Tap water	7.1	400	1.9	ND	455.7	ND	126	ND	65.4	1.8	5.6	0.55	1.4	E	7.4	ND	0.3	ND
Pond water	6.5	200	6.6	17.4	206.1	20	48	0.12	21	2.8	0.85	1.05	0.05	13.6	18.6	ND	0.46	3
Pond water	6.4	720	6.9	26.2	206.1	20	12	0.24	15.6	6.5	0.79	0.105	0.1	10.9	8.7	ND	ND	1
Chitradurga District																		
Borwell water	7.5	400	4.5	ND	325.5	18	90	ND	89.4	0.51	2.4	0.5	1.15	E	4.6	ND	0.31	1
Pond water	7.3	200	3.7	ND	216	20	20.4	ND	475	ND	2.05	E	1.15	E	2.3	ND	ND	ND
Open well water	8.1	400	5	ND	23	29	101	ND	124	0.17	0.75	0.57	1.41	E	0.3	ND	0.47	ND
Dharward District																		
Potable water	7.6	360	4.6	8.72	217	ND	72	0.03	30.3	ND	0.65	0.4	1.4	16.6	1.9	ND	0.21	3
Tap water	7	160	3.2	17.4	80.5	10	12.6	ND	20.6	ND	1.3	1.45	1.2	0.5	0.5	0.03	0.31	1
Hassan District																		
Tap water	7.5	440	3.4	17.4	46	18	160	ND	296	1.3	4.91	2.4	1.8	5.8	0.2	ND	0.16	1
Pond water	7.9	200	4.2	69.8	130.2	15	24	0.04	18.5	81	0.8	ND	2.32	9.8	11.2	ND	0.081	7*
Borewell water	7	200	2.6	8.72	17.4	35	69	ND	32.9	ND	0.02	ND	1.85	E	2.6	ND	0.061	ND

Note: All Parameters are Expressed in mg/L, except pH and colonies.

ND: Not detected; E: Exceeded permissible limit.

*: Highly contaminated water samples (unsuitable for human consumption).

Table 36.2: Physico-chemical and Microbiological Analysis of Water Samples

Location	pH	TDS	DO	COD	Alkalinity	Acidity	Ca Hardness	Fe^{2+}	Cl^-	SO_4^{2-}	Nitrate	Nitrite	Fluoride	Na^+	K^+	Chromium	Zinc	No. of Colonies
Karwar District																		
Tap water	6.8	400	4	17.4	448	14	126	ND	117	ND	2.3	1.15	1.4	E	E	ND	0.1	ND
Well water	6.0	160	4.6	ND	ND	10	4.2	0.05	13.7	ND	0.5	1	1.6	6.4	1.3	ND	ND	ND
Mandya District																		
Open well water	7.3	360	2.2	34.9	271	30	180	0.03	198	ND	0.75	ND	1.5	E	7.4	ND	ND	1
Pond water	8.2	320	5.1	ND	241	15	33.6	ND	41.2	0.45	0.4	1.25	1.6	E	1.2	ND	ND	ND
Borewell water	7.8	480	4.3	8.72	356	10	50.4	ND	61.8	0.35	1.8	ND	1.4	E	2.1	ND	0.2	ND
Mangalore District																		
Tap water	7.0	400	5.9	17.4	173	15	36	ND	16.5	2.05	0.3	0.65	1.4	0.6	5.7	ND	0.2	1
Open well water	6.1	440	5.6	8.72	141	20	36	ND	20.6	ND	1.25	1	1.5	2.4	1.4	ND	0.4	ND
Tap water	7.2	ND	3.2	17.4	92	35	16.8	ND	13.7	ND	0.35	1.15	1.6	0.4	0.5	ND	0.3	ND
Medikere District																		
Tap water	7.0	120	3.4	ND	195	15	75.6	ND	55	0.5	2.8	1.37	1	E	2.8	ND	0.2	ND
Pond water	6.6	ND	5	8.72	151	20	42	ND	15.1	ND	0.25	0.15	1.6	4.7	1.8	ND	ND	3
Well water	6.7	200	3.5	17.4	151	15	48	ND	20.2	ND	0.9	ND	1.5	4.4	0.7	ND	ND	1
Mysore District																		
Tap water	7.6	320	3	ND	338	10	12.6	ND	93.5	1.9	1.65	ND	0.6	E	2.4	ND	ND	ND
Channel water	8.2	40	6.6	87.2	23	16	24	ND	32.9	1.6	0.6	2.05	1.6	7.6	4.6	ND	ND	10*
Well water	7.1	ND	4	ND	187	15	69	ND	162	2.7	0.01	0.4	1.0	18	6	ND	ND	1
Shimoga District																		
Hanambi lake	7.9	ND	5.9	ND	86	20	30	ND	17.2	ND	0.4	2.4	1.5	1.6	0.9	0.02	0.1	1
Municipal water	8.3	120	4.6	ND	.108	4.9	20.4	ND	27.5	ND	1.95	1.65	1	1.0	0.4	ND	ND	1
Udupi District																		
Tap water	6.6	200	3.8	17.4	80	16	16.8	ND	20.6	0.1	1.25	1.55	1.5	5.7	1.7	ND	0.2	ND
Tap water	6.6	80	4	ND	89	29	4.2	ND	13.8	ND	1.35	1.65	1.5	1.7	0.2	ND	ND	ND

Note: All Parameters are Expressed in mg/L, except pH and colonies.

ND: Not detected; E: Exceeded permissible limit.

*: Highly contaminated water samples (unsuitable for human consumption).

Alkalinity

The concentration of alkalinity of the water samples ranged between 13.5 to 448 mg/l.

Acidity

Acidity of water samples in the studied area was found to be within permissible limits.

Chloride

Chloride content for studied samples was well below the limit except two samples.

Calcium

The concentration of Calcium in water samples ranged between 4.2 to 180 mg/l. That means a few samples had higher values than the permissible limit. Normally this metal is not problematic but at higher concentration increase total hardness of water.

Iron

The concentration of Iron is nil except for a few samples, which are also within the permissible limits. The primary concern about iron in drinking water is its objectionable taste.

Nitrate

The Nitrate concentration varied between 0.01 to 4.91 mg/l. The concentration of nitrate in all samples were well within the permissible limit.

Nitrite

Nitrite occurs in water as an intermediate product in the biological breakdown of organic nitrogen, being produced either through the oxidation of ammonia or the reduction of nitrate. One the samples from Chitradurga contained very high amount of nitrite than the desirable limit. In the remaining samples, it is within the permissible limit.

Fluoride

The concentration of Fluoride in Hassan District was very high. In the remaining samples, it is within the permissible limit. Due to high concentration of Fluoride mottling of teeth, skeletal fluorosis, forward bending of vertebral column, deformation of knee joints and of body are seen in the persons who live in that area.

Sulphate

The sulphate concentrations were well below the recommended limit in all the studied samples.

Sodium

Sodium ranks sixth among the element in order of abundance; therefore, it is present in most natural waters, the levels may vary from negligible to appreciable. Relatively high concentration may be found in brines and hard waters softened by the sodium exchange process. Because of this, most of samples contains higher than the permissible limit.

Potassium

The concentration of potassium in water samples ranges from 0.1 to 22.5 mg/l. One of the water sample was higher than the permissible limit.

Dissolved Oxygen (DO)

Dissolved oxygen is a measure of the amount of oxygen freely available in water. All samples were well within permissible limits.

Chromium

Only two samples were found to have 0.02 and 0.03 ppm in all the analysed water samples.

Zinc

Only few samples contained Zinc, but these are well within the permissible limits.

Coliforms

After incubator of EMB agar plates at 37°C, pink coloured colonies were observed growing on the filter discs. Some samples were found to be highly contaminated with coliforms while some other samples were suitable for human consumption. Water is considered safe for drinking if it contains four or lesser colonies per 100 ml of water.

Conclusion

The above data on physico-chemical and microbiological parameters of the study area clearly showed that some water samples are not suitable for drinking and irrigation purpose. But most of the water samples are safe for drinking purpose and not polluted as the values for different parameters evaluated fall far below the acceptable limit. However, it is always advisable to drink water after proper boiling and filteration.

References

Ambika, S.R. and P.C. Shreedharan. 2004. Assessment of water quality of Vrshabhavathi stream loaded with factory effluents and sewage. *Indian J. Environ. and Ecoplan.*, 8(3): 593–596.

Aneja, K.R. *Experiments in Microbiology Plant Pathology Tissue Culture and Mushroom Production Technology*, 3rd Edition. New Age International (P) Publishers.

APHA, AWWA, WPCF. 1975. *Standard Method for the Examination of Water and Wastewater*, 14th Edition.

Bhargava, D.S. 1985. Water quality variation and control technology of Yamuna River. *J. Environ. Pollution Series*, England, 37: 335–376.

Chakraborty, R.D., Ray, P. and Singh, S.B.I. 1959. A quantitative study of the plankton and physico-chemical condition of river Yamuna at Allahabad. *Indian J. Fish.*, 6: 186–203.

Dhankar, R. and Sushila, S. 2004. Water quality assessment from different region of Mahendergarh, Hariyana. *J. Ecotoxicol. Environ. Monit.*, 14(1): 15–22.

Gopalan, R. and Amitha Anand. 1990. *Environmental Chemistry Laboratory Manual*. Reprint.

Khadsan, R.E. and Mangesh V. Kadu. 2004. Drinking water quality analysis of some borewell water of Chikhili town, Maharashtra. 20(1): 31–36

Meenakumari, H.R. and Hosmani, S.P. 2003. Physico-chemical and biological quality of groundwater in Mysore City, Karnataka, India. *Indian J. Environ. and Ecoplan.*, 7(1): 79–82.

Mitchell, Mark K. and Stapp William B. 1996. *Field Manual for Water Quality Monitoring*, 10th Edition.

Pontius, Frederick W. 1990. *Water Quality and Treatment*. Mc Graw-Hill, New York.

Sanjay Gupta, Kumar A., Ojha, C.K. and Gita Seth. 2004. Chemical analysis of groundwater of Sanganer area, Jaipur in Rajasthan. *Journal of Environ. Science and Engg.*, 46(1): 74–78.

Sharma, B.K. and Kaur, H. 1994. *Environmental Chemistry*, 1st Edition. Krishna Prakashan Mandir.

Trivedi, R.K. and Goel, P.K. 1986. *Chemical and Biological Methods for Water Pollution Studies.* Enviro Media Publications, Karad, pp. 209.

Chapter 37

Hydrological Studies for Dams in Ungauged Small Catchments: A Case Study

*Er. Sarbjit Singh Sooch**#, *Er. Deepak Gupta** *and N.K. Khullar***

**Assistant Professor, **Professor-cum-Head,*

Department of Civil Engineering, College of Agricultural Engineering, PAU, Ludhiana

ABSTRACT

For the design of dams, spillways, location of outlets and to determine the life of reservoirs, it is necessary to make detailed hydrological calculations. These calculations will involve determination of design flood and rate of sedimentation and distribution of sediment in the reservoir. Dholbaha Dam and Maili Dam (earthen rockfill dams) in the Kandi area has been considered as representative cases for study as the hydrological, topographical, climatic and geological conditions are more or less indentical in the region. The conclusions derived from these dams for hydrological studies can be applied to other similar dams being provided in the area.

Keywords: *Hydrological studies, Design flood, Sedimentation, Ungauged catchments, Flood control.*

Introduction

In the North-Eastern parts of Punjab State in Ropar, Gurdaspur and Hoshiarpur districts, there are number of flashy torrents called 'choes' which cause heavy damage to the adjoining areas during monsoons. After emerging from the hills, these choes fan out into several small torrents resulting in delta like formation in the plains. The abrupt and flashy discharges in these choes with steep bed slopes cause heavy damage to standing crops and property, whereas the flood during the falling river stages from these choes spread sandy deposits on the fertile lands. Numberous dam sites for storage of excess rain water exist in this region. By constructing a permanent structure of earth or masonry or a combination of both at suitable location in these ungauged catchments, it is not only possible to conserve the rain water and provide supplemental irrigation to the drought–stricken areas but also prevent soil erosion (Singh, 1991).

Hydraulic structures are designed for a flood which is likely to exceed once during the life of the structure. In gauged catchments, the design flood can be computed from the analysis of the flood data where as in the ungauged catchments this has to be computed by other methods.

Sedimentation studies have been carried out to determine the rate of sedimentation and the pattern of silt distribution in the reservoir in order to determine the safe level for fixing the irrigation outlet pipe and to assess the effective storage capacity available for irrigation supplies and flood control during various stages of the life of the reservoir. In this study average rate of sedimentation has been considered (Life of Reservoir, 1977; Mohammed, 1979 and Mutreja).

Methods and Materials

Various methods have been used for estimating design floods *viz.*, Empirical formula, Enveloping curves, Rational method, CWC short term report method, Standard Project Flood method, Synthetic hydrograph method, Comparison with meteorological homogenous catchments (Transposition) method and Flood frequency studies (Log-Normal distribution, Log-Pearson Type III distribution, Gumbel's method, Least Square method (Ven Te Chow method), Partial Duration Series method and Hazen's method) (Allen, 1975; Garg, 1977; Mccuen, 1977 and Mutreja).

Borland and Miller have presented six different methods for predicting the probable distribution of sediment in a reservoir. Of these methods, the empirical area reduction method and area incremental method seem to be more rational (Mohammed, 1979 and Singh, 1991).

From the sedimentation studies carried out on the basis of silt observations made at Dholbaha Dam and Maili Dam site, the annual rate of silt deposition in the reservoirs has been taken to be 8.0 hect-m and 7.34 hect-m respectively. This gives a rate of siltation of 0.1423 hect-m/year/km^2 of catchment for Dholbaha Dam and 0.43 hect-m/year/km^2 for Maili Dam.

Results and Discussion

A summary of results in respect of design flood and sedimentation distribution carried out for the Dholbaha and Maili Dams is given in Tables 37.1 and 37.2.

Table 37.1: Estimated Values of Design Flood Calculated by Different Methods

Sl.No.	*Method*	*Design Flood (cumecs)*	
		Dholbaha Dam	*Maili Dam*
1.	Empirical Flood formulae	1394	617
2.	Rational method	N.A.	635
3.	CWC Short Term Report method	1336	577
4.	Standard Project Flood method	2076	1120
5.	Synthetic Unit hydrograph	1621	611
6.	Transposition method	2261	918

Table 37.2: Peak Discharges

Year	Dholbaha Dam		Year	Maili Dam	
	Annual Peak Discharge (cumecs) X	X^2		Annual Peak Discharge (cumecs) X	X^2
1959	198	38206	1975	201.90	40763.61
1960	100	10000	1976	76.90	5913.61
1961	322	108684	1977	184.85	34169.52
1975	472	222784	1978	220.17	48474.82
1976	222	49284	1979	94.35	8901.92
1977	421	177241	1980	228.95	52418.10
1978	334	111556	1981	47.27	2234.45
1979	47	2209	1982	138.70	19237.69

Design flood calculated by using different frequency analysis methods on the basis of available data of rain for the Dholbaha and Maili sites (Table 37.2) is given in Table 37.3.

Table 37.3: Design Flood by Frequency Analysis

Method	Peak Flood Values of Return Period (cumecs)									
	10		20		50		100		1000	
	Maili	Dholbaha	Maili	Dholbaha	Maili	Dholbaha	Maili	Dholbaha	Maili	Dholbaha
Log-Normal	245	635	330	734	365	832	403	926	542	1204
Log-Pearson III	215	518	309	599	354	673	389	735	–	–
Gumbel	263	398	305	564	361	667	396	747	486	1010
New Gumbel	200	367	216	399	236	440	252	470	303	572
Ven Te Chow's	270	514	335	650	384	750	432	849	591	917
Partial Duration Series	–	504	–	615	–	705	–	791	–	1078
Hazen's	232	461	256	508	279	582	282	604	295	637

From the perusal of the Table 37.1, it is found that the higher magnitude of floods peak is obtained from 'Standard Project Flood' method in case of Maili and Dholbaha Dams. In case of reservoir with storage less than 6000 hect-m, Standard Project Flood (SDF) method of a 100 years flood, whichever is higher is to be adopted as design flood. From the results of frequency analysis summarized in Table 37.3 it is seen that for Maili dam a 100 years flood has a magnitude of 432 cumecs. From Table 37.1 it is found that Standard Project Flood has a magnitude of 1120 cumecs. Therefore a design flood for determining the spillway capacity for the Maili dam may be taken as 1120 cumecs. Since the flood peaks determined from Standard Project Flood method are greatly dependent on the critical sequence of rainfall and the precise storm characteristics not available for the flood, it is desirable that the values of peak flood obtained by this method be cross checked. This cross-checking has been possible in this case by making use of the data available on Dholbaha catchment which has characteristics similar to Maili catchment. By transposition method the flood peak value for Maili dam is obtained as 918 cumecs (Table 37.1).

Keeping in view the above facts and average of values of flood peaks obtained by Standard Project Flood and Transposition method, 1019 cumecs is adopted as design flood value for the Maili dam. Similarly peak flood of 2168 cumecs is adopted as design flood value for the Dholbaha Dam.

Storage capacities calculated by both the methods *viz.*, Area incremental method and Empirical Area Reduction method for Maili dam and Dholbaha dam at the end of 15, 20 and 40 years and 15, 25 and 50 years respectively are not much of difference. Thus any of two methods may be adopted for the purpose of determining sediment distribution in a reservoir. Estimated useful life of Maili dam and Dholbaha dam has been computed as 37 years and 92.50 years respectively.

The elevations of sediment deposits and reduced capacities available in the reservoir resulting from sediment deposition at 15, 20 and 40 years age of the Maili reservoir and at 15, 25 and 50 years age of Dholbaha reservoir are indicated in the Table 37.4.

Table 37.4: Reduced Storage Capacities in Reservoirs after Sediment Deposition

Dam	*Period (Years)*	*Zero Level (metre)*	*Storage Capacities (hect-m)*		
			By Empirical Area Reduction Method	*By Area Incremental Method*	*Average*
Maili	Original	355.40	489.75	489.75	489.75
	After 15 years	359.00	415.334	416.45	415.89
	After 20 years	360.00	407.50	399.02	403.28
	After 40 years	364.00	328.84	331.47	330.15
Dholbaha	Original	391.67	1338.40	1338.40	1338.40
	After 15 years	396.85	1218.50	1216.46	1217.48
	After 25 years	398.70	1138.00	1133.22	1135.61
	After 50 years	402.19	937.50	935.46	936.48

Conclusion

On the basis of sedimentation studies carried out for Dholbaha and Maili Dams, the reservoirs are expected to silt upto elevations 396.85, 398.70 and 402.19 in 15, 25 and 50 years respectively for Dholbaha Dam and elevations 359.0, 360.0 and 364.0 in 15, 20 and 40 years respectively for Maili Dam. The intake of the irrigation outlet pipe is therefore proposed to be provided at EL 403.50 for Dholbaha Dam, and at EL 365.00 for Maili Dam. A temporary intake (to be pludged later on) is proposed at EL 398.00 to enable utilization of additional live storage in the first 15 years life of the Dholbaha Dam when the level of sediment deposits is low. Similarly for the Maili Dam the temporary intake of the irrigation outlet pipe is proposed to be provided at EL 359.0 to enable utilization of maximum quantities of water in early phase of life of dam.

The live storage capacities available for storage of irrigation supplies and for flood absorption at various stages in the life of dams are given in Table 37.5.

Design flood of Maili Dam has been adopted as 1020 cumecs whereas the full reservoir level and irrigation outlet level has been recommended as EL 373.00 and EL 365.00 respectively. Similarly design flood for Dholbaha Dam has been adopted as 2170 cumecs whereas full reservoir level and irrigation outlet level has been recommended as EL 417.00 and EL 403.50 respectively. The useful life for Maili and Dholbaha reservoir has been determined as 37 years and 92.50 years respectively.

Table 37.5: Available Live Storage Capacity of Reservoir

Dam	*Life of Dam (yrs)*	*Irrigation Supplied*	
		Between Elevations (m)	*Storage Available (hect-m)*
Maili	15	359.00 to 373.00	415.89
	20	365.00 to 373.00	369.00
	40	365.00 to 373.00	300.25
Dholbaha	15	398.00 to 417.00	1200
	25	403.50 to 417.00	1035
	50	403.50 to 417.00	920

References

Allen, T., Hjemfelt J.R. and Cassidy, J.J. 1975. *Hydrology for Engineers and Planners.* Iowa Sate University Press, U.S.A.

Garg, Santosh Kumar. 1977. *Water Resources and Hydrology,* 2nd Edition. Khanna Publishers, New Delhi.

Gill, Mohammed Akram. 1979. Sedimentation and useful life of reservoirs. *J. Hydrology,* 44: 89–95.

Hydrological Studies for the Maili Dam. Irrigation Department, Punjab, Head Office, Chandigarh.

Hydrological Studies for the Dholbaha Dam. Irrigation Department, Puniab, Head Office, Chandigarh.

Life of Reservoir. 1977. Technical Report No.19, Central Board of Irrigation and Power, New Delhi.

Mccuen, R.H., Rawls, W.J., Fisher, G.L. and Power, R. 1977. Flood flow frequency for ungauged watersheds: A literature evaluation. *Agricultural Research Service,* Betsvile, Maryland, ARS-NE-8.6, pp. 140.

Mutreja, K.N. *Applied Hydrology.* Tata McGraw Hill Publishing Company Limited, New Delhi.

Singh, Sarbjit. 1991. Hydrological Studies for Dams in ungauged small water shed. *M.Tech Thesis,* Punjab Agricultural University, Ludhiana, India.

Chapter 38

Impact of Tourist Influx on the Courtallam Water Quality Index

G. Gitanjali and A. Kumaresan

ICAR Centre, Sri Parasakthi College for Women, Courtallam – 627 802, Tamil Nadu

ABSTRACT

Water Quality Index (WQI) of Courtallam water was studied after analysing nine hydrochemical parameters. Water samples were collected in triplicate at five sites. Samples collected in main, five and old Courtallam falls before and after anthrogpogenic activity and after were designated as site I and II respectively. Down stream samples were collected at Melagaram, Yanaipalam, Tenkasi outer and they were referred as site III, IV and V respectively. Dissolved Oxygen, biochemical oxygen demand, pH, total alkalinity, calcium, magnesium, sulphate, chloride and nitrate were analysed in the samples collected from all the five sites. WQI calculated as per the method of Kaur *et al.* (2001) giving appropriate unit weights and ratings. The water quality was monitored for five year period considering the tourist influx and cumulative rainfall. The WQI decreased from site I to V and the reduction was drastic from site III onwards. Though the rainfall and flow of water in the falls were more during the season (June to September), the WQI decreased on account of heavy tourist influx. Hence, proper surveillance is needed to maintain the quality of Courtallam water.

***Keywords**: Water quality index, Courtallam water, Tourist influx, Hydrochemical quality.*

Introduction

Water is essential for survival but today about 200 million people in India do not have access to safe drinking water. Most of our water sources are polluted with untreated or partially treated wastes from industry, domestic sewage and fertilizer / pesticide runoff from agricultural fields. The quantity and quality of available water is decreasing and aggravating the already serious situation. When water quality deteriorates, water loses its economic value. Tourism and environment are intricately and inseparably interlinked with many positive relationships as well as several irreversible negative impacts. Tourists are attracted to destinations of high environmental quality. Courtallam is one of the most popular tourist spot in South India known for its water falls and ancient Siva temple (Kurtallanathar). Courtallam falls water is used as drinking water in most villages around. The total population of residents of Courtallam is around 3982 in a total area of 7.56 km^2. According to the report given by Tamil Nadu Tourism Development Corporation, the tourist inflow is approximately 50,000 per day during season (June to September). Nearly 2 lakh Iyyappa devotees visit Courtallam in the months of December and January. Tourism can cause large increase in the number of temporary residents of a tourist area and resulting surge of garbage wastes. Tourists often use far more packaged and disposable goods than average resident and the local disposal system frequently can not handle all the extra trash. Increasing number of tourists year after year with simultaneous reduction in rain fall accelerates the pollution load of Courtallam water thereby reduces the water quality considerably. Therefore, a detailed study was undertaken to assess the water quality index in five sites in relation to tourist influx for a period of five years.

Materials and Methods

Water samples were collected from main, five and old Courtallam falls where 90 per cent of the tourist take bath. Water samples collected before human activities were pooled together and designated as site I. Samples collected after anthropogenic activities were referred as site II. Samples collected at Melagaram, Yanaipalam (entry point of Tenkasi Town) and Tenkasi (exit point of Tenkasi town) were indicated as site III, IV and V.

The samples were collected in triplicate and five such sampling was carried out in one year period. The study was continued for a period of five years to compare the water quality index year after year. The hydrochemical parameters namely pH, alkalinity, dissolved oxygen, biochemical oxygen demand, calcium, magnesium, chloride, sulphate and nitrate were determined in all samples as per the methods shown in APHA (1985). Water quality index was calculated as per the calculation of Kaur *et al.* (2001).

Results and Discussion

Water Quality Index

Water quality index (WQI) may be defined as the rating that reflects the composite influence of a number of water quality factors on overall quality of water. It is one of the most effective ways to communicate information on water quality trends to policy makers to shape sound public policy and implement the water quality improvement programmes efficiently. Water quality index first formulated by Horton (1965) and then by Ott (1978). The mathematical designs given by Horton (1965) and Ott (1978) were reexamined and modified by Kaur *et al.* (2001).

Weighting and Rating Scale

The word weighting implies relative significance of each of the factor in overall water quality and is dependent on the permissible levels in drinking water as suggested by ICMR. Factors having low permissible limits can harm the quality of water to a large extent even on a slight increase. Such factors have high weightings. On the other hand, factors which have higher permissible limit are less harmful and have low weightings. Each chemical factor has been assigned a rating value (Vr) to calculate WQI. The values fall between 0 to 100. The rating Vr = 0 implies that chemical factor exceeds standard maximum permissible limits and water is highly polluted while Vr = 100 implies that chemical factor has the most desirable value and water is absolutely clean.

To calculate water quality index, the products of unit weights (Wi) and rating (Vr) of nine chemical parameters were summated as per Kaur *et al.* (2001).

$$\mathrm{WQI} = \mathrm{Wi} + \mathrm{Vr}$$

where,

$$\mathrm{Wi} \times \mathrm{Vr} = \mathrm{Wi}(\mathrm{DO}) \times \mathrm{Vr}(\mathrm{DO}) + \mathrm{Wi}(\mathrm{BOD}) \times \mathrm{Vr}(\mathrm{BOD}) + \mathrm{Wi}(\mathrm{pH}) \times \mathrm{Vr}(\mathrm{PH}) + \mathrm{Wi}(\mathrm{T\ Alk}) \times \mathrm{Vr}(\mathrm{T\ Alk}) + \mathrm{Wi}(\mathrm{Ca}) \times \mathrm{Vr}(\mathrm{Ca}) + \mathrm{Wi}(\mathrm{Mg}) \times \mathrm{Vr}(\mathrm{Mg}) + \mathrm{Wi}(\mathrm{SO_4}) \times \mathrm{Vr}(\mathrm{SO_4}) + \mathrm{Wi}(\mathrm{Cl}) + \mathrm{Vr}(\mathrm{Cl}) + \mathrm{Wi}(\mathrm{NO_3}) \times \mathrm{Vr}(\mathrm{NO_3})$$

Table 38.1: Classification of Water Quality (According to Kaur *et al.*, 2001)

WQI	*Water Quality*	*Designation*
100	Absolutely clean	Class I
80–99.99	Slightly polluted	
60–79.99	Moderately polluted	Class II
40–59.99	Excessively polluted	Class III
0–39.99	Severely polluted	

Maximum water quality index was observed in site I which gradually decreased in site II, III, IV and V (Table 38.2). Considerable variation in the WQI between site I and II was noticed. Such variation also existed between other sites namely III, IV and V. Minimum WQI was recorded in site V which was not different from samples of sites III and IV. Therefore, there was significant difference ($P < 0.01$) in the water quality of samples collected at various sites. Though the quality of water decreased between sites I to V, no significant difference was recorded between the periods of study (Table 38.1 and Figure 38.1).

Table 38.2: Water Quality Index of Courtallam Water in Different Sites and Periods

Period of Study	*Sampling Sites*				
	Site I *n = 15*	*Site II* *n = 15*	*Site III* *n = 5*	*Site IV* *n = 5*	*Site V* *n = 5*
First year	68.58	60.69	51.93	52.65	52.23
Second year	70.69	61.54	59.41	52.65	46.29
Third year	80.59	67.47	65.68	65.68	52.98
Fourth year	68.15	67.89	58.79	58.35	58.35
Fifth year	85.21	67.47	59.63	46.29	46.51

Impact of tourists influx on water quality index during season and off-season are shown in Figure 38.2. The water quality index decreased in site I, II and III due to heavy tourist influx during season. Impact of tourists influx on water quality index during season and off-season for various sites

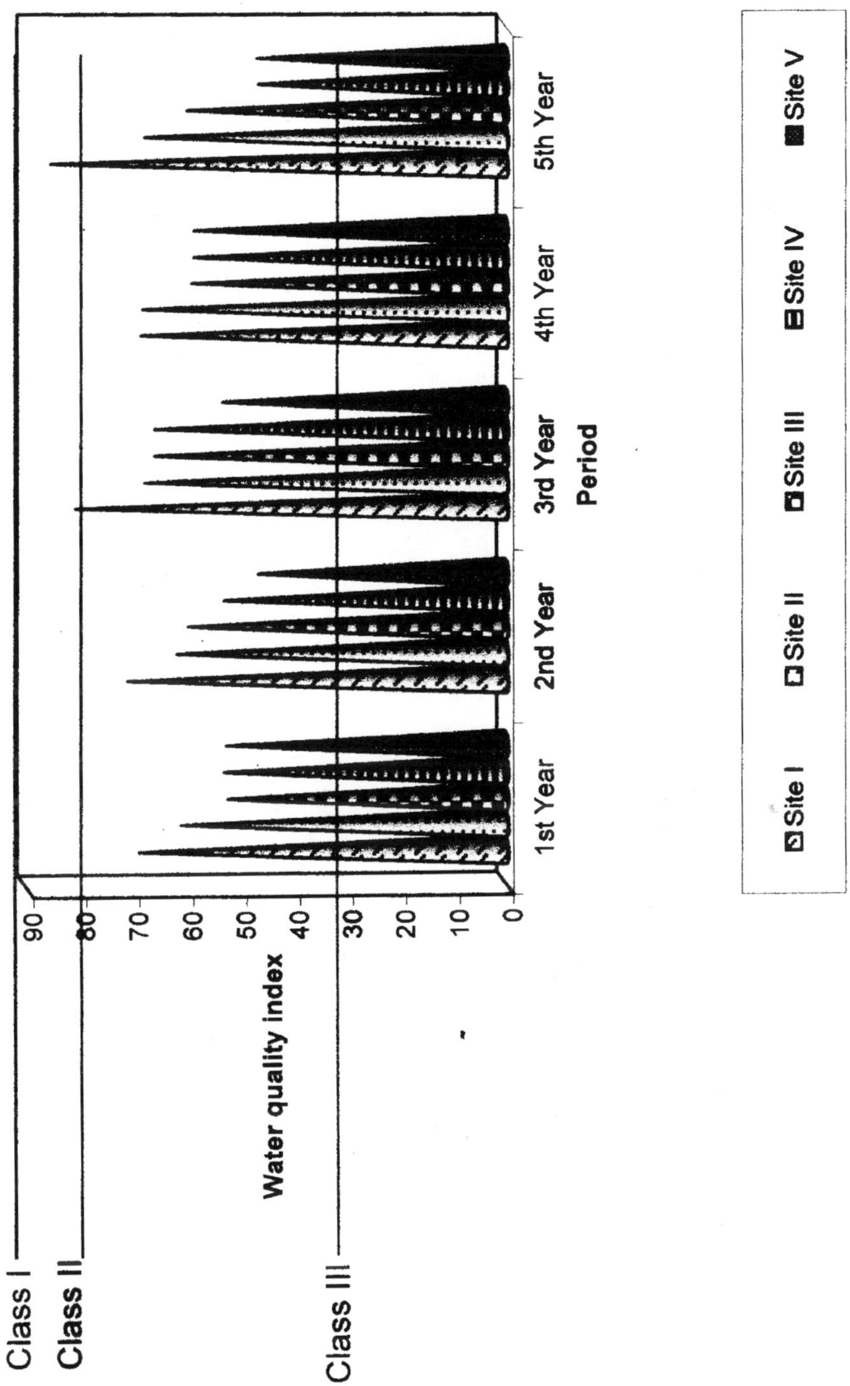

Figure 38.1: Courtallam Water Quality Index Calculated for Various Sites During the Study Period

are shown in Figure 38.2. Cumulative rainfall during various seasons for the period of study is given in Figure 38.3. Normally highest rainfall is recorded in the months of July, August, September and October of every year. January to April is particularly rainless but in May and June fair amount of rain may occur. More than half of the annual rainfall is brought by the south-west monsoon while the rest occurs in October and November. Comparing the rainfall of five year period, maximum was recorded in first two years which reduced to one half in subsequent three year period (Figure 38.3).

Total, floating and accommodated tourists arriving Courtallam and predicted numbers for future years are presented in Figure 38.4. Tremendous increase in tourist influx is observed during the season (June to September) of every year.

Impact of Tourists Influx on Water Quality Index

Travel and tourism creates jobs and wealth and has tremendous potential to contribute to economically, environmentally and socially sustainable development in both developed countries and emerging nations. If travel and tourism are managed badly, they can have a detrimental effect and can damage the small fragile environment. Because so many tourists are attracted to natural, unspoiled destinations, the environmental degradation caused by tourism can hurt the tourism industry. Tourism produces a variety of unwanted byproducts that are disposed, intentionally and unintentionally to the environment and modify the same; they are collectively called as the "externalities". Thus,

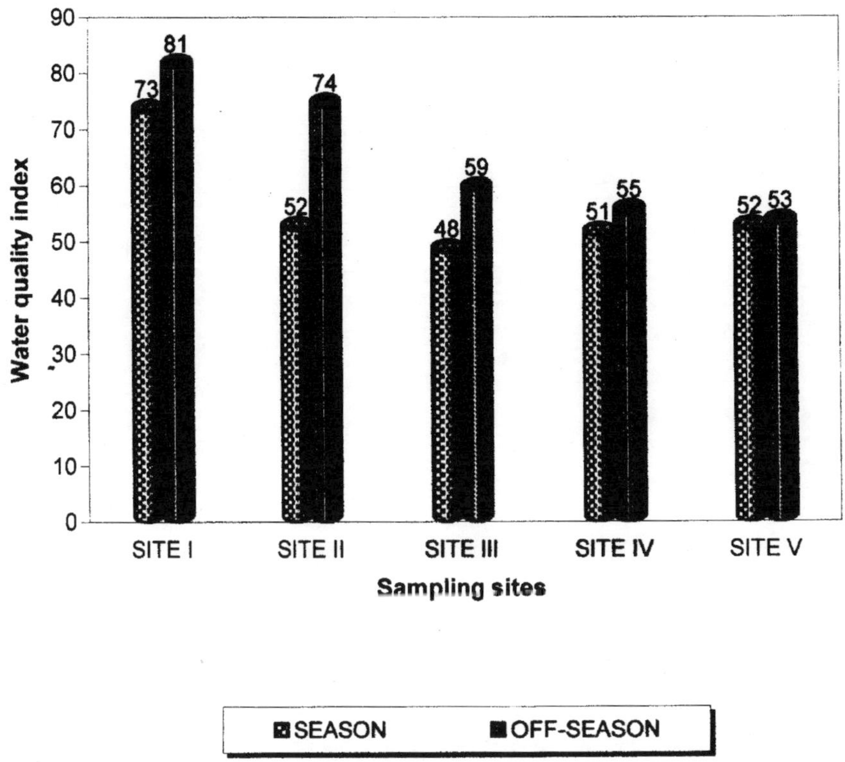

Figure 38.2: Impact of Tourist Influx on Courtallam Water Quality Index During Season and Off-season

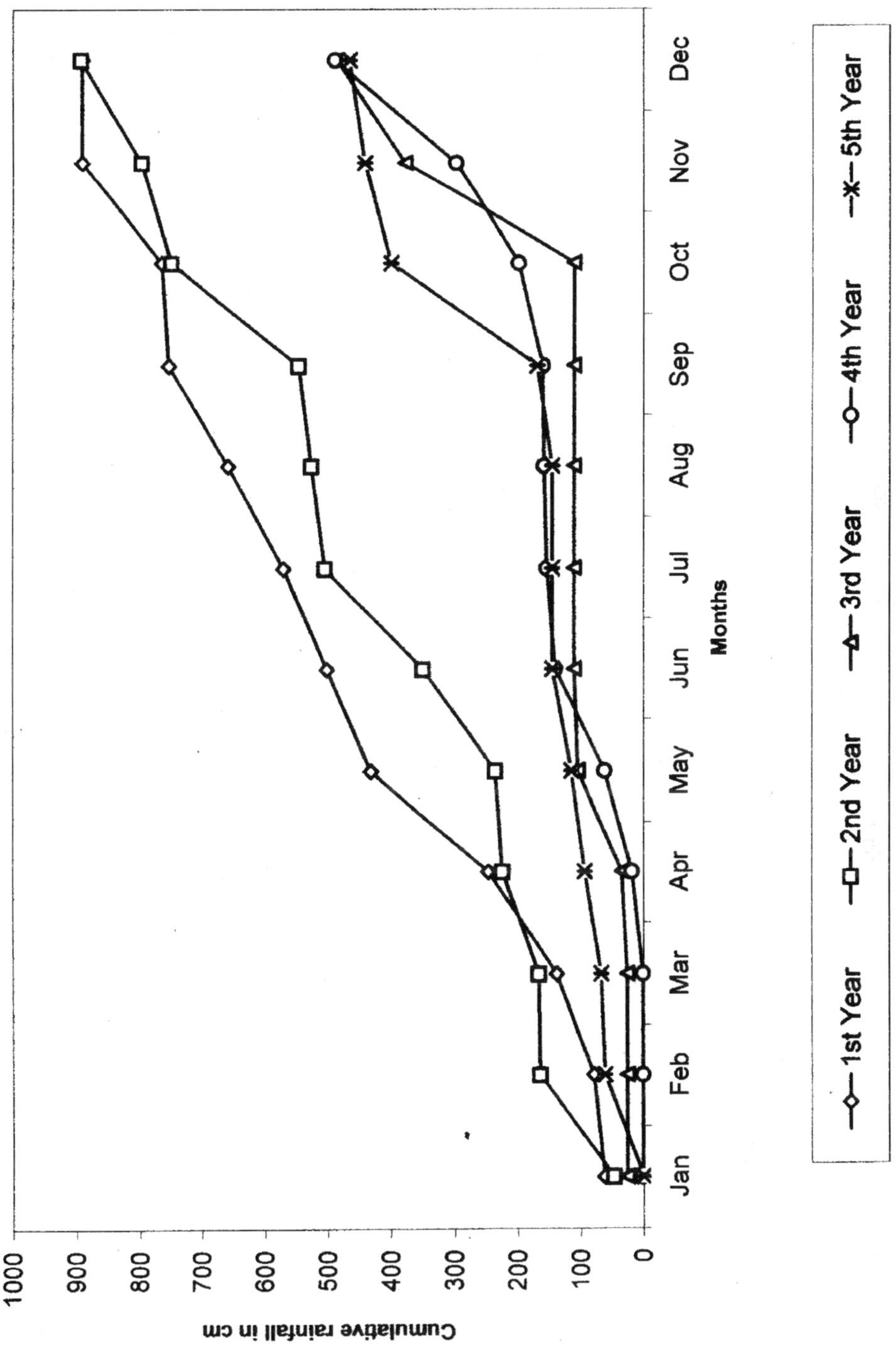

Figure 38.3: Cumulative Rainfall of Courtallam During the Study Period

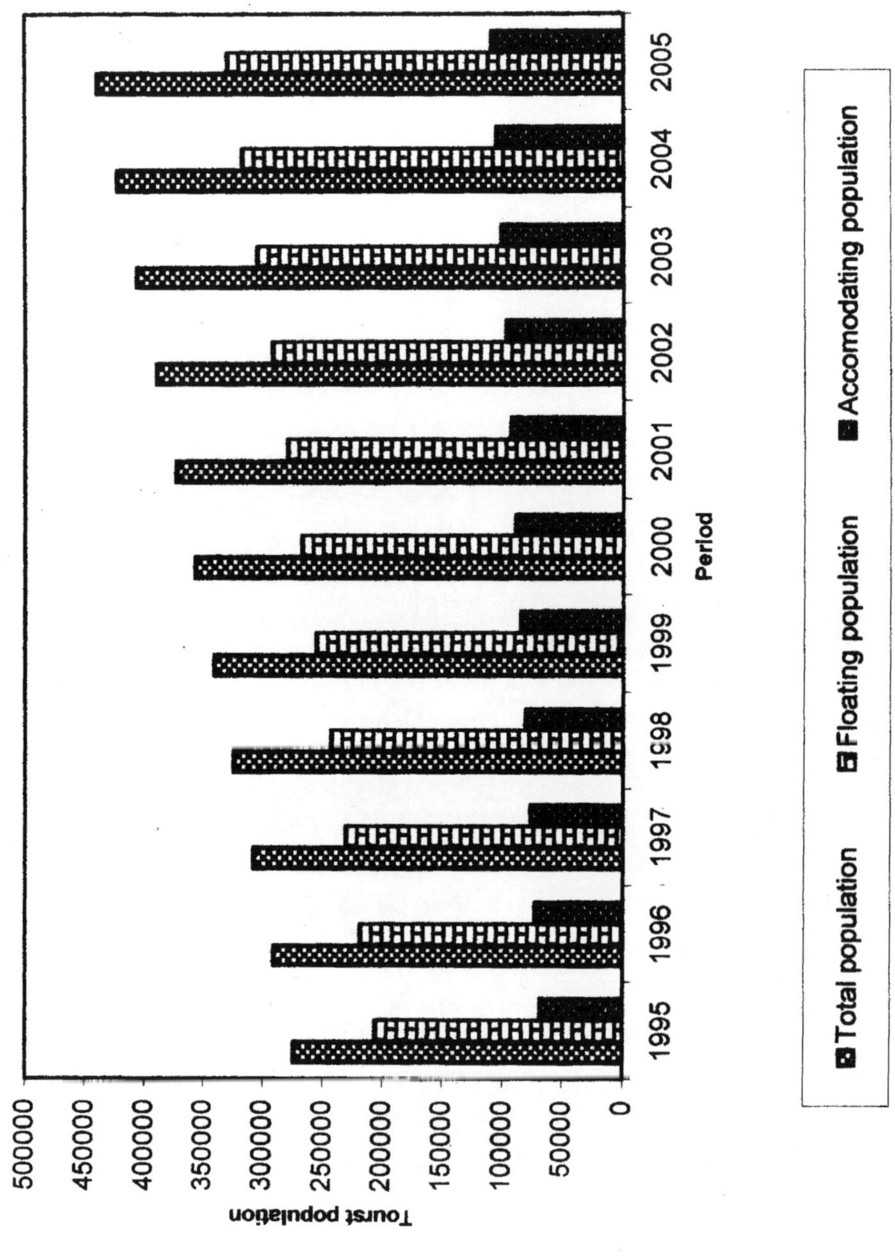

Figure 38.4: Recorded and Predicted Tourist Inflow to Courtallam

sustainable tourism depends in many important ways on the proper handling of the relationship between tourism and the environment (Inskeep, 1991).

Tourism industry should take appropriate measures to protect the environment and the health of local people. There is no facility available to receive the large number of tourists and pilgrims during various periods. Most of the hydrochemical parameters also increased during this period due to human activities. Hence, tourism industry should take remedial measures to save the ecosystem, natural environment and maintain the quality of Courtallam water. This include

1. The construction of more number of hotels to accommodate increasing tourist influx.
2. Sanitary conditions of the town must be increased by constructing urinals and toilets to relieve the natural urge in confined environments.
3. Dust bins must be kept at various points to collect waste dumpes by tourists.
4. Prior information may be published in newspapers to regulate the influx of tourists.
5. Natural, unspoiled scenery, mountains, valleys, water falls, ancient monuments, traditional and picturesque beauty of Courtallam town must be maintained to have vivid natural beauty and attract tourists.

Conclusion

Water Quality Index (WQI) of Courtallam water was calculated using nine hydrochemical parameters. The influence of tourist influx on the water quality index was determined at five locations on the course of the stream. The index decreased from site I to V due to heavy human activities. The water was excessively polluted from site III onwards. The authorities must take remedial measures to save the water quality and natural environment of Courtallam.

References

APHA. 1985 *Standard Methods for the Examination of Water and Wastewaters*, 16th ed. American Public Health Association, Washington DC.

Inskeep, E. 1991. *Tourism Planning: An Integrated and Sustainable Development Approach.* John Wiley and Sons Ltd., London, pp. 252.

Horton, R.R. 1965. An index number system for rating water quality. *J. Wat. Pollu. Cnt. Fed.*, 37: 300.

Kaur, H., Syal, J. and Dhillon, S.S. 2001. Water quality index of the river Satluj. *Poll Res.*, 20(2): 199–204.

Ott, W.R. 1978. *Environmental Indices: Theory and Practice.* Ann. Arbor. Sci. Publis., Inc., p. 202–213.

Chapter 39

Fluoride Concentrations of Water in Vrishabhavathi River Basin, Bangalore District, Karnataka

Y.M. Muniraju, H.C. Vajrappa** and J.M. Neelakantarama****

Department of Geology, Bangalore University, Bangalore – 560 056

ABSTRACT

The Fluoride levels in 100 water samples in different source of the study area were determined using spectrophotometer. The highest fluoride concentration in surface and subsurface water samples is 0.49 and 0.72 ppm respectively. The mean fluoride concentration is 0.42 ppm. Iso-concentration maps almost indicate uniformity of concentration of all samples, however, they show very low content of fluoride below 0.60 ppm. According Bureau of Indian Standards (BIS) 10 subsurface samples are suitable for the drinking purpose.

Keywords: *Fluoride, Iso-concentration map, Spectrophotometer.*

Introduction

The Vrishabhavathi is a 6th order river basin, lies between north latitude 12°44′ to 13°03′ and east longitude 77°23′56″ to 77°43′ with an area of 650 sq.km^2 covering parts of Bangalore district. The western parts of the basin is occupied by prominent hill ranges, the south-eastern and northern parts

E-mail: *munirajuym@rediffmail.com, **hcv123@rediffmail.com, ***neelakantaramajm@rediffmail.com of the basin are made up of undulating terrain and hilly tract. The remaining part of the basin is plain land. The over all slope of the basin is towards north-south. The basin exhibits dendritic pattern of drainage, which is typical of hard homogeneous and crystalline rocks. The drainage segments in the north-western parts of the basin are long straight and are controlled by lineaments in the basin.

Geology of Basin

The main lithological units of the Vrishabhavathi river basin are banded gneisses, granites, amphibolites, and migmatites. Banded gneisses cover major part of the basin. The granites occur with in the gneisses are highly imperceptible and seen as patches on the western flanks of the basin. Migmatites occur as prominent outcrops only certain parts of the basin. Amphibolites occur as small prominent ridges having trends of NNW-SSE.

Fluoride Environment

Fluoride is the common element in earth's crust as component of the rock and minerals. The presence of fluoride, in excess of limits is a serious matter of concern from a public health point of view. Deficiency of fluoride leads to dental caries and higher concentration leads to dental and skeletal fluorosis (Choubisa, 2001 and Faalul Hoque, 2003). Scientists working at national institute of nutrition, Hyderabad found new form of fluorosis characterized by genu-valgum in south India (Prabavathi Nagarajan, 2001). Like any other pollutant the fluoride pollution can also occur due to both natural and man made activities. Fluoride content in groundwater is mainly due to natural contamination but the process of dissolution is still not well understood (Saxena, 2001). Information on the quality of groundwater is an important aspect of any hydrogeological investigation. The chemical quality of groundwater generally varies even at short distances. This variation may be attributed to the variations in the hydrochemical processes. In addition to this the chemical quality of groundwater in an area is influenced by other factors like climate, hydrological conditions, chemical and physical characteristics of soil and geology. Various human activities mainly agricultural practices also contribute to the variation of groundwater quality.

Materials and Methods

A total 100 samples of surface and subsurface water samples have been collected from different parts of the study area for chemical analysis including fluoride. Surface water samples have been collected from the main Vrishabhavathi river and its tributaries. This has been done in order to estimate the effects of various environmental factors, on the parameters of the water. Subsurface water samples have been collected from borewells located in different parts of the basin. Samples were collected in PVC cans of two-litre capacity. In order to get the accurate results, the cans were cleaned with distilled water and rinsed with sample water and then collected the sample. Fluoride is determined by using spectrophotometer.

Results and Discussion

The results of physico-chemical analysis of surface and subsurface water samples are presented in Tables 39.1 and 39.2. The mean pH of the water samples is varies from 6.53 to 8.47 with an average of 7.51 ppm. It is lower than the prescribed limit. The EC is maximum 2.00 and minimum is 0.69 with average of 1.26 Mhos/cm. According to environmental quality standards of ISI the permissible limit is 0.3. These values, indicate, that the water is unsafe for drinking and domestic purposes.

The fluoride content of surface water varies from 0.37 to 0.49 ppm and has an average of 0.43 ppm. According to ISI-10500-1991, the required and desirable limits of fluoride with in surface or

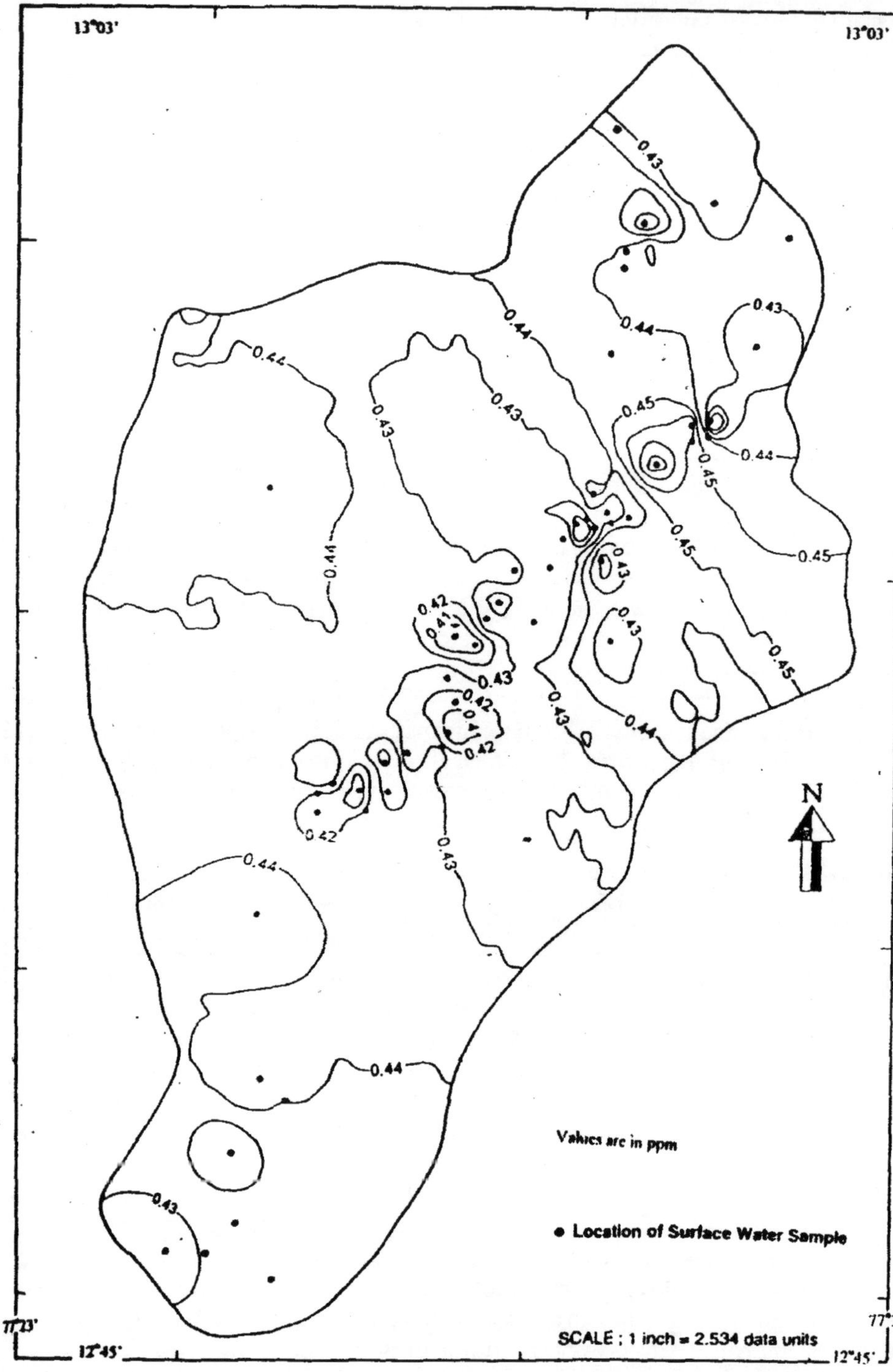

Figure 39.1: ISO Concentration Map of Floride Content of Surface Water

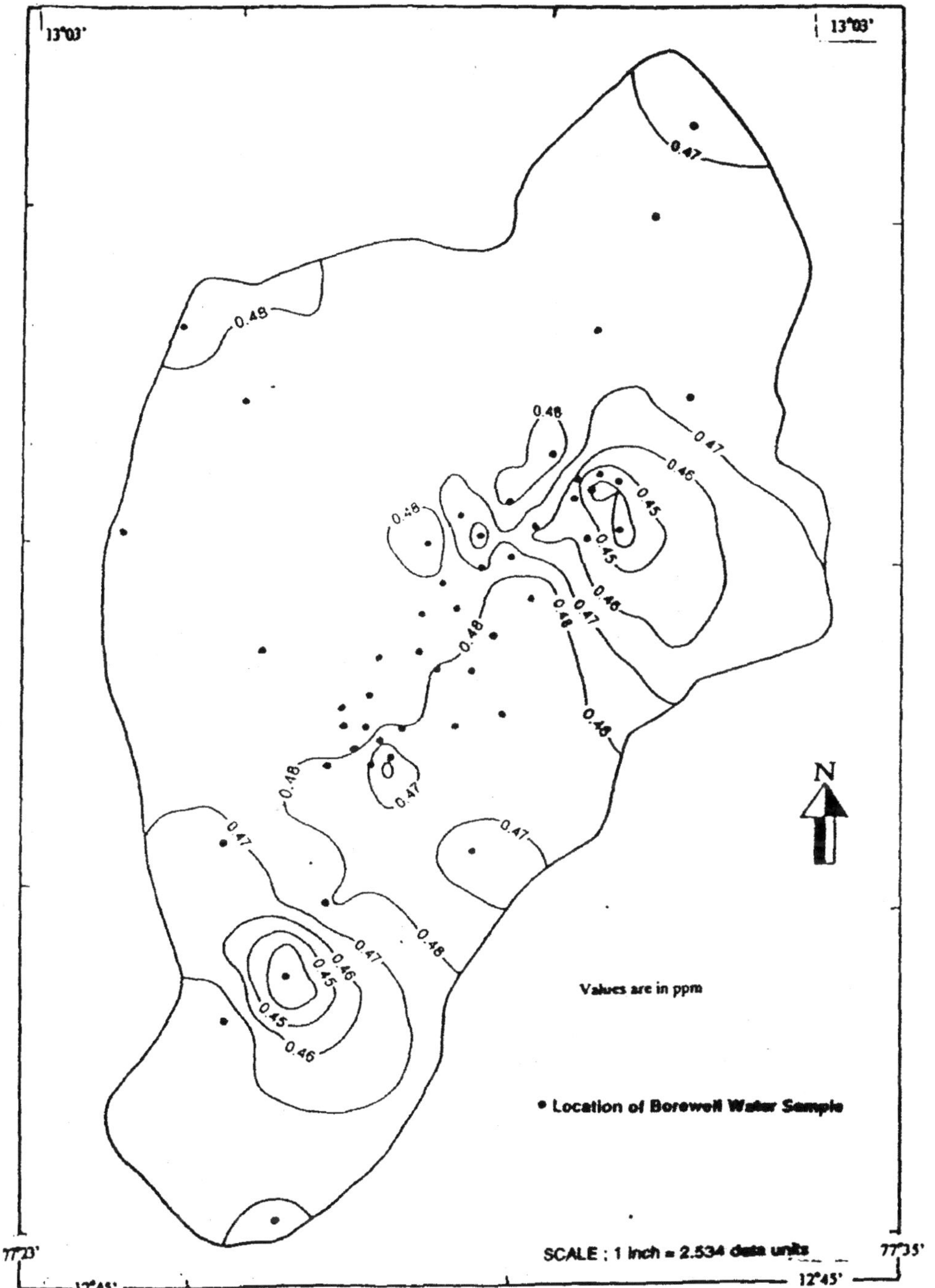

Figure 39.2: ISO Concentration Map of Floride Content of Borewell Water

groundwater is between 0.6 to 1.5, fluoride concentration of all surface water samples are not upto permissible limit. Iso-concentration map (Figure 39.1) also shows that the fluoride concentration is very low in surface water. Low fluoride concentration creates the dental problems so that all surface water samples are not suitable for drinking purpose.

Table 39.1: Fluoride Concentration of Subsurface and Surface Water Samples

Range of Fluoride Concentration (ppm)	Number of Subsurface Water Samples (%)	Number of Surface Water Samples (%)
0–0.5	36 (72)	50 (100)
0.6–1.0	14 (28)	–
1.1–1.5	–	–

Table 39.2: Concentration of the Physical and Chemical Parameters in Subsurface and Surface Water Samples

Parameters	Subsurface Water	Surface Water
pH	7.42	7.61
Ec	1.26	1.37
Ca	73.48	37.33
Mg	41.61	37.57
Na	125.38	101
K	31.6	36
Cl	–	154
SO_4	–	42.42
TH	–	257

Fluoride content of subsurface water varies from a 0.06 ppm to 0.72 ppm and has an average of 0.42 ppm. Out of 50 samples only 10 (20 per cent) samples are ideal for drinking purposes and free from hazards, remaining samples are unfit for the drinking purpose. Iso-concentration maps almost indicate uniformity of concentration of subsurface water (Figure 39.2). However, they show very low content of fluoride below 0.60 ppm.

Conclusion

The surface water is not only lower in fluoride content and it is also highly polluted and therefore it cannot be used for domestic purpose. Where as the subsurface water before being used for domestic purposes, requires fluoridation.

References

BIS. 1991. Specification for Drinking Water, IS: 10500:1991. Bureau of India Standards (BIS), New Delhi.

Choubisa, Sl. 2001. Endemic fluorosis in southern Rajasthan, India. *Fluoride*, 34: 61–70.

Faalul Hoque, A.K.M. *et al.* 2003. Fluoride levels in different drinking water sources in Bangladesh. *Fluoride*, 36: 38–44.

Prabavathi Nagarajan, T., Ramachandramoorthy, Edison Raja, B. Kavitha, R. Lakshmi and C. Sivaraj. 2001. *Indian Journal Environment and Ecoplanning*, 7(2): 311–314.

Saxena, V.K. and Ahmed, S. 2001. Dissolution of fluoride in groundwater: A water-rock interaction study. *Environmental Geology*, 40(9): 1084–1087.

Chapter 40

Assessment of Water Quality of Harishankar Spring: A Pilgrimage Place of Western Orissa

S. Naik and R.K. Patel***

**Teaching Assistant, **Senior Lecturer*
Department of Chemistry, N.I.T., Rourkela – 769 008

ABSTRACT

This article deals with water quality status of Harishankar spring, a pilgrimage place of western Orissa for variables like temperature, turbidity, total dissolved solids, pH, dissolved oxygen (DO), biochemical oxygen demand (BOD), hardness, iron, chloride, sulphate, sodium, potassium etc. The study was conducted during May 2004 to December 2004.The alkalinity of the three sampling sites is found little higher than the normal value.

Keywords: *Harishankar spring, Dissolved oxygen.*

Introduction

Harishankar is a place of pilgrimage in the Patnagarh subdivision of Bolangir district of western Orissa. Geographically the place is located at latitude of 20°50′N and longitude of 82°12′E vide topo sheet no. 64 L/13. It is around 81 km from Bolangir and 471 km from the state capital Bhubaneswar. It is situated on the southern slope of Gandhamardhan hills, famous for its scenic charms and the presence of Hari and Shankar, the dual deities of the Vaishnav and Shaiva sections of Hinduism. A perennial stream gushes forth onto its granite bed forming cascades of small falls at different stages. There is an existing diversion weir-project namely Harishankar MIP at the temple site intercepting the perennial nala. The project has a catchement area of 1.20 sq. mile and ayacut area of 167.42 hectre in Khariff as well as121.37 hectare in Rabi having a well demarcated distribution system at the left side. Thousands of pilgrims visit the place everyday. As the spring water is used to fulfil drinking, bathing and irrigation purposes of the local people the importance of water quality is a matter of great concern.

Materials and Methods

The water samples were collected from three sampling stations situated at three different places of the water fall in three different season *i.e.*, winter, summer and rainy season. Parameters like temperature, pH, electrical conductivity and TDS were measured using portable water analysis kit. The parameters like turbidity, BOD, DO, COD, TSS, Alkalinity, total hardness, Chloride, Sulphate, Iron were analysed as per standard methods (APHA, 1986; Trivedi and Goyal, 1989). Na and K were determined by flame photometer.

Results and Discussion

The analysis results of different physico-chemical parameters of the spring during the year 2003–2004 are given in Table 40.1.

Table 40.1: Physico-chemical Parameters of the Harishankar Spring in the Month of April–2004, July–2004 and December–2004

Parameters	*April–2004*			*July–2004*			*December–2004*		
	S_1	S_2	S_3	S_1	S_2	S_3	S_1	S_2	S_3
Temperature	28	30	30	21	22	22	13	14	13
pH	7.3	7.4	7.2	7.1	7.2	7.1	7.1	7.2	7.5
Ec	271	282	268	270	291	282	280	279	265
Turbidity	34	38	35	45	48	44	34	39	36
TS	342	351	368	380	399	387	358	360	351
TDS	232	225	241	225	245	231	235	249	220
SS	110	126	127	155	154	156	123	111	131
Cl^-	62	61	63	71	70	69	71	70	68
Alkalinity	132	134	136	134	132	138	131	133	134
TH	128	133	131	129	132	131	128	132	131
Ca-Hardness	99	102	102	97	100	96	92	94	91
Mg-Hardness	29	31	29	32	32	35	36	38	41
DO	5	6	6	8	7	7	6	7	6
BOD	2.1	2.3	2.3	1.5	1.6	1.3	1.2	1.2	1.1
COD	15	13	15	15	14	18	12	11	13
SO_4^{-2}	18	16	19	17.1	15.3	6.8	6	5.2	7.1
Fe^{+3}	0.692	0.785	0.882	0.658	0.799	0.813	0.798	0.889	0.882
Na	24	22	21	20	21	23	23	24	22
K	12	8	9.2	10.1	6.8	7.9	9.3	11.1	8.9

Note: All the parameters except Temperature (°C), pH, Ec (μmhos/cm), Turbidity (NTU) are expressed in ppm.

The temperature of the water in the spring varied from 13 to 30°C. The pH values varied from 7.1 to 7.5. It indicates the water is slightly alkaline in nature but within the permissible limit of ICMR for drinking purposes. The electrical conductivity (EC) ranged from 265 to 291 μmhos/cm. Turbidity of the water samples ranged from 34 to 48 NTU. The higher values are due to presence of suspended matters like clay, silt, colloidal particles, planktons and other organisms. In the present study total solids, total dissolved solids and suspended solids are in the ranges 342–399, 225–249 and 110–156 ppm respectively. Total solids contain different types of nutrients which determines the suitability of the drinking water which is well within the permissible limits. The chloride content ranged from 61 to 71 ppm. This may be due to dissolution of salts and soil erosion. Total hardness due to carbonates and bicarbonates ranged from 132 to 138 ppm in the present study (total alkalinity). Total hardness due to Ca- and Mg- ranged from 128 to 133 ppm, the calcium hardness in the range of 91–102 and magnesium hardness in the range of 29 to 38 ppm which is also within the permissible limit prescribed by ICMR. The dissolved oxygen ranged from 5 to 8 ppm in the present study. Depletion in DO in water during summer is due to high temperature and increased microbial activities. The BOD and COD are found in the range of 1.2 to 2.9 and 12 to 18 ppm. The higher BOD values observed in the summer season is due to higher microbial activities. The sulphate content was found to be in the range of 15.2 to 19 ppm, which is well within the permissible limit.

The iron, sodium and potassium, content of the samples are found to be in the range of 0.658 to 0.889 ppm, 21 to 24 ppm and 6.8 to 12 ppm respectively.

Conclusion

The present investigation suggested that Harishankar spring water quality parameters do not exceed the permissible limits except slight alkalinity, which may be due to the presence of basic salts of sodium and potassium in addition to calcium and magnesium. The higher values of BOD may be due to stagnation of water at various points of tourist interest, which deteriorates the water quality. Further the undesirable higher values of turbidity may be due to heavy soil erosion resulting from large scale deforestation and unhygienic social practices such as dumping bio-wastes such like food products, flowers, leaf-plates and other non-biodegradable wastes, especially near the weir adjoining the temple, which may be minimised by periodical monitoring, cleaning and raising public awareness about the quality of spring water.

References

APHA. 1986. *Standard Methods for the Examination of Water and Wastewater*, 16th edn. American Public Health Association, New York.

ICMR. 1975. *Manual of Standard of Quality for Drinking Water Supplies*. Indian Council of Medical Research, Spl. Rep. S. No. 44.

Manivaskaram. 1984. *Physico-chemical Examination of Water, Sewage and Industrial Effluents*. Pragati Prakashan, Meerut, pp. 42

NEERI. 1986. *Manual on Water and Wastewater Analysis.* National Environmental Engineering Research Institute, Nagpur, pp. 340.

Sharma, M.R. and Sharma, R.N. 2002. Water quality of springs of Bilaspur area of lower Himalayas. *Journal of IWWA*, 34(3): 225–229.

Trivedi, R.K. and Goel, P.K. 1986. *Chemical and Biological Methods for Water Pollution Studies.* Environmental Publication, Karad.

Chapter 41

Toxicological Evaluation of Sugar Mill Solid Waste (Leachate) Using Duckweed (*Lemna minor*) Bioassay

Prabhakar Pratap Singh, Vinod Kumar Chaudhary and B.C. Srivastva
Department of Botany, Shiv Pati P.G. College Shohratgarh, Siddharthnagar – 272 205, U.P., India

ABSTRACT

The present investigation has been carried out in order to assess the effect of sugar factory solid waste on aquatic macrophyte duckweed (*Lemna minor*). The study was based on growth of the plant, the frond count, change in biomass, dry weight and chlorophyll content. The effect of different concentration (10 per cent, 25 per cent, 50 per cent, 75 per cent and 100 per cent) of leachate at 7 days experiments showed that 25 per cent concentration was maximum growth supporting with significant increase in chlorophyll ($p < 0.001$). Growth inhibition was measured as a reduction in frond count and chlorosis. The study suggested that duckweed could serve as promising bio-indicator as it is highly sensitive to pollutant.

Keywords: *Chlorophyll content, Chlorosis and Sugar mill solid waste (leachate).*

Introduction

Solid waste generated from domestic, agricultural and industrial sources, pose a serious threat to the environment because the waste materials remains in the environment for relatively longer periods unless removed, burned washed away or other wise destroyed. They are responsible for causing adverse effect to nature, once spread in environment and may be converted in to other form, which may be toxic than the parent compounds. Their toxicity is linked with the ultimate fate of these substances that are likely or unlikely associated in cycling processes of the nature. Sugar factory is among one of the major industry in India. In the process of sugar production, solid waste is generated, which affect the soil and under groundwater through the process of leaching. Therefore, there is a need for simple, sensitive, economical and rapid screening test for detecting the biological damage to the various environmental agents (Verma *et al.*, 1999). In this context an attempt has been made to study the impact sugar factory leachate on the growth of an aquatic plant, duckweed (*Lemna minor*). The *Lemna minor* is a floating, fast growing, vascular plant, with flat, oval, leaf like fronds. The plant have several characteristics that makes it a useful species for laboratory testing including small size, easy to culture, vegetative reproduction, cost effective and world wide distribution in pond and lakes (Hillman, 1976). Several reports are available on duckweed bioassay (Wang and William, 1988; Telsseeire *et al.*, 1998 and Bucky *et al.*, 1996) for the suitability, utility and assessing the environmental impact of chemicals. Hillman (1976) reported that each frond of the *Lemna minor* absorb nutrients from water body through the general body surface and not through the central root system as in the other higher plant. The major objective of this study was to find out the effect of sugar factory leachate on growth and chlorophyll content using *Lemna minor*, a member of family Lemneaceae, commonly called as duckweed or water lentil.

Materials and Methods

100 individuals of *Lemna minor* (L) from laboratory culture with two fronds in a series of 250 ml conical flask with sugar factory leachate were used in the present study. Sample of solid waste were collected from dumping site of Sugar factory situated at Balrampur, Uttar Pradesh.

Leachate Preparation

Percolation of water through the solid waste dumping site generates leachate, which are highly toxic concentrate semi-liquid mixture of chemicals. For leachate preparation from industrial solid waste, samples were dried and crushed with the help of pestal and mortar, to reduce particle size. Powdered samples were taken and 10 per cent leachate was prepared using distilled water as diluents, (USEPA, 1986). *Lemna minor* plants were exposed to 10 per cent, 25 per cent, 50 per cent, 75 per cent and 100 per cent leachate concentration of sugar factory from 1st to 7th days. All the flasks were capped with cotton plug. Each treatment including control was performed in triplicate and for every flask 100 *Lemna minor* used. Every flask was kept in an incubator and was maintained under standard aseptic physiological conditions at 25±0.5°C temperature, 16 hours light (1600 Ft.C. fluorescent tubes, Philips), and eight hours dark cycle. The duckweed (*L. minor*) plants were collected from freshwater pond, washed repeatedly with distilled water, and were maintained in 3 per cent Hoaglands solution (EPA, 1975), for the development of second-generation for experiments. The numbers of duckweed in each triplicate test in conical flask were recorded from 1st to 7th day. The increase in frond count indicated the duckweed growth and the difference between treated and control group were used to elaborate the toxic effect. Addition of new frond with any edge parent plant was counted with change in biomass, dry weight and chlorophyll content. The chlorophyll content was estimated following the method of Arnon (1994) using UV–VIS Spectrophotometer-118 model at end of 7th day exposure.

Results

The results of analysis of number of fronds, chlorosis, change in biomass and dry weight, chlorophyll content are shown in Tables 41.1–41.3. The maximum numbers of fronds were recorded at 25 per cent concentration of sugar factory leachate and showing minimum chlorosis, while at 100 per cent concentration of sugar factory leachate minimum number of fronds and maximum chlorosis was observed in Table 41.l. Similar trend was also observed in respect to dry weight recorded at 1st to 7th day's time interval in Table 41.2 The sugar factory leachate exposed *Lemna minor* plants have maximum change in biomass at 25 per cent of leachate concentration was 30 per cent in respect to the initial biomass, while at 100 per cent concentration, change in biomass was nearly 17 per cent of the initial biomass. The chlorophyll content (a, b, and total) were also maximum at 25 per cent leachate concentration and is highly significant ($P < 0.001$), while at 100 per cent leachate, concentration a significant decline was ($P < 0.01$) observed as compared to control in Table 41.3

Table 41.1: Changes in Number and Chlorosis of Fronds of *Lemna minor* Plants Exposed with Leachate of Sugar Factory Solid Waste

Sl.No.	*Concentration of Leachate*	*Total No. of Fronds*		*Chlorosis*
		At 1st Day	*At 7th Day*	*At 7th Day*
1.	Control	100	120	Not visible
2.	10 per cent	100	103	Visible on 20 plants
3.	25 per cent	100	115	Visible on 11 plants
4.	50 per cent	100	107	Visible on 35 plants
5.	75 per cent	100	105	Visible on 57 plants
6.	100 per cent	100	101	Visible on 72 plants

Table 41.2: Changes in Biomass and Dry Weight of *Limna minor* Exposed with Sugar Factory Solid Waste (Leachate)

Sl.No.	*Concentration of Leachate*	*Change in Fresh Weight*	*Dry Weight*
1.	Control	35 per cent	0.065±0.001
2.	10 per cent	20 per cent	0.040±0.003
3.	25 per cent	30 per cent	0.058±0.001
4.	50 per cent	24 per cent	0.047±0.005
5.	75 per cent	21 per cent	0.043±0.002
6.	100 per cent	17 per cent	0.036±0.005

Table 41.3: Chlorophyll Content (a, b and total) of *Lemna minor* at 7th Days Exposed Sugar Factory Solid Waste (Leachate)

Sl.No.	*Concentration of Leachate*	*Chlorophyll a*	*Chlorophyll b*	*Chlorophyll Total*
1.	Control	0.865±0.002	0.347±0.001	1.120±0.003
2.	10 per cent	0.721±0.015	0.27±0.003	0.981±0.013
3.	25 per cent	0.841±0.001**	0.311±0.002**	1.11±0.002**
4.	50 per cent	0.745±0.005	0.285±0.005	1.012±0.015
5.	75 per cent	0.638±0.003	0.22±0.015	0.837±0.001
6.	100 per cent	0.447±0.002*	0.147±0.004*	0.538±0.003*

Values are arithmetic mean ± SE of three replicates.

*: $P < 0.01$; ** $P < 0.001$.

In this range finding static experiment after 25 per cent leachate concentration, the result shows adverse impact such as alteration in morphological appearance of *Lemna minor*, shrinking of frond size and gradual change in colour from green to whitish as compared to natural death, which occurs in control. The result indicate that at 25 per cent concentration of sugar factory leachate, the possible presence of nutrients such as phosphate, sulphate, chloride, hardness, and heavy metals like Pb, Cu, Fe, Mn, and Zn and other fertilizing minerals optimal for growth and chlorophyll content while at higher concentrations of leachate accumulation of metal create toxicity and have harmful effect due to their binding with enzyme thereby lowering enzyme activity.

Discussion

The rapid industrialization, urbanization and increase in population during last few decades have created several environmental problems. It is found that hazardous and non-hazardous solid wastes are being disposed off indiscriminately on agricultural or low-lying areas. Toxicity affects due to direct and indirect interaction of potentially hazardous solid wastes generated during industrial activity are serious and pressing for the need of the safe managements. Duckweed has been shown to effectively remove nutrients in water treatment wetland system. According to Korner *et al.* (1998) the ability of *Lemna gibbs* to remove N and P from domestic water waste and found that the duckweed did aid in the removal of these essential elements. *Lemna gibbs* also increased the degradation of organic material. However, Boniardi *et al.* (1999), showed that certain dissolve metals, like copper could decrease the effectiveness of *Lemna gibbs* to remove organic pollutants from wastewater. Despite the development in chronic and acute toxicity test, the range finding test is widely used in eco-toxicological studies by environmental agencies in order to assess the potential impact on ecological system (Calow, 1992). Discoloration of fronds, or chlorosis, is linked to loss of chlorophyll pigments and leads to white, red-brown or yellowish fronds (Wang, 1990).

Acknowledgement

The authors wish to express his heartfelt thanks to Shri Manoj Kumar Dubey for their valuable suggestions in all respect.

References

Arnon, D.I. 1949. Copper enzyme in isolated chloroplasts, polyohenol oxidase in *Beta valgaris. Plant. Physiol.*, 24: 1–15.

Boniardi, N., Rota, R. and Nano, G. 1999. Effect of dissolve metals on the organic load removal efficiency of *Lemna gibba. Water Res.*, 33: 530–538.

Calow, P. 1992. The three criteria of ecotoxicology. *Func. Ecol.*, 6: 617–619.

Carpentier, S., Garnier, J. and Flaugnatli, R. 1987. Toxicity and bioaccumulation of cadmium in experimental culture of duckweed, *Lemna Plyrrhiza* (L). *Bull Environ. Contam. Toxical.*, 38: 1055–1060.

Environmental Protection Agency (EPA) 1975. Test methods for assessing the effects of chemicals on plants, (Eds.) R. Rubinstein *et al.* EPA–5600–17–75–008. Final report. U.S. Environmental Agency, Washington, DC.

Hillman, W.S. 1976. Calibrating duckweed: Light, clocks, metabolism flowering. *Science*, 59: 219–225.

Hillman, W.S. 1976. Morphology and uses. *Science*, 193: 453–458.

Korner, S., Iyatuu, G.B. and Vermar, J.E 1998. The influence of *Lemna gibba* (L.) on the degradation of organic material in duckweed covered systems. *Water Res.*, 32: 3092–3098.

Telsseire, H., Couderchet, M. and Vernet, G. 1998. Toxic responses and catalase activity of *Lenma minor* (L.). exposed to toipet, copper and their combinations. *Ecotox. Environ. Safety*, 40: 194–200.

United States Environmental Protection Agency (USEPA). 1986. Test method for evaluating solid waste: Physical/chemical methods. Report S W–846. US Environmental Protection Agency, Office of Solid Waste and Emergency Response, September, Washington, DC.

Verma, Y., Hargan, M.C., Ruparelia, S.G. and Kulkarni, P.K. 1999. Toxicity testing of Tannery effluents using duckweed (*Lemna minor*) bioassay. *Poll. Res.*, 18(4): 497–500.

Wang, W. and Williams, J. 1988. Screening and biomonitoring of industrial effluent using phytotoxicity test. *Environ. Toxical. Chem.*, 7: 645–653.

Wang, W. 1990. Toxicity assessment of pretreated industrial wastewater using higher plants. *Res. J. WPCF*, 62(7): 853–860.

Chapter 42

Seasonal Fluctuation of Primary Production in Bonal Reservoir, Gulbarga District, Karnataka

H. Anjinappa and K. Vijaykumar

Environmental Biology Research Unit, Department of Zoology, Gulbarga University, Gulbarga – 585 106, Karnataka

ABSTRACT

This article deals with biannual variability of primary productivity depends upon the seasonal cycles from October 1999 to September 2001 has been observed. The grass primary productivity is increased in summer seasons followed by northeast monsoon seasons due to variation of environmental factors like the high temperature, low alkalinity, low pH and excessive growth of aquatic vegetation in the reservoir. The productivity and pH photosynthetic efficiency were higher in summer seasons and winter as in compare to rainy seasons. Statistical analysis of primary production in relation to important environmental factors indicates that the effect of pH, dissolved oxygen, total alkalinity, chloride, total hardness, total dissolved solids, turbidity, calcium and magnesium.

Keywords: *Reservoir, Primary production, Environmental factors, Seasonal cycle.*

Introduction

Primary productivity is the most important biological phenomenon in nature, which involves the trapping of radiant energy of the sun and its transformation into high potential biochemical energy by the process of photosynthesis. The production of organic matter in aquatic environment has held the attention of the hydrobiologists since long. The synthesis of basic food *i.e.*, transformation of inorganic substance in to organic form is similar to that of terrestrial environment. The primary productivity relates to the amount of organic matter synthesized in a certain space per unit time. Some of the studies on the primary productivity of Indian freshwaters include those of Sreenivasan (1963, 1964a, 1964b and 1972) from different tanks of Tamil Nadu.

According to Raymont, productivity in a broad sense is a concept of organic matter, synthesis potential which measures ability of an area to support a biological population and sustain a level of growth and respiration. Nasar and Nasar (1976), Gopal *et al.* (1978) have added to Primary production studies of different types of reservoir from Himachal Pradesh. Khan and Zutshi (1980) gave an account on primary productivity and trophic status of a Kashmir Lake. Singh and Desai (1980) observed primary productivity of Rihand reservoir. Ayyappan and Gupta (1985) have opined that the production in Ramasamudra tank from coastal Karnataka was controlled by several hydrographical factors. Sinha *et al.* (1990) reported on the seasonal study of primary productivity of phytoplankton communities in three different types of freshwater bodies from Bihar state. Vijaykumar and Paul (1991) reported on the primary productivity of different water bodies in Dharwad, Karnataka State. Vijaykumar *et al.* (2000), studied on seasonal variation in the primary productivity of Gobbur tank, Gulbarga. Jindal and Kaur (2000) and Shukla and Pawar (2001) investigated on heavy metal toxicity on the production of freshwater ecosystem at Khuda village, Chandigarh and Govingarh lake, Rewa (Madhya Pradesh) and observed minimum rate of primary production during monsoon season. The objective of the present study is to understand the seasonal fluctuations in the grass and net primary productivity rate with respect to variation of environmental factors in the reservoir.

Materials and Methods

The Study Area

Bonal reservoir is shallow water bodies in the northern part of Gulbarga District, Karnataka state, falls under 77°–45′E longitude and 16°–33′N latitude. Which has natural hillocks that forms a natural wall. The total catchement area of the reservoir is 40.96 sq. miles and maximum depth 12 ft and minimum 4 ft. This is perennial nature and water level start rising in premonsoon period and the level drop in summer seasons. The investigation is carried out for a period of 24 months from October 1999 to September 2001. The sampling was done in first week of every month at seven stations were selected to study the variations in these stations. Primary production involving the chaemo-autotriphic processes, forms the base of energy flow in the ecosystems. An understanding of primary production becomes more essential in the evaluation of the capacity of any ecosystem including that of standing water bodies. Thus, this aspect has drawn the attention of numerous limnologists. Primary production *'in situ'* was estimated by Gaarder and Gran (1927). "Light and Dark Bottle Method". A set of one light and dark bottles along with one more bottle (for collection of sample for determining initial dissolved oxygen) was suspended using floats at each station so as to maintain the bottles 10 cm below the water surface. These bottles were filled with surface waters and closed tightly and then the dark bottle (Painted with black emulsion paint) was covered with a black cotton bag of the size of the bottle. The primary productivity study was always carried out after 1000 hours. The time of exposure (incubation period) in the present studies was for a period of 4 hours. The initial dissolved oxygen was determined using the sample from the third bottle. Photosynthesis quotient of 0.375 was used to convert oxygen values. After the completion of incubation period, dissolved oxygen fixed by manganous sulphate and alkali-iodide at the site of the study and transferred to the laboratory for further estimations using Winkler's method (APHA,1998). The rate of grass and net production and consumption of oxygen organic matter during respiration of plankton were arrived by using the oxygen values. The available evidences from the previous works reveal that the productivity profile varies from one region to other Gopal *et al.* (1978), Khatri (1984), Vijaykumar (1991 and 1994), which is influenced by environmental factors and they also opined that the conditions which are generally mentioned to influence the production are the amount of available light, temperature and nutrient levels.

Results and Discussion

The seasonal fluctuation of gross primary productivity at different seasons during from October 1999 to September 2001 (Table 42.1 and Figure 42.1). The minimum value of gross primary productivity from 0.287 gC/m^3/hr to 0.305 gC/m^3/hr recorded during southwest monsoon seasons and maximum from 0.667 gC/m^3/hr to 0.742 gC/m^3/hr were noticed in summer seasons. In present investigation the Bonal reservoir more or less similar trend of fluctuate at same seasons of both the years.

Table 42.1: Seasonal Variations of Primary Production in Bonal Reservoir from October 1999 to September 2001

Primary Production in gC/m^3/hr	*October 1999 to September 2000*			*October 2000 to September 2001*		
	NEM	*S*	*SWM*	*NEM*	*S*	*SWM*
GPP	0.479	0.742	0.287	0.543	0.667	0.305
NPP	0.792	1.195	0.438	0.793	0.959	0.283
CR	0.313	0.451	0.159	0.241	0.283	0.129

Primary productivity can be assessed as gross and net values, while gross primary productivity is the total rate of photosynthesis including the organic matter used up during respiration in a given time, net primary productivity is the total amount of chemical energy left after it has been utilized by

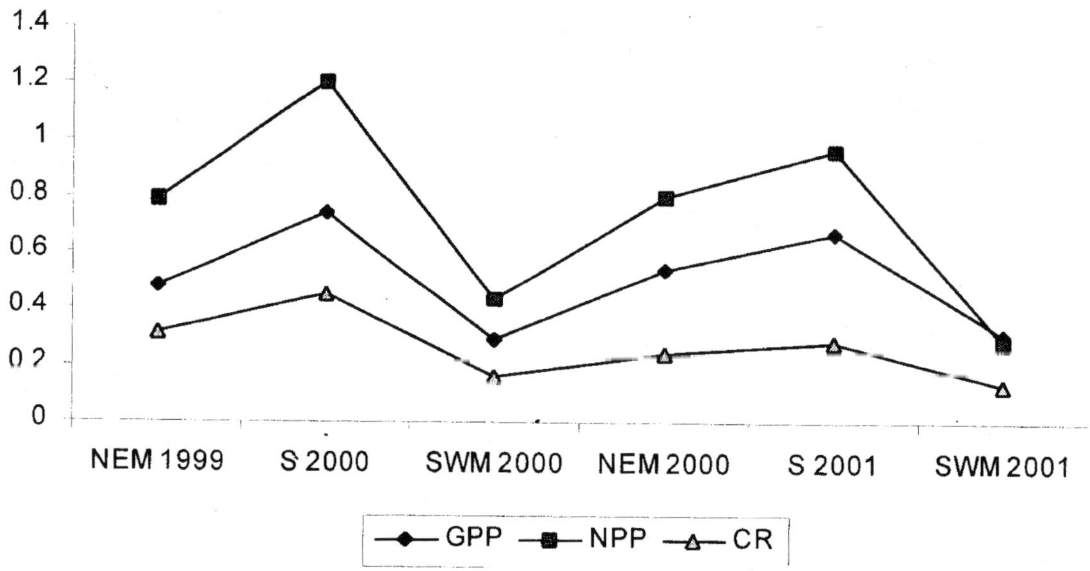

Figure 42.1: Seasonal Variations of Primary Production in Bonal Reservior from October 1999 to September 2001

plants for respiration, primary productivity has been used as potential index of productivity for many diverse ecosystem of the world (Wetzel, 2001). Primary productivity is concerned with the evaluation of the capacity of an ecosystem, to the synthesis of organic matter of high potential chemical.

One of the important aspects of inland waters is the study of regional biological productivity. Voluminous data is available on such studies of temperature waters but there are only a few for Indian reservoirs, lakes and ponds. The available evidences from the previous works reveal that the productivity profile varies from one region to other, Gopal *et al.* (1978), Khatri, (1984), Vijaykumar (1991 and 1994), which is influenced by environmental factors and they also opined that the conditions which are generally mentioned to influence the production are the amount of available light, temperature and nutrient levels.

In the present investigation, the gross primary productivity of the Bonal reservoir followed a more or less similar trend of fluctuations at all the seven stations and also exhibited an increasing trend through the northeast monsoon seasons towards summer season with a distinct peaks and lower values observed during southeast monsoon seasons. Because of influx of the turbid water in to reservoir. However during summer seasons. A distinct rise could be observed which was the highest of the values recorded. Because of influx of the turbid water in to reservoir and high temperature, high transparency, high nutrients, low pH and excessive growth of aquatic vegetation. Gross primary productivity showed positive correlation with pH ($p < 0.01$; $r = 0.87$), dissolved oxygen ($p < 0.01$; $r = 0.97$), total alkalinity ($p < 0.01$; $r = 0.85$), chloride ($p < 0.01$ $r = 0.86$), total hardness ($p < 0.05$; $r = 0.62$), total dissolved solids ($p < 0.01$; $r = 0.75$), biochemical oxygen demand ($p < 0.01$; $r = 0.89$) net primary productivity ($p < 0.01$; $r = 0.98$) and community respiration ($p < 0.01$; $r = 1.0$) respectively.

Net Primary Productivity estimates of the Bonal reservoir followed a rather similar trend of oscillations lower values 0.283 $gC/m^3/hr$ to 0.438 $gC/m^3/hr$ during southwest monsoon seasons due to enters the turbidity of waters and other high environmental factors and an increasing trend 0.959 $gC/m^3/hr$ to 1.195 $gC/m^3/hr$ has been observed during summer seasons. The reservoir followed slight similar trend oscillations at same seasons of the two years. Net primary productivity shows the positive correlationship with pH ($p < 0.01$; $r = 0.82$), dissolved oxygen ($p < 0.01$; $r = 0.96$), total alkalinity ($p < 0.01$; $r = 0.82$), chloride ($p < 0.01$; $r = 0.83$), total hardness ($p < 0.05$; $r = 0.61$), total dissolved solids ($p < 0.01$; $r = 0.73$), turbidity ($p < 0.01$; $r = 0.86$), gross primary productivity ($p < 0.01$; $r = 0.98$) and community respiration ($p < 0.01$; $r = 0.98$).

Community Respiration observations reveal an increase from 0.283 $gC/m^3/hr$ to 0.451 $gC/m^3/hr$ trend towards summer seasons. While decreasing trend observed during southwest monsoon seasons of the two years. In the present study community respiration showed positive correlation with pH ($p < 0.01$; $r = 0.81$), dissolved oxygen ($p < 0.01$; $r = 0.95$), total alkalinity ($p < 0.01$; $r = 0.97$), chloride ($p < 0.01$; $r = 0.90$), total hardness ($p < 0.01$; $r = 0.81$), calcium ($p < 0.05$; $r = 0.69$), magnesium ($p < 0.01$; $r = 0.71$), total dissolved solids ($p < 0.01$; $r = 0.88$), biochemical oxygen demand ($p < 0.05$; $r = 0.65$), turbidity ($p < 0.01$; $r = 0.93$), gross primary productivity ($p < 0.01$; $r = 1.0$) and net primary productivity ($p < 0.01$; $r = 0.99$) respectively.

These reports of quite a significant number and few workers on the Primary production of the standing water bodies from India. Sreenivasan (1963) reported on Primary production in three upland lakes from Nilagiries in Tamil Nadu with one of the lake in Ooty recording high production values, further the author opined that primary productivity correlated well with physico-chemical factors and biological condition. Even though, these observations pertain to waters of comparatively large dimensions, such a correlation could be observed in the present studies. This fact is well substantiated by the extensive contributions of the same author, Sreenivasan (1964b, 1968a, 1968b, 1972 and 1976) from ponds and tanks distributed in Tamil Nadu, where he reported that the varying ranges could be attributed to different factors *viz.*, low oxygen tension, low free carbondioxide and processes such as self shading. Khan and Siddiqui (1971) observed high gross primary productivity between the period from March and May and lower values for the rest of the period from a tropical fish pond at Aligarh and further opined that there is a direct relationship between production, transparency, temperature and phosphate phosphorus amongst nutrients.

However, in the present investigations, comparatively high values were observed during high temperature period. Ayyappan and Gupta (1985) recorded gross primary productivity from a tank in Coastal Karnataka ranging upto 264. 27 mgC/m^3/hr and they are of the view that the higher rates during March to July coincided with low pH, high free carbondioxide, low total alkalinity and high specific conductivity. During present investigation excessive growth of aquatic vegetation were encountered especially towards northeast monsoon and summer seasons which could have caused increased trend in primary productivity during summer seasons of both the years.

References

APHA. 1998. *Standard Methods for the Examination of Water and Wastewater*, 20th Edition, New York.

Ayyappan, S. and Gupta, T.R.C. 1985. Limnology of Ramasamudra tank. Primary production. *Bull. Bot. Soc., Sagar*, 32: 82–88.

Gaarder, T. and Gran, H.H. 1927. Investigations of the production of plankton in the Oslo Fiord. Rapp. P-V. Reun. Common. *Int. Explor. Scient. Mer. Mediterr.*, 42: 1–48.

Gopal, B., Sharma, K.P. and Trivedy, R.K. 1978. Studies on ecology and production in Indian freshwater ecosystem at primary producer level with emphasis on macrophytes. In: *Glimpses of Ecology*, (Eds.) J.S. Singh and Gopal, B. Int. Sci. Pub., Jaipur, p. 349–376.

Jindal, R. and Bharleen Kaur. 2000. Effect of heavy metal toxicity on the productivity of a freshwater ecosystem. *Indian. J. Ecol.*, 27(1): 27–32.

Khan, A.A. and Siddiqui, A.R. 1971. Primary production in a tropical fish pond at Aligarh, India. *Hydrobiologia*, 37: 447–456.

Khan, M.A. and Zutshi, D.P. 1980. Contribution to high attitude Limnology and primary productivity of plankton community of Nilnag lake, Kashmir. *Hydrobiologia*, 75: 103–112.

Khatri, T.C. 1984. Seasonal variation in primary production in relation to some limnological features in Lakhotia lake (India). *Proc. Indian. Acad. Sci. (Anim. Sci.)*., 93(7): 697–702.

Nasar, S.A.K. and Nasar, S.K.T. 1976. An unusual record of oxygen production in a freshwater pond. *Curr. Sci.*, 45: 676–677.

Singh, F.D. 1986. Relationship between primary productivity and environmental parameters of tropical lakes: statistical analysis. *Poll. Res.*, 5(3&4): 103–110.

Singh, H.P. 1998. Studies on primary production in Gobindsagar reservoir, Himachal Pradesh. *J. Environ. Biol.*, 19(2): 167–170.

Singh, R.K. and Desai, V.R. 1980. Limnological observations on Rihand reservoir. Part III. Primary productivity. *J. Inland Fish. Soc., India*, 12(2): 63–68.

Sreenivasan, A. 1968a. The Hydrological features, primary production and fisheries of stanely reservoir, metturdam for the year 1958 to 1965 Chennai. *J. Fish.*, 5: 1–17.

Sreenivasan, A. 1968b. The limnology of fish production in to two ponds in Chingleput (Madras). *Hydrobiologia,* 32(1–2): 131.

Sreenivasan, A. 1972. Limnology of a seepage type impoundment odathurai tank. *J. Inland. Fish Soc., India,* 4; 162–168.

Sreenivasan, A. 1976. Limnological studies and primary production in temple pond. *Hydrobiologia,* 48(2): 117–123.

Vijaykumar, K. 1994. Seasonal variations in the primary productivity of a tropical pond. *J. Ecobiol.,* 6 (3): 207–211.

Vijaykumar, K. and Paul, R. 1991. Primary production of Nuggikeri tank at Dharwad. *J. Nat. Cons.,* 3(1): 33–41.

Vijaykumar, K. 1994. Seasonal variations in the primary productivity of a tropical pond. *J. Eco. Biol.,* 6(3): 207–211.

Vijaykumar, K., Anjinappa, H. Virupanna and Padmavathi, C. 2000. Seasonal variation in the primary productivity of Gobbur Tank, Gulbarga. *Proc. Acad. Environ. Biol.,* 9 (2): 119–123.

Wetzel, G. 2001. *Limnology: Lake and River Ecosystem,* 3rd Edn. Academic Press, UK.

Chapter 43

Computer Models and Water Resources Engineering

Nirmalya Bayal

Department of Mechanical Engineering, Saroj Mohan Institute of Technology, Guptipara, Hooghly, West Bengal

Introduction

Computer model is defined as any simplified representation of a real world system. The discussions here deal with representing water related systems with mathematical formulations that are solved using a computer. Computer modeling of natural and man made water resources systems is a central focus in water resources engineering.

Different types of software tools are now available for building models. There are different types of software are now available:

1. General purpose commercial products.
2. Conventional programming language.
3. Generalized water resources engineering models.

Several alternative types of software may be used for particular applications. Different tools have certain advantages and disadvantages in various situations. Choice of software is also largely dependent on personal preferences.

General Purpose Commercial Software

Water resources engineering continue to discover new used of old software. Several general-purpose software packages are now used and along with examples of popular products are listed in Table 43.1

Table 43.1: Categories of Computer Software

General Purpose Commercial Software	
Spreadsheet-based packages	Excel
	Lotus 1-2-3
Mathematical modeling environment	MATLAB, MathCAD, Mathematica
Geographical Information System	ArcInfo, ArcCiew, ArcGIS
Computer aided drafting and design	AutoCAD
High Level Programming Language	
FORTRAN, C, C++, BASIC	

Spreadsheet programme are used routinely by millions of businesses professionals, universities and industries through out the world. Water resources engineering have recognized the usefulness of electronic spreadsheets since the early 1980's when they were first marketed. Spreadsheet packages also are commonly used as preprocessors for preparing an organizing input data for complex models and as postprocessors for summarizing, plotting and analyzing simulation results.

Water resources engineering models are based on sets of algebraic or differential equations representing governing principals such as conservation of mass, momentum and energy. For fairly simple applications, model developed may consist of formulating the appropriate equations and then solving those using mathematical modeling environments. These software products provide capabilities for solving algebraic and differential equations; performing differentiation and integration, matrix operations, displaying results graphs.

A geographic information systems (GIS) is a set of computer-based tools for storing, processing, combining, manipulating, analyzing and displaying data that are referenced to the earth. Many types of water resources engineering models are developed within GIS environment or interconnected with GIS.

Computer aided drafting and design (CADD) software is used for a variety of graphics applications in various fields, including water resources engineering. CADD programs provide drawing and specialized graphics capabilities used for mapping and general technical purposes.

Generalized Water Resources Engineering Models

Generalized means the computer model is designed for applications to a range of problems dealing with systems of various configurations and locations rather than being developed to address a particular problem. The widely used software packages developed are listed in Table 43.2.

Table 43.2: Generalized Water Resources Engineering Models

Generalized Model	*Model Developer*
KYPIPE Pipe Network Analysis	University of Knetucky
MODFLOW Moduler Groundwater Models	U.S. Geological Survey
WRAP Water Rights Analysis Packages	Texas A&M University
MODSIM River Basin Network Models	Colorado State University

Generalized models are used to manage data, perform computations to simulate real word processes and display results for various analysis applications. Design processes are typically based on iterative execution of simulation models. Computer models also play an important role in documenting and transferring knowledge. Much like textbooks and other reference book and manuals, generalized modeling packages serve to organize record and pass on state-of-the-art knowledge.

Conclusion

Water resources engineering models are often components of decision support systems. The models are even more often applied in other planning, design and resource management situations. The relationship between analyses, design and other decision making processes and modeling systems vary, depending on the particular water resources engineering application.

Chapter 44

Physico-chemical Studies on Pollution in River Sengar at District Etawah (U.P.)

K.K. Saxena*, Raj Narayan and Yogesh Babu Dixit

P.G. Department of Zoology, Janta Mahavidyalaya, Ajitmal, Auraiya (U.P.)

ABSTRACT

The river Sengar at District (Etawah) is an important source of water supply in the last few years. This study of physico-chemical characteristics of water of the river Sengar revealed that the water temperature always followed the atmospheric temperature with a minimum of 18.5°C during winter and the maximum being 33.0°C in summer. The transparency fluctuated between 19.0 to 50.9 cm. The colour of the water was green in summer and winter but muddy in rainy season. The minimum value of pH was 7.0 during rainy season and elevated upto 8.7 during winter season. The minimum DO recorded was 4.0 mg/l while the maximum value of 8.5 mg/l was recorded during winter season. The CO_2 was recorded between 8.3.to 14.0 mg/l. The BOD were recorded minimum value was 4.3 mg/l in winter season and the maximum value was recorded during summer while the COD were found between 26.0 to 44.1 mg/l.

Keywords: *Physico-chemical, Pollution, River Sengar, Etawah (U.P.).*

Introduction

The water quality in terms of physico-chemical characteristics was studies to assess the river water suitability for drinking purpose. Most of the Indian rivers and their tributaries are being used as sites for disposal of domestic sewage and industrial waste leading to the grave problem of water pollution.

* Postal Address: Param Hans Kuti (Bari), Street No. 8, Ashok Nagar, Etawah – 206001, (U.P.), India.

The river sengar originates from and flows through agricultural lands known as gantic plains and confluences with the river Yamuna. The river has its stretch of about 300 km. As per geographical preview, Shikohabad and several villages discharge their domestic and industrial waste and thus make this river highly polluted. The river reduces to a slow spill channel during summers. Moreover, pollution load is increasing at a fast rate due rapid industrialization and urbanisation.

This investigations on the river Sengar were conducted fortnightly at five sampling stations (A to E) it showed the effect of pollution on the physico-chemical aspects of the river water at district Etawah (U.P.) in summer, rainy and winter seasons (2003). A few physico-chemical parameters were studied in the field and rest were done in the laboratory.

Materials and Methods

The river Sengar passes through varied topographical and meteorological conditions, hence it provides heterogeneous ecosystem, there are definite interrelationships between plankton and the physico-chemical factors of the river Sengar and an analysis of data reveals the fluctuations there of and their bearing on the plankton population.

Five sampling stations A to E were selected on the basis of varied ecological condition of the exploration zone due to domestic and industrial wastes. Station 'A' was setup near the entry point in district Etawah at Dhanua village. Station 'B' was setup near the confluence of its tributary 'Sirsa' river at Chakwa which is about 30 km. away from the first sampling station, while station 'C' near the village Morhi which is about 20 km. ahead from sampling station 'B' and the station 'D' is located near village Jua. The distance covered upto this station is about 15 km. from station 'C'. The sampling site 'E' was setup near Dibiyapur. It is about 15 lm. away from the sampling station 'D'.

The water samples were collected fortnightly by immersing a wide mouthed bottle at the sub surface level. Samples were collected during morning hours in between 8.30 to 10.30 A.M. Brown sample bottles having 250 ml. capacity were used for collecting samples for dissolved oxygen. The sample of the water for physico-chemical characteristics were analysed following Standard Methods (APHA *et al.*, 1995) using HACH model DR-EL/4 Kit with a digital pH meter.

Results and Discussion

The growth reproduction and the development of biota is influenced by physico-chemical factors, *i.e.*, temperature. Transparency, pH, dissolved oxygen, free carbon dioxide.

During the present investigation period, the water temperature of the river stretch explored ranged between 18.5°C to 33.0°C. The minimum was recorded during winter and the maximum during summer season. The minimum and maximum temperature of the river shows a clear correlation with atmospheric temperature. Chacko and Ganapati (1949) reported about the water temperature which followed the atmospheric temperature being hottest during May and June and lowest during December and January. Transparency is an important limiting factor in the population and distribution of freshwater plankton. Transparency in the river Sengar was low in rainy season due to heavy rains while it increased in winter and summer at station 'B' and 'D' respectively. Joshi *et al.* (1973b) found that the transparency of the water of the river Bhagirathi at Garwal Himalaya did not show wider annual variations. Colour of the Sengar river water was muddy during rainy season while green colour was in winter and summer seasons. Green colour indicated fish transparency and correlated with increased plankton production.

The pH plays an important role on the biotic life of a river system. As has been suggested by Verma *et al.* (1977); Rai and Munshi (1979) while Chakraborty *et al.* (1959) reported that pH value of the river Yamuna at Allahabad during 1994–95. The value of Hydrogen ion concentration was low (7.0) in monsoon season and highest value was (8.7) in winter season.

Table 44.1: Physico-chemical Characteristics of Water of River Sengar at Various Sampling Stations in Different Seasons (2003)

Parameters	*Seasons*	*Sampling stations*				
		A	*B*	*C*	*D*	*E*
Temperature (°C)	S	29.8	31.0	30.0	29.5	33.0
	R	22.0	23.0	22.8	22.4	24.0
	W	18.5	19.1	18.8	19.0	20.0
Transparency (cm)	S	37.0	35.5	36.0	38.0	35.5
	R	21.0	20.5	20.5	19.0	17.5
	W	50.0	50.5	49.0	49.5	47.5
Colour	S	Green	Green	Green	Green	Green
	R	Muddy	Muddy	Muddy	Muddy	Muddy
	W	Green	Green	Green	Green	Green
pH	S	7.6	7.3	7.5	7.2	7.6
	R	7.0	7.5	7.8	7.2	7.1
	W	8.0	8.7	8.3	8.1	8.2
Dissolved oxygen (mg/l)	S	4.9	5.0	4.8	4.2	4.0.
	R	7.6	5.0	5.3	6.1	5.3
	W	8.5	6.6	6.0	6.5	7.4
Carbon dioxide (mg/l)	S	14.0	9.8	13.6	10.9	14.1
	R	9.8	13.2	11.9	10.3	12.6
	W	8.3	13.0	10.8	10.5	10.2
BOD (mg/l)	S	10.6	10.2	10.8	16.5	14.5
	R	7.7	9.9	7.6	8.5	8.2
	W	5.5	5.2	4.4	4.3	4.3
COD (mg/l)	S	44.1	41.5	40.8	42.1	39.8
	R	26.0	38,5	35.2	37.5	38.8
	W	32.0	30.0	33.2	33.0	31.0

S: Summer; R: Rainy; W: Winter.

Dissolved oxygen plays a vital role in supporting aquatic life in the running water, which is susceptible to the slight environmental changes in D.O. concentration in the river water may be due to polluted affluent, which lowers the dissolved oxygen. In the present investigation the D.O. concentration varied from minimum of (4.0 mg/l) to a maximum of (8.5. mg/l). The high value of D.O. was recorded during the winter season where temperature was low, and low value of D.O. was observed during the summer season when the temperature was high and river flow was also low. This finding about D.O. are in conformity with George *et al.* (1966) and Saxena *et al.* (1966). Carbon dioxide is one of the essential constituents of an aquatic ecosystem and the abundance of it exerts certain specific effects upon aquatic bioata. The river Sengar exhibited maximum carbon dioxide as in summer but low concentration were recorded during winter season. In the present investigation carbon dioxide was found to be inversely proportional to the dissolved oxygen. The same correlation between carbon dioxide and dissolved oxygen was reported earlier by Ray *et al.* (1966) and Prakash *et al.* (1978).

The BOD and COD are the significant parameters of pollution (Odum, 1971). The higher load of them produces obnoxious conditions (Verma *et al.*, 1977) and in the open drains they are directly concerning the productivity (Bewtra and Radhacharan, 1968). The value of BOD was minimum in winter and maximum in summer at all the five stations.

The chemical oxygen demand in the river Sengar was higher than that of BOD. The maximum values were recorded in summer and minimum were in rainy season. In river Yamuna COD was also found higher than BOD by Joshi and Bisht (1993) from which it may be concluded that same degree of biodegradable oxygen demanding pollutants are present in the water.

Acknowledgement

Authors are greatly indebted to Dr. R.R.S. Chauhan, Reader, Department of Zoology, Janta Mahavidyalaya, Ajitmal (Auraiya) for his constant guidance, suggestions and necessary help to this work. Our sincere appreciation goes to Arvind Kumar Dubey, Head of the Department for providing necessary laboratory facilities.

References

APHA, AWWA, WPCF. 1995. *Standard Methods for the Examination of Water and Wastewater*, 19th edn. American Public Health Society Pub. Washington, D.C. pp. 874.

Bewtra, J.K. and Radhacharan. 1968. Further observations on progression of biochemical demand. *Environ. Hlth.*, 10: 51–60.

Chacko, P.I. and Ganapati, S.V. 1949. Some observations on the Adyar river with special reference to its hydrobiological conditions. *Indian Geogr. J.*, 24(3): 25–49.

Chakrabarty, R.D., Roy, P. and Singh, S.B. 1959. A quantitative study of the physico-chemical conditions of the river Yamuna at Allahabad in 1954–55. *Indian J. Fish*, 6(1): 186–203.

George, M.G., Qasim, S.Z. and Siddiqui, A.Q. 1966. A limnological survey of the river Kali with special reference on fish mortality. *Environ. Hlth.*, 8(4): 262–269.

Joshi, B.D., Pathak, J.K., Singh Y.N., Bisht, R.C.S. and Joshi, N. 1993b. Phytoplankton production in the snow bed river Bhagirathi in the Garhwal Himalaya. *Him. J. Env., Zoo.*, 7: 60–63.

Joshi, B.D. and Bisht, R.C.S. 1963. Some aspects of physico-chemical characteristic of western Ganga canal near Jawalpur at Haradwar. *Him. J. Env. Zool.*, 7: 76–82

Odum, E.P. 1971. *Fundamentals of Ecology*, 3rd edn. W.B. Saunders Co., Philadelphia. pp. 574.

Prakash, C., Rawat, D.C. and Grover, P.P. 1978. Ecological study of river Yamuna (Poiya Ghat to Jawahar Bridge) IAWPC. *TECH. Ann.*, 5: 32–45.

Rai, D.N. and Munshi, J.S.D. 1979. Observation on diurnal changes of some physico-chemical factors of three tropical swamps of Darbhanga (North Bihar) India. *Comp. Physiol. Ecol.*, 4(2): 52–55.

Ray, P., Singh, S.B. and Sehgal, K.L. 1966. A study of some aspects of ecology of the river Ganga and

Yamuna at Allahabad (U.P.) in 1958–59. *Proc. Nat. Acad. Soc. India,* (B), 36(3): 235–272.

Saxena, K.L., Chakrabarty, R.N., Khan, A.Q., Chattopadhya, S.N. and Harish Chandra. 1966. Pollution studies of the river Ganga, near Kanpur. *Environmental Health,* 8: 270–285.

Verma, S.R., Tayagi, A.K. and Dalela, R.C. 1977. Studies on the characteristic and disposal problem of industrial effluents with reference to ISI Standards (Part II). *Ind. J. Environ. Hlth.,* 19 (3): 165–175.

Chapter 45

Distribution of Nutrients at Different Seasons in Tharangambadi–Vanjur Coasts, South East Coast of India

P. Martin Deva Prasath and T. Hidayathullakhan***

**Post Graduate Department of Chemistry, TBML College, Porayar – 609 307, Nagapattinam District, Tamil Nadu, South India*

***Post Graduate Department of Chemistry, Khadir Mohideen College, Adiramapattinam, Tamil Nadu, South India*

ABSTRACT

Investigations on the physico-chemical factors at two different stations namely, Tharangambadi (Station 1) and Vanjur (Station 2), lying along the southeast coast of India showed spatial and temporal differences for all the parameters observed. Of the two stations studied, Station 1 recorded higher values of surface water temperature, pH and salinity than Station 2. Station 2 recorded higher concentrations in nutrients of NO_2, SiO_3 whereas Station 1 recorded higher concentrations in nutrients of PO_4, NO_3.

Keywords: *Nutrients, Tharangambadi, Vanjur, India, Marine water quality.*

Introduction

Information on the coastal hydrography of Nagapattinam coast is essential to know the pollution status and consequent productivity of the ecosystem. Little information is available on the physico-chemical aspects of the Nagapattinam coast in southeast coast of India. There is no detail and long-term (continuous monitoring) study in this area. The present communication reports on the seasonal distribution of some hydrographical parameters and micronutrients in selected two stations like Tranquebar and Vanjur coast in Nagapattinam district, southeast coast of India. The studies were carried out for the period of two years from January 2000 to December 2001. Since Nagapattinam is one of the main fishing centers of (South) Tamil Nadu, it is essential to know the hydrographical characters of the coast. For this would provide valuable information on the nature of the coastal waters.

Materials and Methods

Two stations were selected for the present study. Station 1 is Tranquebar coast (Lat. 11°02′N; Long. 79°49′ E) and station 2 is Vanjur coast (Lat. 10°2′ N; Long. 79°59′ E) in Nagapattinam district of Tamil Nadu state. The distance between the stations is 20 km. Both the stations are occupied with sandy shore beach and receive freshwater from the river Uppanar and Bravadayan river. Station 1 is clean control site where there is no industrial pollution. Station 2 is crowded with 36 small chemical industries, which came up in a short span of 10 years.

Surface water samples were collected at monthly intervals from the study areas for the estimation of various physico-chemical parameters for the period of two years from January 2000 to December 2001. Temperature (air and surface water) was measured using a standard centigrade thermometer. Salinity was estimated with the help of a Salinometer model E-2 and pH was measured by using a Elico pH meter. The nutrients like dissolved inorganic phosphate, reactive silicate, nitrate and nitrite was measured by adopting the method described by Strickland and Parsons (1972). Rainfall data were collected from statistical department of Nagapattinam.

For the sake of convenience and easy interpretation a calendar year (of study) was divided into four seasons *viz.*, Postmonsoon (January–March), Summer (April–June), Premonsoon (July–September), Monsoon (October–December) seasons, which centers round the north east monsoon. During this monsoon only, Tamil Nadu part of India (where the present study areas are situated) gets the bulk rainfall.

Results

Monthly variation in the physico-chemcial parameters like rainfall, air and surface water temperature, salinity, pH and nutrients in station 1 and station 2 are shown in Figures 45.1 and 45.2.

Rainfall

Annual total rainfall during first and second year was 1043.3 mm and 1397 mm. There was no rainfall during February and March 2002 of the second year. Maximum rainfall was recorded in November 2000 (234.8 mm) and December 2001 (389.6 mm) for the first and second year respectively. The monthly mean rainfall for the first and second years was 86.94 mm and 116.4 mm respectively.

Air Temperature

Air temperature varied from 28.2°C–37.1°C and 29.3°C–37.9°C during first and second year respectively at Station 1. At Station 2, it varied from 28.2°C–37.9°C and 29.2°C–37.8°C during first and second year respectively. The minimum temperature was recorded in December in both the years and stations.

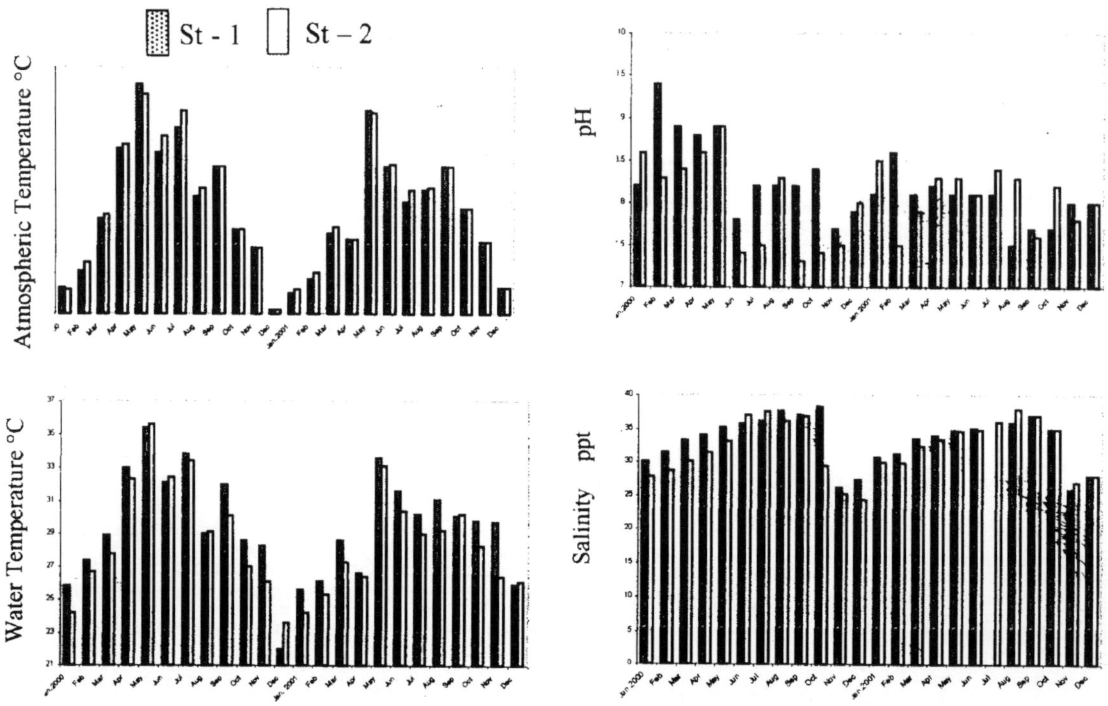

Figure 45.1: Monthly Variations in Temperature (Air and Surface Water), pH and Salinity

Surface Water Temperature

Surface water temperature varied from 22°C–35.4°C and 25.6–33.6° C during first and second year respectively at Station 1. At Station 2, it varied from 23.6°C–35.6°C and 24.2°C–33.1°C during first and second year respectively.

Table 45.1: Simple Correlation Co-Efficient (R) for Physico-chemical Parameters at Tharangambadi (Station 1) and Vanjur (Station 2)

Parameters	*'r' Value (Station 1)*	*r' Value (Station 2)*
Surface water temperature × Atmospheric temperature	0.9575***	0.9836***
Surface water temperature × pH	0.0757	– 0.3385
Atmospheric temperature × pH	0.0346	0.3314
Salinity × Atmospheric temperature	0.6868***	0.4480**
Salinity × Surface water temperature	0.5910***	0.4651**
Salinity × pH	0.0436	– 0.1095
Salinity × Rainfall	– 0.7557	– 0.8562

***: 0.001 level; **: 0.01 level; *: 0.05 level.

pH

pH ranges from 7.7–9.4 and 7.7–8.6 at Station 1 and at Station 2, it varied from 7.3–8.9 and 7.5–8.5 during first and second year respectively.

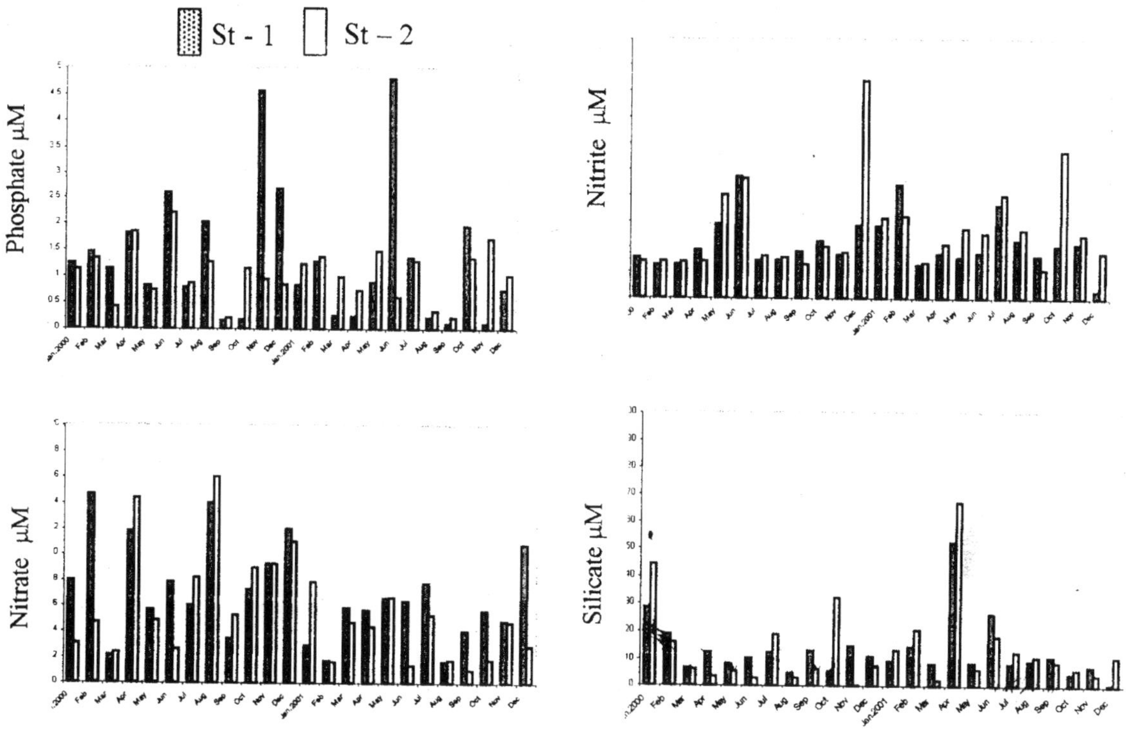

Figure 45.2: Monthly Variations in Nutrients Concentration

Salinity

Salinity varied from 26.2–38.2 per cent and 26–36 per cent at Station 1 during first and second year respectively. At Station 2, it varied from 24.4–36.9 per cent and 27–36.1 per cent during first and second year of the study.

Phosphate

Concentration of inorganic phosphate varied from 0.16–4.58 μM and 0.11–4.82 μM at Station 1 during first and second year respectively. At Station 2, it varied from 0.21–2.22 μM and 0.23–1.37 μM during first and second year respectively.

Nitrate

Nitrate concentration varied from 2.19–14.72 μM and 1.69–10.83 μM at station 1, during first and second year respectively. At Station 2, it varied from 2.67–10.04 μM and 1.01–6.67 μM during first and second year respectively.

Nitrite

Concentration of nitrite varied from 0.64–2.37 μM and 0.14–2.21 μM at Station 1 and 0.65–4.21μM and 0.69–2.86 μM at Station 2 during first and second year respectively at both the stations.

Silicate

Concentration of silicate varied from 4.54–28.52 μM and 0.64–52.57 μM at Station 1 and 0.13–44.24 μM and 4.01–20.69 μM at Station 2 during first and second year respectively at both the stations.

Discussion

Monsoon season recorded more rainfall *i.e.*, 39.75 per cent during first year and 72.3 per cent during second year study period, owing to influence of the northeast monsoon which can bring about profound changes in the hydrography due to floods and runway water (Qasim, 1990). Temperature is one of the important factor in the coastal environs which influx the physico-chemical features and the distribution and abundance of flora and fauna. Prevalence of high temperature during the summer and premonsoon season and lower temperature during the monsoon season could ascribed to the fluctuation in air temperature caused by the seasonal changes in the incoming solar radiation (Bhattathiri, 1987). Surface water temperature was higher during the summer season while it was lower during the monsoon season at both the stations due to intensity of solar radiation; evaporation and freshwater influence the surface water temperature and caused seasonal variation. Further, seawater and air temperature are directly proportional to one another as evidenced by a significant positive correlation obtained from this two (r = 0.9575 at Station 1, r = 0.9836 at Station 1), (Kannan and Kannan, 1996 and Rajaram, 2001).

Higher salinity recorded during summer and premonsoon season was due to low rainfall, decrease of freshwater inflow, land drainage and rise in temperature, non-homogenous mixing and incursion of low-density water (Nair *et al.*, 1983 and Mathew and Pillai, 1990). Lower salinity during monsoon season at both the station was due to high dilution of coastal water by massive freshwater input through river as evident by obtaining a negative correlation between salinity and rainfall (r = – 0.7557 at Station 1, r = – 0.8562 at Station 2).

Higher pH at both the station in post monsoon season might have resulted as a cause of redox changes in the sediments and water column apart from influx of the freshwater (Kumaraguru, 1980) and uptake of CO_2 by photosynthesizing organisms (Ananthan, 1994).The lower pH observed during the monsoon season was due to the influence of freshwater influx resulting in the dilution of seawater, low temperature and organic matter decomposition (Zingde, *et al.*, 1987; Rajasegar, 1999 and Rajaram, 2001).

Higher concentration of nutrients like phosphate, nitrate, nitrite and silicate was observed during premonsoon and monsoon season at both the stations due to the land runoff, following rains and flushing of domestic effluents and influx of river water, land runoff from irrigation channels and release of minerals from sediments (Chandran and Ramamoorthy, 1984). Higher phosphate concentrations are found in both the coasts because of the strong wind stirring action, which blows from the phosphate rich sediments. Seasonal variation and higher concentration of nitrate are due to phytoplankton excretion, oxidation of ammonia, reduction of nitrite (Kannan and Kannan, 1996) and air-sea interaction for exchange of chemical elements (Mathew and Pillai, 1990). Higher level of silicate during monsoon season at both the coasts is due to heavy influx of freshwater derived from land drainage carrying silicate leached out from rock, shrimp farm discharges in Station 1 and industrial effluent discharges in Station 2 (Ramaraju *et al.*, 1987; Zingde *et al.*, 1987 and Sarma *et al.*, 1988).

The lower level of nutrients noted during postmonsoon and summer season at both the coasts are due to decreased land runoff, higher salinity, high pH, less freshwater inputs, utilization of phytoplankton (high density), bacterial decomposition and nitrifying bacterial activity (Mathew and Pillai, 1990).

References

Ananthan, G. 1994. Plankton ecology and heavy metal studies in the marine environs of Pondicherry, India. *Ph.D. Thesis*, A.U., India, pp. 125.

Bhattathiri, P.M.A. 1987. Laccadive Sea: Its environmental characteristics. In: *Contribution in Marine Sciences,* Dr. S.Z. Qasim, Sixtieth birthday felicitation volume, p. 105–115.

Chandran, R. and Ramamoorthy, K. 1984. Hydrobiological studies in the gradient zone of the Vellar estuary: II. Nutrients. *Mahasagar-Bull. Natl. Inst. Oceanoger.,* 17(3): 133–140.

Kannan, R. and L. Kanna. 1996. Physico-chemical characteristics of seaweed beds of Palk Bay, Southeast coast of India. *Indian J. Mar. Sci.,* 25: 358–362.

Kumaraguru, A.K. 1980. Studies on the chemical and biological transport of heavy metal pollutants copper, zinc and mercury in Vellar estuary and the toxicity of three pollutants to some estuarine fish and shellfish. *Ph.D. Thesis,* Annamalai University, Parangipettai.

Mathew, L. and Pillai, V.N. 1990. Chemical characteristics of the waters around Andamans during late winter. *Proc. First Workshop Scient. Result.* FORV Sagar Sampada, 5–8 June, 1989, p. 15–18.

Nair, V.R., S. N. Gajbhiye and Sayed, F.Y. 1983. Organic carbon content of zooplankton from the near shore water of Bombay, India. *J. Mar. Sci.,* 12: 183–185.

Qasim, S.Z. 1990. Adaptation in phytoplankton to changing condition in tropical estuaries. *Mahasagar-Bull. Natl. Inst. Oceangogr.,* 13(2): 117–145.

Rajaram, R. 2001. Studies on hydrobiology and mercury content in Uppanar estuary, Cuddalore, Southeast coast of India. *Ph.D. Thesis,* A.U, India, pp. 128.

Rajasegar. M. 1998. Environmental inventory of Vellar estuary (Southeast coast of India) in relation to shrimp farming. *Ph.D. Thesis,* A.U., India, pp. 95.

Ramaraju, V.S., Sarma V.V., Narasimha Rao, T.V and Vijayakumar, R. 1987. Variation of physico-chemical characteristics with tide in Visakhapatnam harbour water, east coast of India. *Indian J. Mar. Sci.,* 16: 218–222.

Sarma, V.V., Narasimha Rao T.V., Ramaraju V.S., Vijayakumar R. and Suguna, C. 1988. Interrelationship and distribution of hydro-chemical parameters in coastal waters off Vishkhapatnam, east coast of India. *Mahasagar-Bull. Inst. Oceanogr.,* 21(4): 197–207.

Strickland, J.D.H. and T.R. Parson, 1972. A practical handbook of seawater analysis. *Bull. Fish. Res. Bd., Canada,* 167: 310.

Zingde, M.D., S.A.H. Abidi, P. Sarma and M.A. Rokade, 1987. Base water quality of Tha. In: *Contribution in Marine Sciences.* Dr. S.Z. Qasim, Sixtieth birthday felicitation volume, p. 307–318.

Chapter 46

Diversity of Phyto and Zooplankton with Reference to Pollution Status of Kalavam Bazaar Lake, Arcot, Vellore District

V. Indra, V. Prabakaran and R. Balachandar*

**Department of Zoology, Thiruvalluvar University, Fort Campus, Vellore – 632 004*

ABSTRACT

To study the impact of waste disposal on plankton community structure in different seasons of the Kalavam Bazaar lake. The qualitative and quantitative analysis of Phyto- and Zooplankton was carried out. A total of 15 Zooplankton and 23 Phytoplankton taxa were identified from 4 sampling sites. The occurrence of diatom flora at all sites were recorded. More number of Phytoplankton species in three sites indicted the organic enrichment of this habitat. This also reveals that algae are sources for monitoring the lake water quality. The plankton communities in the discharge point indicate the nature of pollution in the lake is discussed.

Keywords: *Seasonal variation, Pollution status, Zooplankton, Phytoplankton and Physico-chemical parameters.*

Introduction

The Kalavam Bazaar lake in Arcot, Vellore District receives effluent from various sources of multiple nature. The addition of these materials not only influences the microflora of freshwater but also favours the development of variety of new biotic species rendering it unfit for human consumption, like fishes, crab etc, consumed by common people in and around Arcot town. The water of the Kalavam Bazaar lake is also used for irrigation, washing and drinking purpose.

Phytoplankton is the fundamental component of aquatic ecosystem as they are the major source of biologically important organic compounds, located at the base of food chain. The magnitude and dynamic of Phytoplankton population has become an essential factor to assess the general health of an aquatic ecosystem (Gaikwad *et al.*, 2004).

Phytoplanktons of freshwater ecosystem have been studied extensively in India (Bilgram's,1989; Nandan and Patel, 1992 and Thorat *et al.*, 2000). The importance of lake water has been increased now due to the provision of aquatic animals, therefore, the present study was carried out to assess the physico-chemical parameters and various forms of bacteria, Phytoplankton and Zooplankton of Kalavam Bazaar lake, based on various seasons. Such studies have also been conducted by Annon (1998) and Biswar and Konar (2000). There is no information available on the ecology of Zoo and Phytoplankton in Kalavam Bazaar lake. Hence, this study has been focused to study the Phytoplankton, Zooplankton, various forms of bacteria and characteristics of the Kalavam Bazaar lake water and its extent of pollution.

Materials and Methods

A total of 8 water samples were collected from October 2004 to March 2005 from 4 sites which comprised four each for rainy (October–December) and postmonsoon period (January–March), respectively. Standard methods (APHA, 2000) were used during the collection, preservation and estimation of physico-chemical, bacteriological analysis and plankton sample analysis. The planktons were collected from all the sites, were fixed in small vials with 5 per cent formalin and few drops of glycerin was added. Samples were centrifuged and concentrated to 10ml in the laboratory. Qualitative and quantitative estimation of Phytoplankton was done with the atlas of Desikachary (1959), Baltish (1992) and Agarkar *et al.* (1994) and Zooplankton identification was done using standard keys of Edmondson (1959), Ruttner-Kolisko (1974), Ranga Reddy (1984), Sharma (1988) and APHA (2000).

Results and Discussion

The physico-chemical parameters analysed in the present study are illustrated in Table 46.1. Zooplankton was identified at different sites such as Cladocera, Copepoda, Rotifera and Ostracoda were indicated in Table 46.2. Distribution and abundance of Phytoplankton were showed in Table 46.3 and the results of bacterial analysis are expressed in Table 46.4.

The pH data of the water sample showed slightly alkaline nature during rainy season in all the stations. Whereas, during postmonsoon, season all the values of the samples showed above 8. It was ranged from 7.4 to 8.5. The total dissolved solids showed more than permissible limits of (WHO 1994) *i.e.*, ranged from 1020 to 1625 mg/l during postmonsoon and 1230 mg/l to 1360 mg/l during rainy season. The BOD and COD values are quite high and indicate that inorganic and organic pollution in the lake is high. The phosphate content was more than 0.2 mg/l, may be considered as productive nature of water. Natural waters mostly contain < 0.1mg/l of phosphorous, which is permissible limit. However, excess of Soluble Reactive Phosphate (SRP) fraction of phosphate promotes eutrophication (Kodarker, 1995). The whole lake is influenced by major point source of pollutant that may be due to freshwater runoff during rainy season. More or less all the physico-chemical parameters were showed slightly higher values during rainy season than in the postmonsoon season. Though the rainy season dilute the lake, the freshwater runoff may increase the concentration of organic and inorganic pollutant load. The less concentration during postmonsoon season may be due to high temperature (evaporation). The water of this lake was greenish and odour is fishy due to proliferation of Phytoplankton population *i.e.*, indication of eutrophic habitat and occasional release of water from mouth of the mixing and discharge zones.

Table 46.1: Seasonal Variations of Physico-chemical Parameters of Kalavam Bazaar Lake, Arcot, Vellore District

Parameters (mg/l)	*Rainy*				*Postmonsoon*			
	S1	*S2*	*S3*	*S4*	*S1*	*S2*	*S3*	*S4*
pH	7.8	7.4	7.5	7.4	8.61	8.54	8.41	8.5
Temp. °C	27.0	28.0	28.0	27.5	38.0	37.0	38.0	38.0
Colour	Sg	Sg	Sg	Sg	G	G	G	G
Odour	Unobjection-able	Objection-able	Fishy	Objection-able	Objection-able	Objection-able	Fishy	Objection-able
BOD	15.0	17.0	16.0	45.0	94.0	54.0	88.0	90.0
COD	110.0	2200	236.0	180.0	225.0	240.0	225.0	235.0
Total hardness	250	280	260.0	276.0	210	260.0	250.0	240.0
Total solid	1275	1280	1360.0	1230.0	1438	1625	1340	1020
DO	7.60	7.83	6.45	8.32	8.0	8.45	8.46	8.00
Organic nitrogen	0.026	0.029	0.024	0.035	0.046	0.056	0.043	0.056
Nitrate	32.0	34.6	33.45	35.47	40.37	43.86	47.44	46.7
Phosphate	2.31	2.62	2.42	3.10	1.86	1.48	1.40	1.63

Sg: Slightly green; G: Green.

Table 46.2: Seasonal Variation of Zooplankton Diversity in Kalavam Bazaar Lake, Arcot

Zooplankton	*Rainy Season*				*Postmonsoon*			
	S1	*S2*	*S3*	*S4*	*S1*	*S2*	*S3*	*S4*
Daphnia	42	78	40	31	55	64	60	68
Bosmina	13	11	7	9	8	4	11	3
Macrothrix	–	4	–	–	–	2	–	–
Moina	8	8	2	1	4	3	8	8
Ceriodaphnia	14	35	42	11	38	29	27	19
Polyhemus	12	4	9	3	2	8	1	4
Diaptomus	11	13	15	9	8	12	10	6
Cyclops	8	7	11	5	4	–	–	3
Asplanchna	31	42	33	15	18	21	20	9
Brachionus	7	11	4	9	11	6	–	–
Keratella	–	–	2	1	–	–	3	–
Halicyclops	1	7	–	3	5	1	8	–
Mesorchra	1	3	1	2	1	1	–	–
Filinia	3	–	1	1	2	–	1	3
Cypris	1	–	2	1	1	2	–	–

Table 46.3: Seasonal Variation of Phytoplankton Abundance and Diversity in Kalavam Bazaar Lake

Phytoplankton	*Rainy Period*				*Postmonsoon Period*			
	S1	*S2*	*S3*	*S4*	*S1*	*S2*	*S3*	*S4*
Spyrogyria	14	26	23	12	25	27	38	16
Volvox	22	11	14	19	29	20	31	37
Eudorina	–	6	–	4	3	1	5	1
Closterium	–	–	–	8	–	–	6	4
Hydrodictyon	4	3	1	–	1	2	–	–
Pachycladom	2	8	4	3	2	7	6	–
Cyclindrocapsa	–	1	4	1	3	1	2	2
Bumillaria	1	2	–	–	1	3	1	–
Amphora	2	1	1	–	–	–	–	1
Oscillatoria	3	5	–	–	–	–	–	1
Anabaena	18	13	21	24	–	27	–	29
Spirulina	15	17	30	12	11	–	1	15
Navicula	3	8	6	5	4	9	11	8
Melosura	–	–	4	–	1	1	3	2
Acanthes	–	–	–	1	–	–	1	–
Tabellaria	2	7	4	3	1	–	1	–
Fragillaria	2	1	2	1	1	1	–	1
Camphonema	1	1	1	–	–	–	1	–
Nitzschia	2	1	1	–	–	3	5	7
Euglena	11	8	7	–	4	–	4	2
Microcystis	3	2	5	1	4	4	2	8
Auliscus	1	–	1	–	–	–	1	–
Coscinodiscus	3	1	5	1	3	2	1	3

Table 46.4: Total Viable Count of Various Bacteria in Kalavam Bazaar Lake

Species	*Rainy*				*Postmonsoon*			
	S1	*S2*	*S3*	*S4*	*S1*	*S2*	*S3*	*S4*
Escherichia coli	350	368	450	340	345	344	90	250
Staphylococcus aereus	132	102	130	146	157	134	231	200
Pseudomonas sp.	120	110	95	62	84	176	211	160
Micrococcus sp.	90	68	46	35	38	60	64	96
Aerobactor aerogenus	64	55	98	93	110	88	55	38
Clostridium sp.	15	22	44	38	55	37	43	63

It was also found that most of the Phytoplanktons were recorded considerably higher number in different sites, during postmonsoon season, similar findings were recorded by Chakraborty and Konar

(2000), and Gaikwad (2004). The dominant groups like Myxophyceae and Chlorophyceae were recorded. Thus, the appearance of blue green algae was an indication of organic enrichment of these habitats. According to Biswas and Konar (2000), it can be correlated with high concentration of dissolved organic load and was obviously scripted from sewage, dead body burning, damping of animal carcasses and defaecation of the lake. The appearances of tolerant taxa of Chlorophyceae also support the above explanation.

The results of the Zooplankton and bacteriological analysis are given in Tables 46.3 and 46.4. The maximum viable count was observed in rainy season than postmonsoon season. This may be due to addition of land washing and organic matter by rain. The specific bacteriological examination indicates the notable presence of *Eschericia coli,* and *Pseudomonas* sp. Sengar *et al.,* (1996) reported the presence of similar genera in different sources of water.

In the present study the abundance and diversity of Zooplankton and Phytoplankton was found to be dependent on water quality and co-existing biotic communities at given point of time. Due to organic load disturbances from various non-point sources was clearly reflected by shifting of Phytoplankton and Zooplankton at various sites, which highlights the importance of non-point and point sources to cause pollution in the Kalavam Bazaar lake, Arcot.

References

Agarkar, M.S., Goswami, H.K., Kaushick, S., Mishra, S.M., Baipai, A.K and Sharma, U.S. 1994. Biology, conservation and management of Bhojmet land 1. Upper lake ecosystem in Bhopal. *Bionature,* 14: 20–28.

Anon. 1998. The river Dhamodhar and its enrichment. *Bull. Cent. Inland Fish. Res. Inst.,* Barakpore, 79: 32.

APHA. 2000. *Manual for Water and Wastewater Analysis*, Washington, DC.

Battish, S.K. 1992. *Freshwater Zooplankton of India*, Oxford and IBH Publishing Co., New Delhi.

Bilgrams, K.S. 1989. Biomonitoring of water quality of Ganga. In: *The Ganga Project Directorate*. North Book Centre, New Delhi, p. 101–106.

Biswas, B.K and Konar, S.K. 2000. Influence of Nunia Nullah (canal) discharge on plankton abundance and diversity in the river Damodar at Narankuri (Raniganj) in West Bengal. *Indian J. Environ. and Ecoplan.,* 3: 209–217.

Chakrabotry and Konar, S.K. 2000. Studies on the hydrobiology of river Damodar from Raniganj region (West Bengal) with reference to Phytoplankton. *Indian Journal of Environ. and Ecoplan.,* 3: 485–490.

Desikachary, J. V. 1959. *Cynophyta: Monographs on Algae*. ICAR, New Delhi.

Edmondson, W.T. 1959. *Freshwater Biology*, John Wiley and Sons Inc., New York, 2nd, p. 657–663.

Gaikwad, S.R, S.R. Thorat and T.P. Chavan. 2004. Plankton diversity with reference to pollution status of river Tapi in North Maharashtra region. *J. Curr. Sci.,* 5(2): 749–754.

Kodarkar, M.S. 1995. *Conservation of Lakes*, IAAB Publi., No. 2, IAAB, Hyderabad.

Nandan, S.N and Patel, R.J. 1992. *Ecological Studies of Algae in Aquatic Ecology*, Ashish Pub House, New Delhi, p. 70–99.

Rangareddy, Y. and Radhakrishna, Y. 1984. The calanoid and cyclopoid fauna of lake Kolleru, South India. *Hydrobiol.*, 119: 27–48.

Ruttner and Kolisko, A. 1974. Plankton Rotifers. *Biennengeswasser*, 26: 1–146.

Sengar, S.G., Naqshi, A.R and Mir, G.R. 1996. Impact of floating gardens on the limnological features of Dal lake. *Poll. Res.*, 15: 217–221.

Sharma, S.S.S. 1988. New records of rotifer. *Hydrobiol.*, 160: 263–269.

Thorat, S.R. and Massarat, S. 2000. Pollution status of Salim Ali lake, Aurangabad (M.S.). *Poll. Res.*, 19: 307–309.

WHO. 1994. *Guidelines for Drinking Water Quality*. Geneva, 2: 91–96.

Chapter 47

Zooplankton Diversity in the Chennai Coast, Tamil Nadu

V. Indra and R. Ramanibai*

Department of Zoology, University of Madras, Guindy Campus, Chennai – 25

ABSTRACT

Zooplankton community constitute an important component in the faunal composition of the water body. Distribution of major groups of Zooplankton population were conditioned by various behavioural and physiological adaptations of the plankton population to everchanging hydrographical condition. Zooplankton density was maximum during monsoon and minimum during premonsoon season. A variety of Zooplankton taxa numbering 28 were identified from 4 major groups, the cladocera, copepoda, ostracoda and rotifera. Among the 4 groups copepoda was one of the major dominating group of the micro zooplankton fauna is represented by 16 species under 3 sub groups. After the group copepoda, the rotifera was represented by 8 species with most common species Brachionus to rare species like Trichocera. The most striking feature of the group cladocera was represented by one single species of Daphnia. The ostracods were represented by 3 species. Dominant, common, rare taxa were also isolated under each group and compared between the different zones of the coast. Relative occurence showed distinct variations with copepoda and rotifera groups.

Introduction

Diversity refers to the range of variation or differences among some set of entities; zooplankton diversity thus refers to variety within their community. The members of the Zooplankton community are important for their role in the tropic dynamics and in energy transfer in the aquatic ecosystem.

They provide food for fishes in fresh and marine and play a major role in the fish production in the Bay of Bengal. The present study an intensive survey of Zooplankton diversity from 4 different habitat of Chennai coast have been made and the collected data have been analysed to ascertain the distribution pattern in the marine ecosystem. In spite of their great importance, very little work has been done on this group from freshwater wetlands of West Bengal (Venkataraman *et al.,* 2000) Kakinada Bay (Chandramohan *et al.,* 1999).

Distribution of Zooplankton and their variation at different zones of a water body is known to be influenced by physico-chemical parameters of water. Zooplankton study provides a relevant and convenient point of focus for research on the mechanism of eutrophication and its adverse impact on an aquatic ecosystem. Zooplankton constitutes a vital link in food chain and its productivity depends on water quality at a given time. The present study reveals some aspects of Zooplanktonic dynamics and explains with physico-chemical characteristics of water.

The present investigation was conducted in the near shore coastal zone of the Chennai coast which extends between Ennore backwater on the northern side and the Kovalam backwater on the southern side. There are 8 major inlet through which the inland water bodies communicate with the Bay of Bengal. The study area was classified in to 4 zones *viz.,* (*i*) Riverine Zone (*ii*) Estrarine Zone (*iii*) Backwater Zone and (*iv*) Near Shore Zone. The nearshore area was also divided in to southern and central and northern Zone. The plankton samples were collected by passing 100 litres of surface water through a 63μm plankton net. The samples were preserved in 5 per cent formalin at the sampling site itself. Enumeration of the organisms were performed in a two step protocol. Slow moving, generally abundent organisms, such as rotifers, nauplius were counted first. Counts were carried out using sub samples (or) total samples. The sub samples were then divided in to 10–12 drops on a multidepression microscope slide. During the second step, fast swimmers like copepods and cladocerans were enumerated. When rotifer counting was completed, the sample was first fixed with 5 per cent formalin. Counts of Zooplankton forms were affected on fine successive sub samples of the entire sample. Planktons were examined under a light microscope at a magnification of 10 × 10 ×. Systemic identification was done at the level of the highest taxa using standard keys (Edmoson, 1959; Ranga Reddy and Radhakrishna, 1984; Sharma, 1988; APHA, 1995).

The most abundant Zooplankton taxa collected from the surface samples of all the 21 sampling stations were presented in the Table 47.1. Totally 28 Zooplankton taxa were identified from 4 major groups. The cladocera, Copepoda, Ostracoda and Rotifera. Among the four groups enumerated in the study area, the bulk of the biomass was accounted distinctly by few species and their occurrence is given in the (Table 47.1). Dominant, common and rare taxa were isolated under each group and compared between the different zones of the coast.

Plankton exhibit a pronounced seasonality between zones and the seasonal peaks of Zooplankton biomass varied depending on the sampling locations. Classical investigations of Zooplankton have concentrated on describing the dominant species and their seasonal abundance. Highest biomass at the nearshore coastal stations. (S.C.N) was observed during monsoon seasons, while lowest biomass during summer and premonsoon seasons. Copepods were much more abundant than the other three almost all occasions. Out of a total 15 copepod species were present in monsoon season and 12 species were present in the summer season. Comparatively lower number of species was noticed in the riverine zone during monsoon season. This dominane was decreased up to single genus namely *Eucalanus* in the river Adyar and *Bryocamptus* in the river coovum.

Table 47.1: The List of Zooplankters Identified in the Chennai Coast

Group	*Sl.No.*	*Name of the Organism*
Cladocera	1.	Daphnia
		Calanoida
	2.	Diaptomus sp.
Copepoda	3.	Eucalanus
	4.	Epischura
	5.	Eurytemora
	6.	Nauplii
	7.	Oithona
		Cyclopoida
	8.	Cyclops
	9.	Eucyclops
	10.	Halicyclops
	11.	Heterocyclops
	12.	Mesocyclops
	13.	Paracyclops
		Harpacticoida
	14.	Bryocamptus
	15.	Mesochra
	16.	Paradactylopodia
	17.	Stenocaris
Ostracoda	18.	Cypris sp.
	19.	Candona sp.
	20.	Limnocythere sp.
Rotifera	21.	Brachionus
	22.	Filinia
	23.	Keratella
	24.	Lecane
	25.	Monostyla
	26.	Notholca
	27.	Platyias
	28.	Trichocera

Ointhona was not present during pre and postmonsoon seasons in the coastal zones. In the estuarine (16 Nos) and in the backwater zones (12 Nos) the number of copepod taxa enumerated was quiet similar to that of near shorecoastal zone. An overall reduction in the number of copepod taxa was found during pre and postmonsoon seasons. An increased number of organisms were noticed during monsoon and summer season.

Rotifera

The second most abundant taxa was the rotifera and the species enumerated frequently on all the six zones were the *Brachionus, Keratella* and *Filinia*. More or less in all the five zones except in the riverine zone recorded up to seven species. In the river Coovum only the *Brachionus* sp. was present throughout the four season. In the Adyar river 3 taxa were present (*Brachionus, Filinia* and *Keratella*). *Filinia* was absent during summer and monsoon season and was replaced by *Karatella*.

Ostracoda

Much reduced abundance of ostracoda taxa was recorded and composed of cypris. *Candona* and *limnocythere*. All the taxa were present during monsoon in all the five zones except in the riverine zone. In the river Adyar Candona and Cypris were present in monsoon season and this density and number was reduced to the single species (Cypris) during postmonsoon and summer season, and nil species was recorded during premonsoon season. In the river Coovum the ostracod population was completely altered in all the four seasons.

Cladocera

This showed high density during monsoon (Daphnia SP) season, and was absent during pre-postmonsoon and summer in the coastal zone. In the river Coovum daphnia was present in the monsoon, pre and postmonsoon season. Except during summer it was also present in the estuarine zones. In the backwater zone it was absent during premonsoon season.

On the whole the planktonic abundance was more during monsoon season at all the six zones and reduction was noticed during preland postmonsoon seasons. A small increase in the planktonic abundance was noticed during summer season. This fluctuations in the abundance of Zooplankton was clearly depicted by two groups namely the copepods and rotifers.

Zooplankton

Zooplankton community constitute an important component in the, faunal composition of the water body. Plankton data vary on a wide range of spatial and temporal scales and describing this variability is an important problem in plankton ecology. Distribution of major groups of Zooplankton population were conditioned by various behavioural and physiological adaptations of the plankton populations to ever changing hydrographical conditions. Tropical waters are, in general, poorly studied, but the planktonic communities surrounding India have been focus of extensive study over the past decade.

Copepods have generally been considered the most important metazoan, secondary producers in pelagic marine ecosystems, both in terms of abundance and biomass. It has been stated that copepods are the most plentiful multicellular animals on earth 7 out numbering the insects which have more species but fewer individuals. Copepods are important in fisheries, maintaining regenerated primary) productions. The plankton populations were dominated by the copepods in Chennai coast which represented an average of 60 to 65 per cent of total zooplankton. The biomass value and total counts varied significantly between seasons. The present study is in conformity with the works carried out in the coast and estuarine systems of temperate regions. Copepoda dominated the zooplankton population in the Mandovi-Zuari estuarine system in Goa, west coast of India constituting 66.2 per cent of the total counts (Madhupratao, 1999). In the Kakinada Bay, east coast of India copepods were represented by 75–80 per cent of total zooplankton (Chandramohan *et al.*, 1999). Higher levels of abundance were relatively more during March to July than those observed from August to February in

Kakinada bay, east coast of India. Here, copepoda exhibited a decrease in density in low saline waters and many species totally disappeared during flood period. This agrees with the present study where low percentage of copepods present in the riverine zone. In Bahuda river estuary, east coast of India registered a maximum of 40 species of copepods and their peak was found during premonsoon season. The premonsoon and monsoon seasons were characterized by maximum and minimum diversity indices respectively (Sujathamishra and Panigrahy, 1996).

In contrast to the seasonality, where an overall composition of copepods were more or less same throughout the year in the Arabian sea and small filter feeding forms and carnivores dominate (Madhupratao, 1999). Copepod population contributed 72.7 per cent and 79.6 per cent to the total zooplankton count respectively for Versova and Mahim transects of the nearshore waters of Bombay (Gajbhiye *et al.*, 1991). Copepods were abundant and 80 species were identified from coastal waters of Goa, west coast of India (Goswami, 1985) where a minimum copepod diversity was obtained during the month of October. The plankton data presented here, clearly demonstrated that, the abundance of zooplankton peaked during monsoon and decreased during pre and postmonsoon seasons. Furthermore, the abundance of dominant taxa, the copepods also decreased during pre and postmonsoon seasons. In the Nethravati–Gurupur estuary, the copepods and copepodites together formed the bulk of the plankton population. Except for great reduction during peak monsoon months, they were available in sizeable numbers throughout the year in this estuary (Bhat and Gupta, 1983).

The most diverse group within the zooplankton in the lower swan estuary in South West Australia was the copepoda, which comprised of 21-species, the majority of which were holoplanktonic forms. Copepod nauplii were by far the most abundant taxon, contributing 32.9 per cent to the total number of zooplankton in the Wilson islet of western Australia. River zooplankton are often characteristically dominated by small forms such as rotifers, bosminids and juvenile copepods (Shiel *et al.*, 1982; Thorpe *et al.*, 1994). The biomass of river zooplankton can be substantially higher in the tidal freshwater portion than that in the non-tidal portion. On contrary to this, where as in the case of the river Coovum and Adyar the riverine zone of the study area did not register significant zooplankton biomass of either copepoda or the rotifer fauna. The probable reasons could be the poor water quality and the effect of discharge of wastewaters of the city through these rivers.

Rotifera enjoy a cosmopolitan distribution which is not limited to continents. Rotifers are important components of planktonic community seasonality. Furthermore, rotifers influence various interactions within the microbial food web which occur at several trophic levels (Arndt, 1993). Detailed studies dealing only with rotifers in estuarine environments are rare (Edborge and Tawari, 1987; Edborge, 1994; Green, 1995). The Indian freshwater rotifer fauna is well studied with reports from different parts of the Indian sub continent starting with are available.

According to there are only approximately 50 strictly marine forms of rotifers are available most of which are littoral. This may be one of the reason for the sparse literature available on the occurrence of Indian marine rotifer fauna. Reports on rotifer fauna of east and west coast of India is negligible. The rare available report is from a coastal mangrove, east coast. Studies from pichavaram mangrove, south east coast of India also confirms that Brachionus is dominant with more species and abundance (Govindasamy and Kannan, 1991). The earlier view that rotifers are entirely cosmopolitan based on accumulated information on their geographic distribution patterns is likely (Dumant, 1983).

The cladocerans are filter feeders and their association to phytoplankton especially diatoms is fairly well known. Ostracods inhabit a wide variety of environment mostly free living forms (Edmoson, 1959). When compared with the other two groups, cladocerans and ostracods contribute much low percentage composition during the present study. The cladoceran and the ostracod standing stocks represent 15.19 per cent and 0.21 per cent in the coastal waters of Goa, west coast of India (Goswami, 1985). The faunal composition of Mandovi and Zuari estuarine system of Goa, west coast of India also revealed low percentage composition (1.8 per cent) of cladocera (Padmavathi and Goswami, 1996). The planktonic biomass and counts were also, poor in this estuary during monsoon period (June–September). The two groups were seen occasionally and in few numbers in the plankton hauls off Kakinada bay. Large cladocerans such as Daphina have been recorded in many rivers including the river Murray in Australia (Shiel *et al.*, 1982). Daphnia, has reached a maximum density of 0.7 animals l^{-1} in the Nile river, 0.8 animals l^{-1} the Ohio river (Thope *et al.*, 1994) and 2.98 animals l^{-1} is the lower river Rhine (*Dumant*, 1983). But there was no record of Daphina in Hawkesbury Nepean during 1996 (Kobayashi *et al.*, 1998).

The zooplankton of temperate Bay and estuaries were typically higher in abundance and lower in diversity than zooplankton of adjacent neritic waters. The distribution of various plankton groups was also influenced by rainfall to some extent during monsoon period. This study indicates that the zooplankton in the coastal waters of Chennai is dominated by the copepods in both number of species as well as abundance followed by rotifers. The maximum number of copepods are found in the nearshore waters, than in other zones of the coast and this has been the general observation from Indian coast (Madhupratao, 1999; Goswami, 1985; Bhat and Gupta, 1983; Chandramohan *et al.*, 1999). However, the number of species of copepods recorded in this study are only 16 in number of which half of them constitute much less in abundance.

The decline in the major planktonic communities has been predicted by (Sivasamy, 1990) in his report on plankton in relation to costal pollution at Ennore, Madras coast. Though his studies using two way correlation between the pollutants and planktonic fauna yielded statistically no significant relations, he has predicted more deterioration of water quality especially in northern coastal \waters, considering the impact of urbanisation and industrial development. The present study has confirmed the trends of decline.

Gajbhiye *et al.*, 1991 have compared the data on copepod population in the creek system collected during 1990 with samples collected during 1979–1980. Though a consistent pattern could not be observed at different stations, they could record the influence of deteriorating water quality due to industrial and urban wash off from greater Bombay on the incidence of copepod population. It is even suggested that the copepod *Acartia spinicanda,* a predominant species in the heavily sewage polluted areas of the coast as an indicator organism whose presence and abundance can indicate pollution load in the coastal waters.

The cladocerans prefer low saline conditions, unlike copepods which are truly marine, feed generally on diatoms and therefore associated with fluctuations of diatom flora. The diatom population in the Chennai coastal waters are very much reduced in diversity and abundance as reported by since diatoms are most sensitive group to pollution load. In the present study, cladocerans are impoverished and is represented by only one species (Rahukumar and Anil, 2003) established facts that nutrient availability largely determines the diversity of primary producers.

It appears that the Chennai coastal area has more of physically controlled than biologically accommodated communities and are low in diversity of zooplankton fauna unlike tropical/subtropical marine, environment. In the present study the sampling in the nearshore waters have been done in 50 m zone that are likely to be influenced more by monsoon inflows. The high variability of water quality, levels and zooplankton fluctuation may have been due these causes also

References

APHA. 1995. American Public Health Association, Washington D.C.

Arndt. 1993. *Hydrobiologia,* 225/256: 231–246.

Bhat, B.V. and Gupta, T.R.C. 1983. *Ind. J. Mari. Sci.,* 12: 36–42.

Chandramohan, P., Raman, A.V. and Srinivas, N. 1999. *Ind. J. Mari. Sci.,* 28(2): 192–197.

Dumant, H.J. 1983. *Hydrobiol.,* 104: 19–30.

Edborge, A.B.M. 1994. *Hydrobiologia,* 272: 95–104.

Edborge, A.B.N. and Tawari, P. 1987. *J. Plank. Reas.,* 9L 1–13.

Edmoson, W.T. 1959. John Wiley and Sons, Inc. New York, 2nd Edn., p. 657–663.

Gajbhiye, S.N. Rosamma Stephan, Vijayalakshmi, R., Nair and Desai, B.N. 1991. *Ind. J. Mari. Sci.,* 20: 187–194.

Goswami, S.C. 1985. *Ind. J. Mari. Sci.,* 14: 177–180.

Govindasamy, C. and Kannan, L. 1991. *Hydrobiol approach. Mahasagar,* 24: 39–45.

Green, 1995. *J. Hydrobiologia,* 313/314: 47–56.

Kobayashi, T., Shiel, R.J., Gibbs, P. and Dixon, P.I. 1998. *Hydrobiol.,* 377: 133–145.

Madhupratao, M. 1999. *Ind. J. Mar. Sci.,* 28: 146–149.

Padmavathi, G. and Goswami, S.C. 1996. *Ind. J. Mari. Sci.,* 25: 268–273.

Rahukumar, S. and Anil, A.C. 2003. *Curr. Sci.,* 84(7): 884–891.

Ranga Reddy, Y. and Radhakrishna, Y. 1984. *Hydrobiol.,* 119: 27–48.

Segers, H., Sharma, S.S.S., Kattasery, F.K. and Nayar, C.K.G. 1994. *Hydrobiologia,* 287: 251–258.

Sharma, S.S.S. 1988. *Hydrobiol.,* 160: 263–269.

Shiel, R.J., Walker, K.F. and Williams, V.W.D. 1982. *Aust. J. Mari. Sci.,* 33: 301–327.

Sivasamy, S.N. 1990. *Ind. J. Mari. Sci.,* 19: 115–119.

Sujata Mishra and Panigraphy, R.C. 1996. *Ind. J. Mari. Sci.,* 25: 98–102.

Thorpe, J.H., Black, A.R., Haag, K.H. and Wehr, J.D. 1994. *J. Fish. Aqua. Sci.,* 51: 1634–1643.

Van Dijk, G.M. and Vanzantan, B. 1995. *Hydrobiol.,* 304: 29–38.

Venkataraman, K., Das, S.R. and Nandi, N.C. 2000. *J. Aquatic. Biol,* 15: 19–25.

Wiebe, P.H., Davis, C.S. and Greene, C.H. 1992. *Oceanus,* 35: 100–106.

Chapter 48

Biodiversity of Chlorophyceae in Haranbari Dam of Baglan (Maharashtra)

N.H. Aher and S.N. Nandan***

**Department of Botany, Arts, Commerce and Science College, Sakri – 424 304, District Dhule (MS)*
***P.G. Department of Botany, S.S.V.P.S's L.K. Dr. P.R. Ghogrey Science College, Dhule – 424 001 (MS)*

ABSTRACT

A critical study has been made on the diversity of Chlorophyceae in Haranbari dam of Baglan during the period from July 1998 to June 2000. Algal samples were collected at first week of monthly intervals from 3 stations of Haranbari dam during the period of two years of investigation. Algal samples preserved in 4 per cent formalin for taxonomic investigation. Line drawings of Chlorophyeae were made by camera lucida. Taxa of Chlorophyceae were identified with the help of standard monographs and recent literature. In present study this class consists of mainly species of *Gonium, Chlorococcum, Pediastrum, Tetradedron, Westella, Crucigenia, Ulothrix, Oedogonium, Zygnema, Spirogyra* and *Cosmarium.* The population of *Desmids* and *Chlorococcales* was dominant in winter and summer seasons. From Ulothrichales like *Ulothrix* and from Oedogoniales like *Oedononium,* all these filaments forms were observed from all 3 stations. In the present study 25 taxa of belonging to 11 genera of Chlorophyceae were studied taxonomically in algal community.

Keywords: *Biodiversity, Freshwater, Green algae, Dam.*

Introduction

Algae are natural inhabitants of water. It is important group of thallophytes. Algae constitute a part of food chain of aquatic ecosystem. They affect the population of Zooplankton and other aquatic organisms. The number of publications on limnology and ecology of freshwater rivers, dams etc., have been appeared in Maharashtra (Deore, 1978; Barhate and Tarar, 1981; Gunale and Balu, 1981; Jagdale *et al.*, 1987; Bodas, 1991; Ragothaman and Jaiswal, 1995; More, 1997; Ansari Ziya, 1997; Patil and Nandan, 1998; Nandan and More, 2001, 2002). Though many rivers of dam of Maharashtra were observed hydro-biologically very few rivers and dams carried out work with relation to algae. Biodiversity of 4 groups of algae and limnological studies of algae were made by selecting Haranbari dam of Baglan of Nasik district of Maharashtra. In present investigation Biodiversity of Chlorophyceae of Haranbari dam is explained.

Materials and Methods

Haranbari dam is located on the Mosam river. The dam is situated towards the west direction of Taharabad and near the village Haranbari. Geographically Haranbari dam is located at latitude 20°51′2″ North and longitude 74°11′23″ East at 714.52 meters above the MSL. The catchment area of dam is about 5478 hectares. The water storage capacity of it is 34.78 million cubic meter. The Haranbari dam is the earthen dam. The wall length of the dam is 748.60 meters and height of dam wall is 44.44 meters. The dam is mainly constructed for irrigation and drinking water purpose.

Algal samples were collected in acid washed glass bottles and preserved in 4 per cent formalin at monthly intervals from 3 stations of Haranbari dam from July 1998 to June 2000.

For qualitative study of Chlorophyceae algae and identification of Chlorophyceae line drawings of different forms of taxa were made by Camera lucida. The members of Chlorophyceae algae were identified by relevant monographs and recent available literature. (Kant and Gupta, 1998; Ramanathan, 1964; Gonzalves, 1981; Randhawa, 1959; Bernard, 1908; Iyengar and Desikachary, 1981; Philipose, 1967; Nandan and Patel, 1985a).

Taxonomy of Chlorophyceae

Biodiversity of Chlorophyceae genera and species were recorded at 3 stations of Haranbari dam. Genes Gonium 1 taxon, Chlorococcum 1 taxon, Pediastrum 2 taxa, Tetradedron 3 taxa, Westella 1 taxon, Crucigenia 2 taxa, Ulothrix 3 taxa, Oedogonium 2 taxa, Zygnema 1 Taxon, Spirogyra 5 taxa, Cosmarium 4 taxa.

1. ***Gonium Pectorale* Muller (Plate 48.1, Figure 1)**

   Huber. Pestalozzi. G.1961, 16(5):624, F.877

   Colonies square with rounded corners, cells 15.6μ broad and 20.8μ long.

   Habitat: HD–II, October, 1999.

2. ***Chlorococcum infusionum* (Schrank) Meneghini (Plate 48.1, Figure 2)**

   J. Meneghini, 1842, P.27, Pl.II, F.3

   Cells usually spherical and variable dimensions 101.8μ in diameters.

   Habitat: HD–II, November, 1991.

3. ***Pediastrum simplex* Mayen (Plate 48.1, Figure 3)**

   F.I.F. Mayen 1829, P.772, Pl.43, F.1.5

   Colonies circular, cells 8–13μ broad, 19–26μ long.

   Habitat: HD–I, December, 1999.

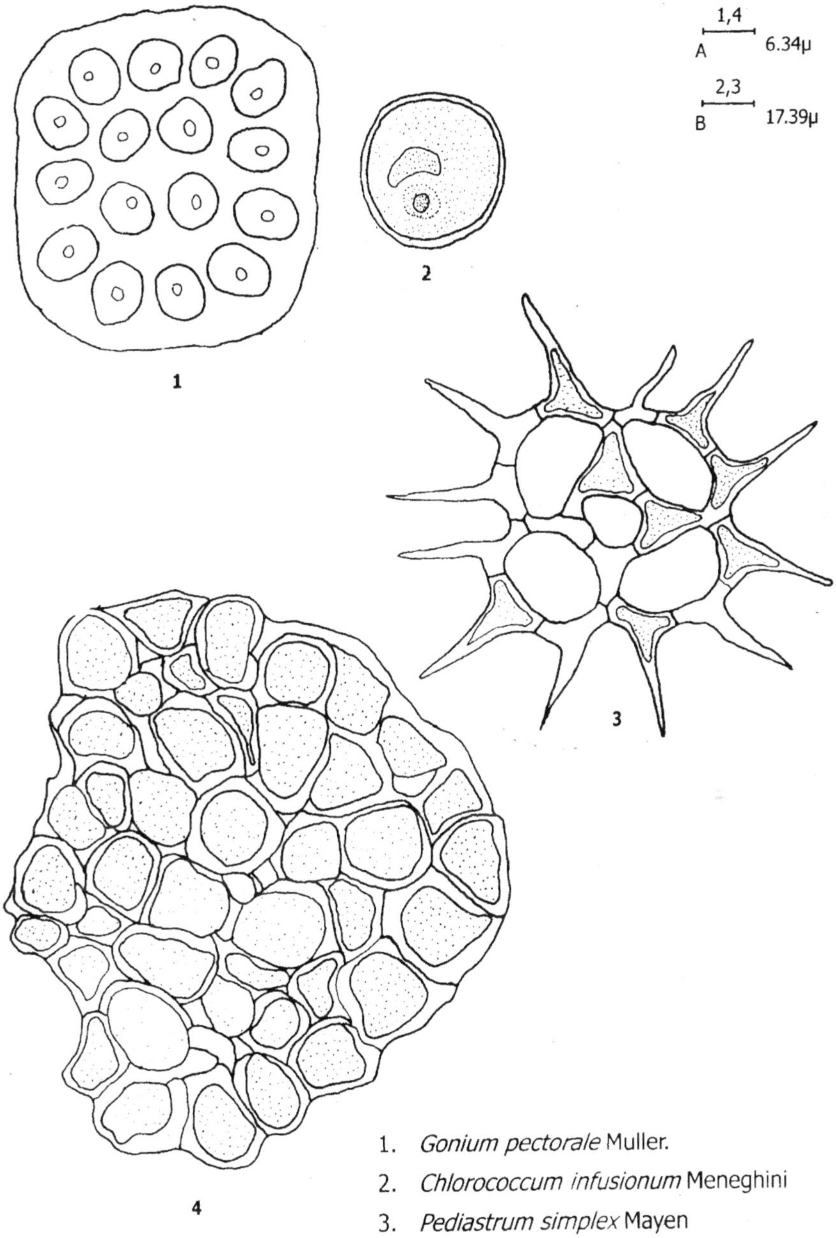

1. *Gonium pectorale* Muller.
2. *Chlorococcum infusionum* Meneghini
3. *Pediastrum simplex* Mayen
4. *Pediastrum tricornutum* Borge

Plate 48.1

4. ***Pediastrum tricornutum* Borge (Plate 48.1, Figure 4)**
Kant and Gupta, 1998, P.80, Pl.23, F.9.
Outer face of Marginal cells with two projections, cells 9μ board and 13μ long.
Habitat: HD–II, December, 1999.

5. ***Tetradedron minimum* (A. Braun.) Hansgirg (Plate 48.2, Figure 1)**
A. Hansgirg, 1988, Pl.131, 1889, P.18.
Cells small and quadrangular with the sides concave and angles rounded. Cells 12μ in diameter.
Habitat: HD–I, August, 1999.

6. ***Tetradedron scrobiculatum* (Lagartheim) De. Toni (Plate 48.2, Figure 2)**
C.B.De Toni, 1989, P.601.
Cell wall coarsely scrobiculate, Cells 10.77μ in diameter.
Habitat: HD–II, September, 1999.

7. ***Tetradedron octaedrium* V. Spinosum (Rernsch) W.et. G.S. West (Plate 48.2, Figure 3)**
W. and G.S. West, 1901a, P.183.
Cells octagonal with eight lateral planes. Cells 13.3μ diameter and the spines 3.17μ long.
Habitat: HD–I, October, 1999.

8. ***Westella botrayoides* (W. West) Dewildeman. (Plate 48.2, Figure 4)**
Dewildeman, 1897, P.532.
Colonies of Irregular shape and about 40–80 cells. Cells 3-8μ in diameter.
Habitat: HD–III, September, 1999.

9. ***Crucigenia fenestrata* (Schmidle) Schmidle (Plate 48.2, Figure 5)**
W. Schmidle, 1900, P.234.
Colonies rectangular 9.51 μ in diameter.
Habitat: HD–III, September, 1999.

10. ***Crucigenia tetrapedia* (Kirchner) W.et.G.S. West (Plate 48.2, Figure 6)**
W. and G.S. West, 1920a, P.62, Pl.1, F.11–12.
Colonies 4 celled, outersides of cells always concave, cells 3.17μ in diameter, four celled colonies 7.6μ diameter.
Habitat: HD–I, July, 1999; HD–II, August, 1999; HD–III, July 1998.

11. ***Ulothrix-aequalis* Kuetzing (Plate 48.2, Figure 7)**
Ramanathan, 1964. P.36, Plate 9, I–L.
Cells generally cylindrical, 14.5μ broad and 1–2 time as long as broad.
Habitat: HD–III, August, 1999.

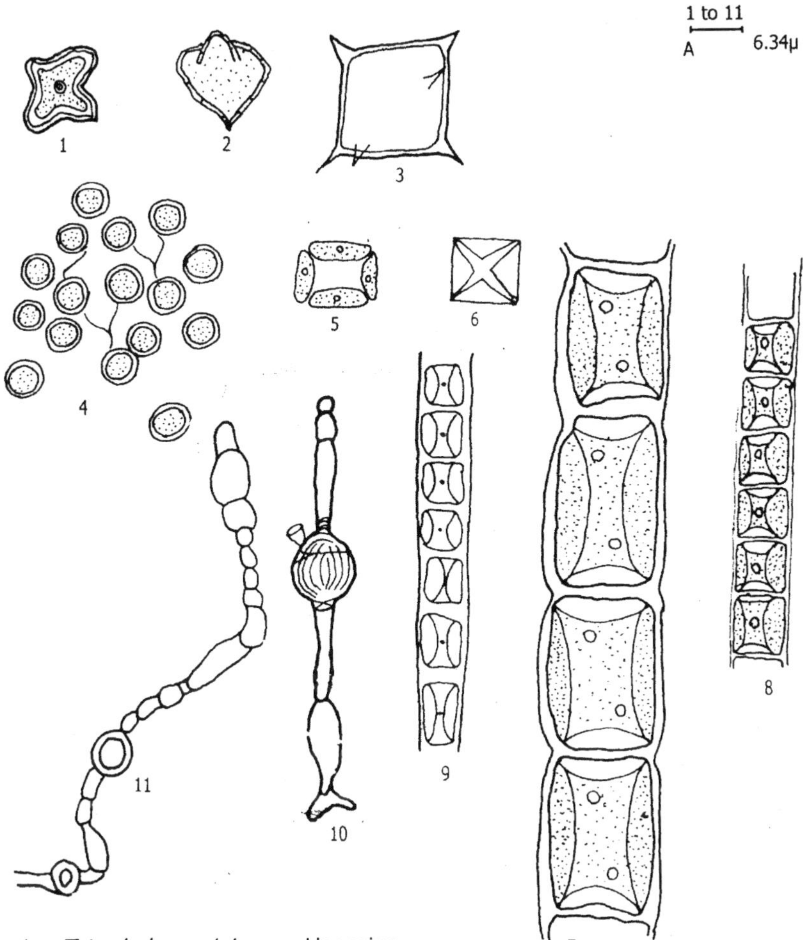

1. *Tetradedron minimum* Hansgins
2. *Tetradedron scorbiculatum* De Toni.
3. *Tetradedron octaedrium* V.Spinosium W.et. G.Wet.
4. *Westella botrayoides* Dewildeman
5. *Crucigenia fenestrata* Schmidle
6. *Crucigenia tetrapedia* G.S.Wet.
7. *Ulothrix aequalis* Kutzing.
8. *Ulothrix tenerrima* Kuetzing
9. *Ulothrix variabilis* Kuetzing
10. *Oedogonium exostriatum* Triff.
11. *Oedogonium concatenatum* Wittr.

Plate 48.2

12. *Ulothrix tenerrima* Kuetzing. (Plate 48.2, Figure 8)

Ramanathan 1964, P.37, Plate 10 A–C.

Cells 7.6µ thick, 2/3-1½ times as long as broad with one pyronoid.

Habitat: HD–I, August, 1999.

13. *Ulothrix variabilis* Kuetzing (Plate 48.2, Figure 9)

Ramanathan, 1964, P.39, Pl.10, D–F.

Filaments sometimes with a pointed basal cells; Cells 5–7µ, ½-1½ times as long as broad.

Habitat: HD–I, November, 1999; HD–II, August, 1999.

14. *Oedogonium exostriatum* Triff. (Plate 48.2, Figure 10)

E.A. Gonzalves, 1981, P.378, F.9–271.

Nannandrous, vegetative capitellate, 10µ in diameter, 39.30µ long, oogonia depressed-globose, 28µ in diameter 21µ long.

Habitat: HD–III, November, 1999.

15. *Oedogonium concatenatum* Wittr. (Plate 48.2, Figure 11)

E.A. Gonzalves, 1981, P.446, F.9.345A.

Nannandrous, gynandrosporous, vegetative cell, Cylindrical, 25–40µ in diameter, 75–400 long, anthridia single or upto 4 seriate, 13–15µ in diameter, 12–25µ long.

Habitat: HD–I, July, 1999.

16. *Zygnema collinsianum* Transeen (Plate 48.3, Figure 1)

Randhawa, 1959, P.249, F.209 a–d.

Vegetative cells 22 × 45µ chloroplast 2 in each cells, reproduction by conjugation.

Habitat: HD–II, March, 1999.

17. *Spirogyra varians* (Hassall) Kuetzing (Plate 48.3, Figure 2)

Randhawa, 1959, P.297, F.256.

Vegetative cells 29 × 45µ end wall plane, Chloroplast 1, Conjugation scalariform, and lateral.

Habitat: HD–II, July, 1999.

18. *Spirogyra sahnii* Randhawa (Plate 48.3, Figure 3)

Randhawa, 1959, P.307, F.274 a–e.

Vegetative cells 60.86 × 70.37µ usually they are broader than long. Conjugation lateral.

Habitat: HD–III, July, 1999.

19. *Spirogyra hyalina* Cleve (Plate 48.3, Figure 4)

Randhawa, 1959, P.318, F.294 a–c.

Vegetative cells, 59.12 × 126.94µ with plane endwalls. Tubes formed by both gametangia.

Habitat: HD–I, January, 2000.

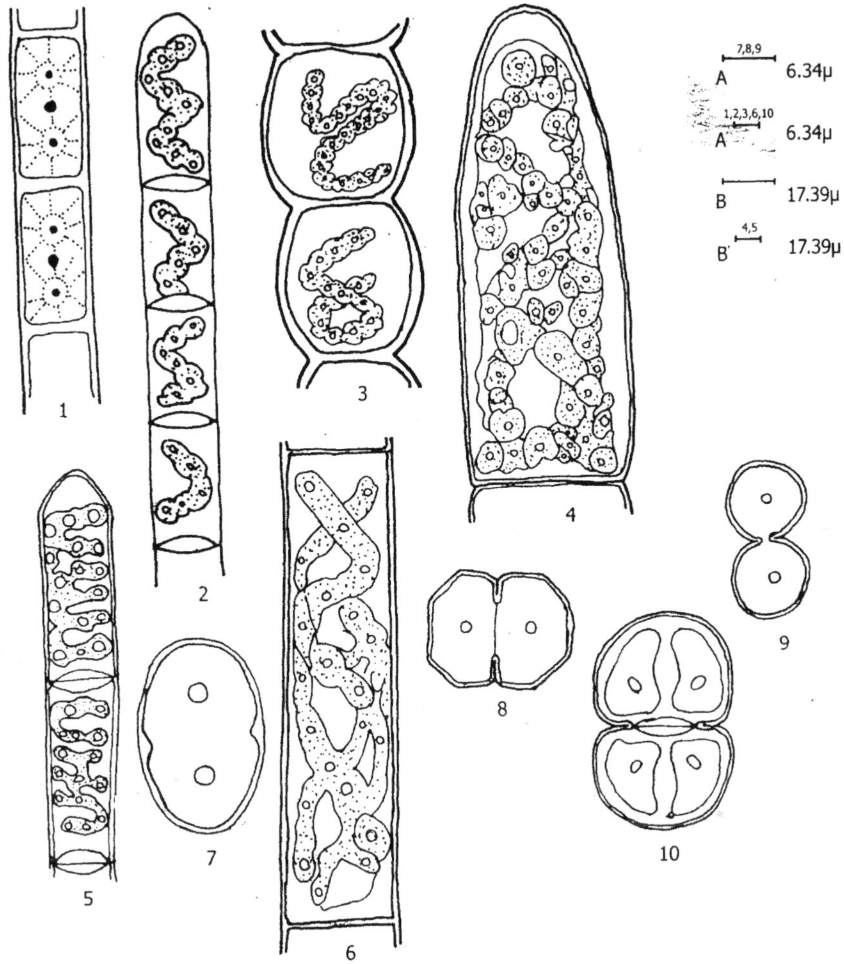

1. *Zygnema collinsianum* Transeen.
2. *Spirogyra varians* Kuetzing.
3. *Spirogyra sahnii* Randhawa.
4. *Spirogyra hyalina* Cleve.
5. *Spirogyra columbiana* Czurd.
6. *Spirogyra frankliniana* Tiff.
7. *Cosmarium moniliforme* Ralfs.
8. *Cosmarium meneghini* Breb.
9. *Cosmarium ellipsoideum* v.*minus* Raclb.
10. *Cosmarium pyramidatum* Breb.

Plate 48.3

20. *Spirogyra columbiana* Czurd (Plate 48.3, Figure 5)

Randhawa, 1959, P.318, F.295.

Vegetative cells 53.9 × 92.16μ with plane end walls, chloroplast 1, conjugation scalariform, tubes formed both gametangia.

Habitat: HD–I, January, 2000.

21. *Spirogyra frankliniana* Tiffany (Plate 48.3, Figure 6)

Randhawa, 1959, P.413, Fig.500.

Vegetative cells 32.96 × 84.95μ with plane end walls. Conjugation scaliform; tube formed by both gamentangia.

Habitat: HD–II, July, 1999.

22. *Cosmarium moniliforme* (Turp) Ralf. (Plate 48.3, Figure 7)

Ch. Bernarad, 1908, 2, P.90, F.58.

Cells 30μ long, 18μ broad, Isthmus 2μ.

Habitat: HD–III, February, 2000.

23. *Cosmarium meneghini* Breb. (Plate 48.3, Figure 8)

Ch. Bernard, 1908, 2. P.90, F.64, 65.

Cells 17μ long, 13μ broad, Isthmus 3μ.

Habitat: HD–I and HD–II, September, 1999.

24. *Cosmarium ellipsoideum* Elfv. v. minus. Raclb. (Plate 48.3, Figure 9)

Ch. Bernard, 1908, 2. P.106, F.65.

Cells 17μ long, 9.51μ broad, Isthus, 2–2½μ.

Habitat: HD–I, September, 1999.

25. *Cosmarium pyramidatum* Breb. (Plate 48.3, Figure 10)

Ch. Bernard, 1908, 2, P.107, F.75–76.

Cells 71μ long, 61μ broad, Isthmus 20–25μ.

Habitat: HD–I and HD–II, September, 1999.

Acknowledgement

We are grateful to Principal, Dr. S.N. Nandan, S.S.V.P.S's L.K. Dr. P.R. Ghogrey Science College, Dhule for providing the facility for investigation.

References

Deore, L.T. 1978. Studies on the freshwater algae of Maharashtra. *PhD. Thesis*, University of Poona.

Eloronta, P. 1976. Species diversity in the phytoplankton some finfish lakes. *An. Bot. Fennicl.* 13: 42–48.

Fritsch, F.E. 1907. The sub-arrial and freshwater algal flora of the tropics. *Ann. Bot.* 20: 236–275.

Gonzalves, E.A. 1981. *Oedogoniales.* ICAR Monograph, New Delhi, p. 1–757.

Gonzalves, E.A. and Joshi, D.B. 1946. Freshwater algae near Bombay–I. The seasonal succession of algae in a tank of Bandra. *J. Bomb. Nat. Hist. Soc.*, 46: 144–176.

Hegde, G.R. 1988. Freshwater algae of Karnataka state–Certain new records, from Dharwad. *Indian. Bot. Reptr.*, 7(1&2): 51–53.

Iyengar, M.O.P. and Desikachary, T.V. 1981. *Volvocales.* ICAR Monograph, New Delhi, p. 1–532.

Jyothi, B.B., Sudhakar, G. and Venkateswarlu, V. 1989. Ecological studies on a desmids bloom. *J. Indian Inst. Sci.*, 69: 285.

Kant, S. and Gupta, P. 1998. *Algal Flora of Ladakh.* Jodhpur, p. 1–341.

Lal, A.K., Sarkar, A., Sarkar, S.K. and Shah, K.L. 1982. Species diversity of periphyton as an index of population of river Ganga. *J. Int. Fish. Soc. Ind.*, 14: 44–51.

Mahajan, S.K. 1987. Algal flora of Pawagiri Oon, M.P. *Phykos*, 26: 61–62.

Nandan, S.N. 1993. Algal flora of fish pond in Dhule, Maharashtra. *Indian Bot. Rept.*, 12(1&2): 61–63.

Nandan, S.N. and R.J. Patel. 1985a. Species diversity of algal flora in Vishwamitri river Baroda. *Pol. Arch. Hydrobiol., Poland*, 32(1): 1–6.

Patel, R.J. and Daniel, J.K. 1990. Some Chloroccales new to India. *Phykos*, 29: 129–135.

Philipose, M.T. 1967. *Chlorococcales.* ICAR, New Delhi, p. 1–356.

Ramanathan, K.R. 1964. *Ulotrichales.* ICAR, New Delhi, p. 1–188.

Randhawa, M.S. 1959. *Zygnemaceae.* ICAR, New Delhi, pp. 478.

Tarar, J.L. and Seema Bodkhe. 1998. Studies on Chlorococcales of Nagpur. *Phykos*, 37(1&2): 107–114.

Chapter 49

Physico-chemical Characteristics of Khari Nadi Water, Agra

Sunita Solanki, K.S. Rana and Arun Kumar Singh
Department of Zoology, Agra College, Agra – 282 002, U.P., India

ABSTRACT

The present article deals with seasonal variations in some important physico-chemical characteristics of Khari Nadi water. Khari Nadi is stretch from Fatehpur Sikri to Kheragarh town in Agra district (U.P.). The physico-chemical characteristics of Khari Nadi water undertaken from 2002–2003 for studying various physico-chemical parameters such as water temperature, transparency, depth, pH, dissolved oxygen and free carbon-dioxide, the overall concentration of the dissolved oxygen, level appeared to be low (6.2 to 8.8 ppm) while the concentration of free carbon-dioxide significantly varied from 2.7 to 4.9 ppm in site 1st and 2.3 to 4.6 ppm in site 2nd, the nature of water is found to be acidic to slightly alkaline (pH 6.12 to 7.97).

Keywords: *Pollution, Physico-chemical characteristics, Khari Nadi.*

Introduction

Though the defilement of water and deterioration of aquatic systems is as old as civilization, however escalating industrialization, urbanization, development and agricultural activities have brought irreversible change in such systems. Water is only the most important source, essential constituents for all animals, plants and other organisms. It is also potable survivality of the mankind in the Biosphere (Sharma, 1987). It is the main source of epidemics since it is universally used and supplied to a large populations from a few centralized sources.

The growth and distribution of the plankton, depends upon several physico-chemical characteristics of water, such as-amount of dissolved gases, temperature and transparency of water, intensity of incident light etc. Potable water is one that is considered safe for drinking and contain an organic or an inorganic substances deleterious for health and no pathogenic organisms (Park and Park, 1994). There is no doubt the physical and chemical characteristics of water can adversely effect the aquatic environment and ecological balance in many different way. Considerable investigations have been done on the physico-chemical characteristics of different types of water bodies in country. Thus the present study aimed to study some important physico-chemical characteristics of two different sites of Khari Nadi in Agra.

Materials and Methods

Physico-chemical characteristics of Nadi water were studied from March, 2002 to February, 2003. Water samples from surface (maximum depth 5.6m) were collected monthly with necessary precautions in plastic bottles from different sites of Nadi. Depth of water was measured with a straight and pointed bomboo-pole while pH was determined with the help of electronic pH meter. Other parameters were done following standard methods given by Michael (1984).

Results and Discussion

Physico-chemical parameters from March, 2002 to February, 2003 is shown in the Table 49.1. Water temperature of the Nadi varied from 19.4 to 33.0°C in site 1 and 18.6 to 31.9°C in site 2. Water temperature influence the aquatic life and concentration of dissolved gases like oxygen, carbon-dioxide and solutes. The water temperature increased during summer period in site 1 and in site 2. The transparency varied from 32.8 to 89.5cm in site 1 and 26.5 to 58.3 cm in site 2. The maximum transparency was observed during the summer period and minimum during the rainy periods. Similar observations were reported by Iqbal and Kataria (1993). The poor transparency might be due to silt brought into the Nadi during rainy season.

Table 49.1: Seasonal Variations of the Physico-chemical Characteristics of Khari Nadi Water in Parameters

Parameters	*Site 1*						*Site 2*					
	Summer		*Rainy*		*Winter*		*Summer*		*Rainy*		*Winter*	
	Min.	*Max.*	*Min.*	*Max.*	*Min.*	*Max.*	*Min.*	*Max.*	*Min.*	*Max.*	*Min.*	*Max.*
Water Temperature (°C)	33.0	33.2	25.4	28.6	19.4	22.1	30.3	31.7	25.2	31.9	18.6	21.3
Transparency (cm)	67.4	89.5	32.8	36.3	58.3	61.0	50.3	58.3	26.5	30.4	42.8	54.2
Water Depth (m)	2.9	3.8	4.8	5.6	3.1	3.6	2.6	2.8	4.6	5.3	3.9	4.4
Hydrogen Ion Concentration (pH)	7.28	7.74	6.12	6.67	6.92	7.14	7.43	7.65	6.12	6.22	7.12	7.97
Dissolved Oxygen (ppm)	7.5	7.9	6.2	6.7	7.6	7.9	6.8	7.2	6.9	7.4	8.2	8.8
Free Carbon-dioxide	2.3	2.5	4.4	4.9	3.9	4.0	2.7	2.9	3.7	4.6	3.6	4.1

The depth of the different sites of Nadi varied in different seasons which was been shown in Table 49.1. The maximum depths 5.6 m in site 1 and 5.3 m in site 2 were observed during rainy season and the minimum depths 2.9 m in site 1 and 2.6 m in site 2 were found during summer in both the study sites. The maximum depths which were observed during rainy season maybe attributed to the supply of rain water while the minimum depths 2.9 m and 2.6 m during summer season may have resulted mainly due to very less rainfall during this period substantiated by excessive loss of water due to high evapa-transpiration rate. Hydrogen ion concentration (pH) also showed in seasonal variations. Highest value occurred during summer periods and the lowest values during rainy periods. The highest values during summer periods may be due to removal of sufficient amount of CO_2 by photosynthetic process of the aquatic macrophytes (Shardendu and Ambasht, 1988).

The quantity of dissolved oxygen in the water body is of prime importance. During the present investigation, the concentration of DO_2, ranged between 6.2 to 7.9 ppm in site 1 and 6.8 to 8.8ppm in site 2 which has been shown in Table 49.1. In the present study the overall DO_2 level appeared to be low 6.2 to 8.8 ppm. The concentration of free CO_2 significantly varied from 2.3 to 4.9 ppm in site 1 and 2.7 to 4.6 ppm in site 2. The lowest concentration 2.3 to 2.7 ppm of free CO_2 during winter periods in both the study may be due to high photosynthetic activity of the aquatic macrophytes. The maximum concentrations 4.9 ppm in site 1 and 4.6 ppm in site 2 were recorded during rainy period which may be due to diffusion of atmospheric CO_2 into the rainwater producing carbonic acid (Chakraborty *et al.*, 1995).

From the studies by comparison with standards, it is revealed that the Nadi water is suitable for drinking after Chlorination and also for agricultural purpose.

References

Chakraborty, R.D., Roy, P. and Singh, S.P. 1995. The quantitative study of the plankton and the physico-chemical conditions of the River Yamuna at Allahabad. *Ind. J. Fish.*, 6: 186–203.

Iqbal, S.A. and Kataria, H.C. 1993. Physico-chemical analysis and water quality assessment of upper lake of Bhopal. In: *17th Intern. Conf. Sci. and Tech.*, New Delhi.

Michael, P. 1984. *Ecological Methods for Field and Laboratory Investigations*. Tata McGraw-Hill Publs. Company Ltd., New Delhi.

Park, J.E. and Park, K. 1994. *Parle's Textbook of Preventive and Social Medicine*, 13th Ed. Banarsidas Bharat Publishers, Jabalpur, p. 367–380.

Shardendu and Ambasht, R.S. 1998. Limnological studies of a rural and urban tropical aquatic ecosystem: Oxygen form and ionic strength. *Trop. Ecol.*, 92: 98–109.

Sharma, K.N. 1987. The limnological study of Chambal River in relation to pollution and its influence on aquatic fauna. *Ph.D. Thesis,* Agra University, Agra.

Chapter 50

Occurrence of a Cyanophycean Bloom in Mallapura Tank Near Chitradurga, Karnataka

A.B. Banakar, B.R. Kiran*, R. Purushothama*, E.T. Puttaiah* and S. Manjappa***

**Department of Environmental Science, Kuvempu University, Shankaraghatta – 577 451, Karnataka*
***University BDT College of Engineering, Davanagere, Karnataka*

ABSTRACT

The present study deals with ecological aspects of cyanophycean bloom in relation to certain physico-chemical factors from Mallapura tank, near Chitradurga of Karnataka during October 2004 to March 2005. Physico-chemical variables on the growth of this bloom both collectively and individually include alkaline pH, higher phosphate, chloride, sulfate, E.C. and temperature. The major bulk of the bloom was represented by the genus *Arthospira.*

Introduction

Algae are involved in water pollution in a number of ways. Phytoplankton, such as *Microcystis, Anabena, Pandorina,* and *Euglena* etc., produce thick massive blooms in inland waters of the tropics (Round 1965; Pandey 1973). The algal blooms can create severe problems in the maintenance of water supplies and in meeting the ever-increasing demand for potable water.

Cyanophycean algae are distributed all over the world and found in marine as well as freshwater habitats. They grow luxuriantly in eutrophic lentic water bodies. Blue green algae have reported to accumulate heavy metals and therefore ameliorate the toxicity of flyash with increased fertility (Rai *et al.*, 2000). Some non-heterocystous blue green algae like Spirulina are used as single cell protein and in wastewater treatment considerable work has been done on blue green algae of fresh water habitats of Uttar Pradesh, India (Prasad and Srinivas, 1965; Tiwari, 1979; Ahmed, 1972; Chaturvedi and Pandey, 1976). Our article discusses the ecological aspects of various cyanophycean blooms and their significance in evaluating the trophic status of the habitats.

Study Area

Mallapura tank is situated at 76°22′19″ North latitude and 14°13′30″ East longitude with a water spread area of 91.84 hectares. The water storage capacity is 189 MCFT (Million Cubic feet). Surface water samples were collected from Mallapura tank during October 2004 to March 2005. The filtered samples were analyzed for various physico-chemical parameters by standard methods (Gilterman and Clymo, 1969; APHA 1998) simultaneously 100 ml of the sample was collected separately and preserved in 4 per cent formaldehyde solution. Using Sedgewick-rafter counter cell did the quantitative estimation of algae and finally, the number of plankton was calculated per litre. The data was statically analyzed by following the methods of Snedecor and Cochran (1967).

Results and Discussion

The data presented in Table 50.1 shows the physico-chemical properties of water from this tank. Table 50.2 shows the occurrence of blue green blooms during this study. In this investigation the species of *Aphanocapsa, Oscillatoria, Merismopedia, Spirulina* and *Nostoc* was occurring rarely in this waterbody. While, *Arthrospira* bloom was dominant in all the months.

Table 50.1: Physico-chemical Complexes of Mallapura Tank

Parameters	*Oct*	*Nov*	*Dec*	*Jan*	*Feb*	*Mar*
pH	8.2	8.8	8.6	8.5	8.2	9.6
Total hardness	216	240	230	242	252	248
E.C	1120	1160	1190	1300	1880	2200
Water temperature	25	26.5	28.5	29	29.5	31.5
Sulfate	48	54	107.30	198.4	255.2	269.7
Calcium	49.699	56.112	42.48	28.05	34.46	38.47
Magnesium	20.93	22.69	28.96	26.47	39.45	100.8
D.O	9.73	13.186	9.986	13.586	5.186	4.8
Iron	0.027	0.021	0.085	0.085	0.116	0.006
Phosphate	0.24	0.62	0.87	0.70	0.85	1.02
Nitrate	0.005	0.007	0.03	0.035	0.015	0.02
Chloride	107.78	326.2	358.82	468.03	496.39	787.15

Note: All the parameters are in mg/l except pH, temperature (°C) E.C. (μmhos/cm).

Table 50.2: Occurrence of Cyanophycean Blooms in Mallapura Tank (Organisms/Litre)

October	*November*	*December*	*January*	*February*	*March*
2.7×10^5	3.575×10^5	3.634×10^5	6.362×10^5	2.162×10^5	5.434×10^5

The water temperature of the sampling sites ranged between 25–31.5°C, EC 1120–2200 µmhos/cm, pH 8.2–9.6, total hardness 216–252 mg/l, Calcium 28.0 56.11 mg/l, Magnesium 20.93–100.8 mg/l, Chloride 107.78–787.15 mg/l, phosphate 0.24–1.02 mg/l, Dissolved oxygen 4.8–13.58 mg/l, in the current study 5 genera of blue greens have been identified of which *Arthrospira* bloom was dominant (90 per cent) as compared to other Cyanophycean bloom.

According to Tucker *et al.* (1984) high temperature supports the luxuriant growth of blue greens. In this study, the temperature fluctuated between 25–31.5°C. The dominance of blue green algae with high temperature support the work of Munawar (1974) who stated that the blue green algal forms develop profusely towards the end of monsoon when the temperature rise or during winter (October) when the temperature was fairly high, ranges between 25–32° C.

In the present waterbody the Cyanophycean blooms as a whole exhibit a direct relationship between dominance of algal species with nitrates. Moreover this tank much harbour having less quantities of nitrate confirming the result of Pearsallis (1932) and Zafar (1967), that cyanophycean are capable of growing rapidly in minimal quantities of nitrate. It is well known that blue green algae occur more abundantly in hard water than in soft water bodies and the major factors that influence their growth are light water temperature, high pH, DO, Chloride, Phosphate and low nitrate.

In general dense blooms occur in nutrient rich water and many blooms are highly concentrated. Blooms develop frequently during certain times of the year in temperate region or almost any time of the year observed in the tropics (Reynolds and Walsby, 1975) the abundance of any Phytoplankton organism in a water body depends not only on such factors as light, temperature and content of nutrient, but also on the changes which the plankton population by its own growth may bring about in its environment. Certain organic substances liberated by the algae themselves may be important for successful growth (Brook, 1957) on the basis of the experimental work, Lam and Silver (1979) concluded that (i) the effective uptake of nutrients by one species may cause it to dominate over other algae or (ii) the liberation of inhibitory substances by one alga may eliminate the other species.

The temporary water bodies such as rainwater pools, the area is very small and if they have high nutrient loading (hypertrophic condition), the species with fastest rates of growth can develop into a bloom for a short period. After a certain period of growth these blooms disappear suddenly from the habitat perhaps, due to the following reasons:

1. Exhaustion of essential nutrients
2. Concentration of certain elements which may become toxic
3. Extracellular substances released by the organism may be of inhibitory nature (autoinhibition)
4. Ion binding effect of certain extra cellular substances which will alter the ionic composition of the medium.

Uhlmanns (1971) models of growth of Phytoplankton in various habitats indicate that small species of algae are capable of dominating at high rates of nutrient loading and short retention times by virtue of their rapid rates of growth.

Like the preceding species, the blue green alga *Arthrospira* sp. also occurs in polluted environment where the medium is alkaline (pH 9.6) and very rich in chloride, total hardness and Electrical conductivity. It is interesting to note that although the habitat shows all the characteristics of eutropic waters, its oxygen content is very high and this could attributed to the vigorous photosynthetic activity of alga at the time of collection which may be further substantiated by the very high pH value during the study.

Conclusion

1. Nutrient loading of water bodies favour bloom formation
2. Algal blooms can develop at any time of the year
3. Species with fastest growth rate in a short span develop into blooms
4. Depending on the type of environment and nature constituents of the medium bloom forming species vary greatly
5. Extracellular substances released by the algae also play a significant role in bloom formation as they have great biological significance in ecology
6. Well oxygenated waters are more favorable to blooms.

References

Ahmed, M.R. 1972. Algae of Allen forest Lake Kanpur India. *Nova Hedwigia,* 23(1): 125–129.

APHA. 1985. *Standard Methods for the Examination of Water and Wastewater*, 16th edition. USA.

Brook, A.J. 1957. *New Biology*, 23: 36–101.

Chaturvedi, V.K. and U.C. Pandey. 1976. A list of blue green and green algae from Rohilkhand division, U.P., India *N. Phykos,* 15(1–2): 127–137.

Golterman, H.L and R.S Clymo. 1069. *Methods for Chemical Analysis of Freshwater.* I.B.P. Hand Book No. 8 Blackwell Scientific Publications, Oxford, London.

Lam, W.Y. Catherine and W.B. Silvester. 1979. *Hydrobiologia,* 63: 135–143.

Munawar, M. 1974. Limnological studies on freshwater ponds of Hyderabad, India, IV. The biocenose, periodicity and species composition of unicellular and colonial phytoplankton in polluted and unpolluted environments. *Hydrobiologia*, 45(1).

Pearsall, W.H. 1932. Phytoplankton in the English lakes II. *J. Ecology,* 20: 241–262.

Prasad, B.N. and P.N. Srivastava. 1965. Some interesting morphological features in *Talypothrix* species. *Hydrobiologia*, 26(3–4): 517–520.

Reynolds, C.S. and A.E. Walsby, 1975. *Biol. Rev.,* 50: 437–481.

Snedecor, G.W. and W.G. Cochran. 1967. *Statistical Methods*, 6th edition. Oxford Publishing Co., New Delhi.

Tiwari, G.L. 1979. A study of the blue green algae from paddy field soils of India IV. Taxonomic consideration of Nostocales and Stigonematotales. *Nova Hedwigia,* 63: 133–159.

Tucker, C.S., Loyed, S.W. and R.L. Bush. 1984. Relationship between Phytoplankton periodicity and concentration of total and unionized ammonia in channel catfish. *Hydrobiologia,* 11(1): 75–79.

Uhlmann, D. 1971. Mitt. der Intern. Ver. Fur Theor. And Angew. Limnol., 19: 190–224.

Zafar, A.R. 1967. On ecology of algae in certain fish ponds of Hyderabad, India II. *Hydrobiologia,* 30(1): 6–112.

Chapter 51

Study of Sitarampur Dam, Saraikela-Kharsawan, Jharkhand to Evolve Strategies for Biodiversity Conservation

Gouri Suresh
Reader in Botany and Co-ordinator, Environment and Water Management, Jamshedpur Women's College, Jamshedpur – 831 001

ABSTRACT

Sitarampur Dam in the state of Jharkhand was studied for evolving long-term strategies for conservation of aquatic biodiversity in Indian wetlands. The present study was carried out as part of a project at the national level initiated by UNDP, MoEF and SACON. One prime aim of the present study was to identify the plant species and their distribution patterns as plants available decide animals that can be sustained. Another aim was to get detailed information on the socio-economic issues related to the wetland. These data will be used along with other required data to determine the methods to be adopted for biological conservation of Indian wetlands.

Keywords: *Sitarampur, Conservation, Wetland.*

Introduction

India is a developing country supporting the second largest population in the world. We have a mainly agrarian economy. Therefore India has a significant impact on all natural resources including wetlands. This shows how essential it is to have an inventory of the important wetlands in India.

There has been no systematic attempt to evolve an action plan for conserving aquatic biodiversity for which, primarily, identification of such reserves should be done.

The Ministry of Environment and Forests, Government of India conducted the first nation-wide survey of wetlands of India in the 1980's. This has been a good beginning. However, much more has to be done. IUCN (International Union for Conservation of Nature and Natural resources) has categorized India as one of the Asian countries needing more data about wetlands than what is available now. This shows that a more serious and systematic inventory is required. Wetlands more than 60 hectares in size have been mapped on the basis of visual interpretation of satellite data. These maps however do not give the exact nature of the flora and fauna and water quality. Moreover, many wetlands less than 60 hectares in size are also biologically important.

Socio-economic values of wetlands are a very important aspect for prioritization of biological conservation sites. The location of these wetlands, present legal conservation status and major use to the local people are important outputs for conservation values.

On realising the significance of the work, United Nations Development Programme (UNDP), Ministry of Environments and Forests (MoEF), Government of India and Salim Ali Centre of Ornithology and Natural history (SACON) came together with a massive project for prioritisation of biological conservation sites in Indian wetlands. Sitarampur Dam, among other dams in the vicinity of Jamshedpur, was studied as apart of this project. The parameters studied were the profiles, status regarding conservation monitoring, socio-economic values and environmental impact.

Materials and Methods

All species of hydrophytes–submerged, floating and emergent–were collected from the dam. Plants growing adjacent to the water-body, from areas that get regularly inundated in the rainy seasons, were also collected. A required quantity of water from the dam was collected too, for studying the phytoplanktons. All these were brought to the laboratory and identified with the help or relevant monographs and other authentic texts. Data sheets were prepared for the aquatic angiosperms for three parameters–density, spread and distribution patterns.

For collecting information on socio-economic values of the wetland, local people were interviewed. The officials of the pumping station provided information regarding the dam itself: its capacity, beneficiaries of water from the dam etc.

The information thus collected and arranged systematically is given below:

Results and Discussion

Sitarampur Dam is located in Saraikela-Kharsawan, one of the twenty-two districts of the newly carved out state of Jharkhand in India. It is a very picturesque spot and consists of a reservoir of water spread over 19.20 square kilometers. There is no river feeding this reservoir. It is the water collected in a low-lying rain fed catchment area. The dam was built in 1964. Maximum height of the dam is 16.77 meters while its length is 2043 meters. Length of the spillway is 35 meters, and gross storage is 882.62 hectare meters.

The soil at the edge of the reservoir is very soft and fertile and is home to a wide variety of plants. Many of these are medicinal and are regularly used by the villagers nearby. The water itself is clean. It was evident that the villagers are aware of the necessity to keep it clean. They were also very co-operative in answering the questions put to them.

Figure 50.1: Sitarampur Dam

The biological as well as the socio-economic information gathered by the present study is arranged below as per a specific plan.

Profile of the Wetland

Name of the wetland: Sitarampur Dam

Location: Adityapur

1. Latitude: 22° 47′ N
2. Longitude: 86° 09′ E
3. Altitude: 500 meters
4. Nearest city/town and its distance in kms: Jamshedpur, 14 kms.
5. Biogeographic region in which the wetland is situated: Gangetic Plain
6. Micro-geographic area in which the wetland is situated: Chotanagpur Plateau, part of Gondwanal and formed in the Pre-Cambrian era.

Water spread area in sq.kms.

1. Maximum: 19.20 sq.m.
2. Minimum: 10 sq.m. (apx)

Status of the Wetland with Reference to Conservation Monitoring

Present legal conservation status of the wetland: Protected fully

1. Year since which the wetland is being protected: 1964
2. Protecting Agency: PHED (Public Health Engineering Department)
3. Protection Status: Fishing regulated

Overall concern of the local people regarding wetland biodiversity and its conservation: Moderately concerned

Level of awareness among local people about wetland degradation: Moderately aware.

Socio-economic Values of the Wetland

Uses of the wetland to the local people: Agriculture, Grazing, Fishing, Irrigation, Domestic, Drinking, Wallowing by cattle, Recreation as a Picnic spot.

Species having socio-economic values other than direct economic utilities (with ecological groups): *Ficus religiosa* and *Ficus bengalensis* are species having religious importance in the locality. Though typically xerophytes, these plants are growing on wet soil near the dam.

Table 51.1: Plant Species having Direct Economic Utilities (with ecological group)

Economic Utilities	*Plant Species*	*Ecological Group*
Food for local communities	*Oryza sativa*	Amphibious hydrophyte
	Different leafy and green vegetables	Mesophytes
Food for commercial purposes	*Oryza sativa*	Amphibious hydrophyte
	Phaseolus radiatus	Mesophyte
	Different Vegetables	Mesophytes
Medicines for local communities	*Phyllanthes niruri*	Helophyte (marshy)
	Euphorbia hirta	Helophyte
	Hydrocotyle asiatica	Helophyte
	Catharanthus roseus	Mesophyte
	Vitis quadrangularis	Mesophyte
Fodder for livestock	*Cynadon dactylon*	Meso/Helophyte
	Different trailers and small sized herbs	Meso/Helophytes

Table 51.2: Abiotic Attributes having Direct and Other Socio-economic Values/Utilities

Abiotic Attribute	*Values*					
	Economic	*Subsistence*	*Religious*	*Scientific*	*Educational*	*Recreational*
Water quality						
Hydrology-related						
Nutrient status						
Ecological process						
Habitat diversity						
Other attributes						Picnic Spot

Environmental Contamination

Overall trend in fish yield over the last 5/6 years: No change

Species-wise particulars: Katla, Rehu, Tengra, Aira, Chingri (small and big), Chui etc.

Any industrial effluent joining the wetland: No

Whether the wetland is subjected to seasonal cultivation: Yes

Whether agricultural activity close-by: Yes

Names of crops cultivated: Paddy (Ghoda dhaan, *i.e.*, large grained rice, in plains and fine quality rice in water logged soils), urad dal and vegetables.

Names of pesticides applied on the crops: 1. For paddy, Thimet for pests and Dhanwan (Chlorpyriphos) for flies in particular; 2. For ladies' finger, Superkiller (Cypermethrin 10 per cent and 25 per cent) and Sulphur powder in later stages. Asatop (Acephate 75 per cent) also used; 3. For brinjal, Fluradan and Superkiller; 4. Roger and Superkiller for all vegetables.

Names of fertilizers used: For paddy and also all vegetables, urea in the initial stages. Suphala for tomato, potato etc. Urea enhances growth; suphala increases yield.

Any incidence of mortality of fish or bird in the wetland in the recent past: No.

The Dam and its Surroundings

Daily water supply from the dam is 50 lakh gallons. Of this, 11 lakhs are supplied to housing colonies: 27 lakhs to industries and the rest to non-organized sector. Water is supplied to Jugsalai, Shastri Nagar, Bagbera, Gamharia, Adityapur, Tayo etc. Total cost for constructing the dam in 1964 was Rs. 43,30,500 only. Common domestic animals in the surroundings are cows, buffaloes and sheep. The area adjoining the dam is neat and clean, away from the city, with fresh air. Stone pitching of the dam is strong. Local people live in neat huts slightly away from the dam. Majority poor. Grieved that rehabilitation work not at all satisfactory.

Table 51.3: Helophytic Herbaceous Angiosperms on the Wetland

Sl.No.	Botanical Name of the Species	Abundance	Sociability
1.	*Boerhavia diffusa*	O	S–1
2.	*Phyllanthus niruri*	O	S–2
3.	*Xanthium strumarium*	O	S–2
4.	*Euphorbia hirta*	O	S–2
5.	*Croton bonplandianum*	A	S–2
6.	*Evolvulus nummularius*	A	S–3
7.	*Lippia nodiflora*	A	S–2
8.	*Cynodon dactylon*	A	S–3

The present study shows that the dam is well protected. Abundance of green algae and the minimal presence of blue green algae is an indication that the water is not polluted. As there has been no fish or bird mortality in the recent past, the pesticides and fertilizers used for cultivation in the vicinity can be taken to be within safe limits. The awareness of the villagers regarding environmental issues, moderate at present, can be enhanced to higher levels, because they were found to be receptive to information.

Figure 51.2: Datasheet for Aquatic Angiosperms

(All data based on approximately two meters wide edge of the water body)

1. EMERGENT

DENSITY				
100				
75				
50	9	2		
25	< FOR 7,10,11,12	3		
SPREAD	25	50	75	100

DISTRIBUTION PATTERN
a-2,7; b-3,10,11,12; c-9

2. FLOATING

DENSITY				
100				
75				
50				
25				<1
SPREAD	25	50	75	100

DISTRIBUTION PATTERN
a-1

3. SUBMERGED

DENSITY				
100				
75		5,6		
50				
25	< FOR 4,8			
SPREAD	25	50	75	100

DISTRIBUTION PATTERN
a - 4,8 ; c - 5,6

4. OPEN WATER

	SPREAD
25	
50	
75	
100	✓ SUBMERGED PLANTS

DISTRIBUTION PATTERN
a

Sign for lesser than (<) used where required. *E.g.* < 25, plant number is given after that sign.

Distribution Patterns

(*a*) Localised;

(*b*) Patchy;

(*c*) Uniform

1. ***Aponogeton natans***
2. ***Eleocharis atropurpurea***
3. ***Scirpus supinus***
4. ***Vallisneria natans***
5. ***Hydrilla verticellata***
6. ***Najas graminea***
7. ***Paspalum distichum***
8. ***Potamogeton crispus***
9. ***Ipomoea carnea***
10. ***Heliotropium indicum***
11. ***Hydrocotyle asiatica***
12. ***Eclipta alba***

Table 51.4: Large Herbs/Shrubs

Sl.No.	*Botanical Name of the Species*	*Abundance*	*Sociability*
1.	*Datura fistuosa*	O	S–1
2.	*Calotropis procera*	O	S–1
3.	*Tephrosea* sps.	F	S–2
4.	*Lantana camara*	O	S–1
5.	*Zizyphus jujuba*	O	S–1
6.	*Crotalaria* sps.	R	S–1

Abundance Classes

R: Rare (very sparse); O: Occasional (sparse); F: Frequent (not numerous); A: Abundant (numerous).

Sociability Classes

S–1: Plants growing singly; S–2: Plants growing in small groups; S–3: Plants growing in small patches.

Abundance Class VA (Very Abundant) and Sociability

Classes S–4 (Plants in large patches); S–5 (Plants in continuous populations) were not observed.

Table 51.5: Algae of the Wetland: (Dominant genera *Chara* and *Nitella*)

Sl.No.	*Generic Name of the Alga*	*Habitat*
1.	*Chara*	Submerged, attached to mud
2.	*Nitella*	Submerged, attached to mud
3.	*Chlorella*	Free floating
4.	*Oedogonium*	Epiphytic on macrophytes
5.	*Oscillatoria*	Floating on damp soil
6.	Navicula	Free floating on bottom mud
7.	Pinnularia	Free floating on bottom mud

Acknowledgement

The author is thankful to SACON for conceiving and initiating the project, to Dr. D.S. Srivastava, State Co-ordinator for arranging the necessary finance and to Dr. Alim for providing the monograph on wetland flora. Thanks are also due to the Principal, Jamshedpur Women's College, Jamshedpur for providing the laboratory facilities.

References

Ambasht, R.S. 1990. *A Textbook of Plant Ecology*. Students' Friends and Co., Varanasi

Chopra, G.L. 1993. *Angiosperms*. Pradeep Publications, Jalandhar

Prasad, V.P., Daniel Mason, Joy, E. Marburger, Ajith Kumar, C.R. 1998. *Illustrated Flora of Keoladeo National Park, Bharatpur, Rajasthan: A General Guide to the Wetland Flora of the Gangetic Plains.*

Subranyam, K. 1974. Aquatic angiosperms: A systematic account of common Indian aquatic angiosperms, *Botanical Monograph*, No. 3.

Chapter 52

Studies on the Ecology of Chlorococcales of Sitarampur Dam, Jharkhand, India

Gouri Suresh and Meena Krishnan*

**Department of Botany, Jamshedpur Women's College, Jamshedpur – 831 001, India*

ABSTRACT

The present communication is a comparative study on the ecology of 12 taxa of Chlorococcales belonging to 8 genera collected from Sitarampur Dam. This dam is located in the Saraikela-Kharsawan District of the state of Jharkhand, India. Water samples from the dam were collected and analyzed for various physico-chemical characteristics and phytoplankton composition for 12 months. As Sitarampur Dam provides drinking water to the residential areas in addition to the industries in the neighbourhood, limnological studies on this water body are most essential.

Keywords: *Sitarampur, Chlorococcales, Systematic description, Physico-chemical properties, Ecology.*

Introduction

Sitarampur Dam is located in Saraikela-Kharsawan, one of the twenty-two districts of the newly formed state of Jharkhand in India. The dam has a catchment area of 19.20 sq. kms. It supplies 50 lakhs gallons of water/day–11 lakhs to housing colonies, 27 lakhs to industries and the rest to non organized sectors in the surrounding areas. There is no river feeding the dam. Rainwater collected in the low-lying stone pitched catchment basin is the source of water for the dam. The dam is located away from the city, and its surroundings are neat and clean. No limnological studies have been carried out in detail on this economically important wetland of Jharkhand, as revealed by a perusal of available literature. It was in this context that the present work was undertaken to find out the phytoplankton composition as well as the physico-chemical characteristics of the water of the Sitarampur Dam. The present article is a comparative study on the taxonomy and ecology of members of Chlorococcales that the authors came across, while studying the algai flora in relation to different parameters and standards of drinking water. Other common algae growing here are *Chara, Nitella* and diatoms beionging to different genera. Blue green algae are comparatively few, though *Merismopedia* was rather common and *Microcystis, Oscillatoria* and *Anabaena,* in that order of occurrence, were observed too.

Materials and Methods

Monthly collections were made from selected sampling points of Sitarampur Dam from January 2003 to December 2003. Algal samples were collected in polyethylene bottles and studied fresh and also after preservation in 4 per cent formalin. Camera Lucida diagrams were drawn from both living and preserved specimens. For identification of Chlorococcalean taxa, the monograph on Chlorococcales by Philipose (1967) was used. The collection has been deposited at the laboratory of Department of Botany, Jamshedpur Women's College, Jamshedpur.

For analyzing the physico-chemical parameters, water samples were brought to the laboratory. Ambient temperature, Water temperature and pH were analyzed on the spot of collection. Standard methods of American Public Health Association (APHA) (1985) and Nagpur Environmental Engineering Research Institute (NEERI) were followed for physico-chemical analysis.

Results and Discussion

Twelve taxa of Chlorococcalean algae, belonging to eight genera were collected and identified in the present study. Their taxonomy is being discussed first, and then based on the months of their occurrence, preferences shown by different taxa towards various physico-chemical characteristics of the water at the time of collection are discussed. The results of this study thus combine taxonomical and ecological aspects of the Chlorococcales of Sitarampur Dam. The commonest Chlorococcales are *Tetraedron* represented by four taxa, and *Scenedesmus* represented by two taxa. Other genera are represented by a single taxon each.

Systematic Enumeration

***Chlorococcum infusionum* (Schrank) Meneghini [Plate 52.1; Figure (A)]**

Cells spherical, solitary, with diameter 22 to 25 µ. Chloroplast a hollow sphere with a notch on one side. Single pyrenoid.

Collected in August from the edge of the water body, intermingled with other algae attached to small, submerged stones and aiso on soft wet mud. The same sample also contains different diatoms, *Merismopedia* and *Cosmarium.*

***Tetraedron trilobulatum* (*Reinsch*) Hansgirg [Plate 52.1; Figure (B)]**

Cells triangular with all sides equal in length. Sides deeply concave. Angles broadly rounded. Cell size 16–22 µ.

Collected in November, from the edge of the dam, when water level was low. Along with *Coelastrum microsporum, T. trigonum, Cosmarium* and *Microcystis.*

***T. trigonum* variety–*verrucosum* Jao [Plate 52.1; Figure (C)]**

Cells flat, triangular with slightly concave sides. Each angle with a stout spine. Cells 22–25µ in diameter. Spines 8–9µ long.

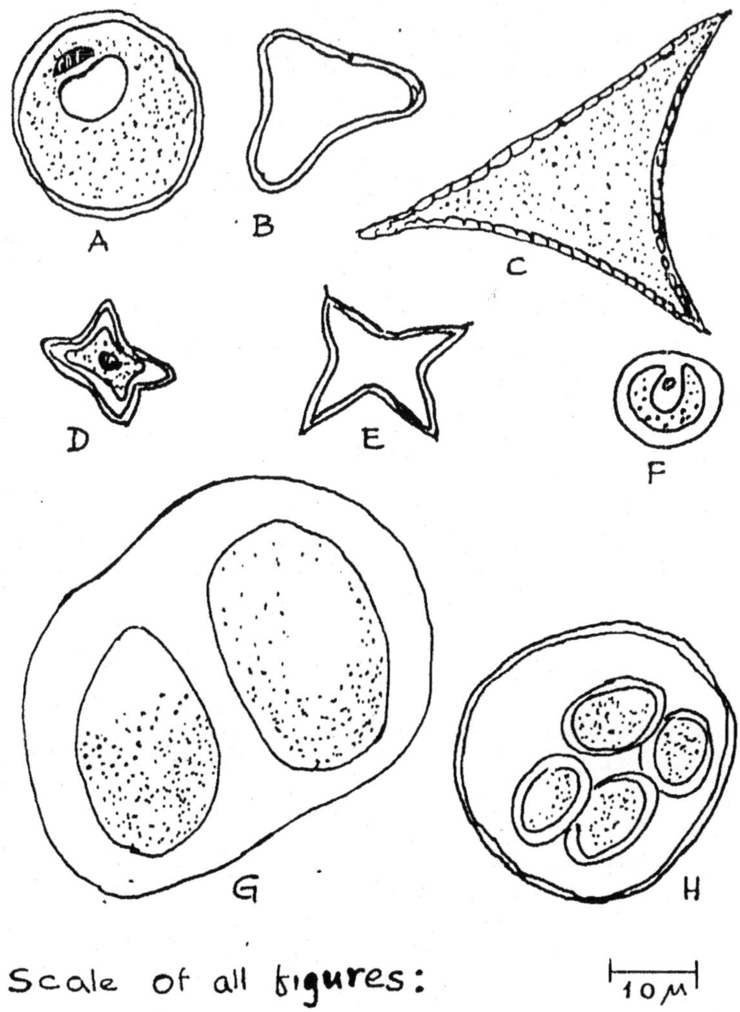

Plate 52.1: (A) *Chlorococcum infusionum*; (B) *Tetraedron trilobulatum*; (C) *Tetraedron trigonum, V. verrucosum*; (D) *Tetraedron minimum*; (E) *Tetraedron regulare V. torsum*; (F) *Chlorella vulgaris*; (G) *Gloeotaenium loitlesbergerianum*; (H) *Oocystis borgei*

Collected in November from water near the edge of the dam along with *T. trilobulatum, Cosmarium* and diatoms.

T. minimum **(A. Braun) Hansgirg [Plate 52.1; Figure (D)]**

Cells small and quadrangular, side-slightly concave and angles broadly rounded. Cell walls smooth. Processes absent. Cells size 13–17µ.

Collected along with *Merismopedia, Microcystis, Chlorella* and diatoms in November, from the edge of the water body.

T. regulare variety torsum (Turner) Brunnthaler [Plate 52.1; Figure (E)]

Cells tetragonal. Each cell has two halves, twisted in a cruciate manner. All four angles have short spines. Cell diameter varies from 17.5–20μ.

Found in November near the edge along with *Meresmopedia, T. minimum* and diatoms.

Chlorella vulgaris **[Plate 52.1; Figure (F)]**

Solitary spherical cells. Cell membrane thin. Cup-shaped parietal chloroplast very prominent. Abundant in the habitat. Cell diameter varies from 7.8–9.3μ. Found in almost all samples collected throughout the year. But in the months of August and November they are more in number and comparatively larger in size.

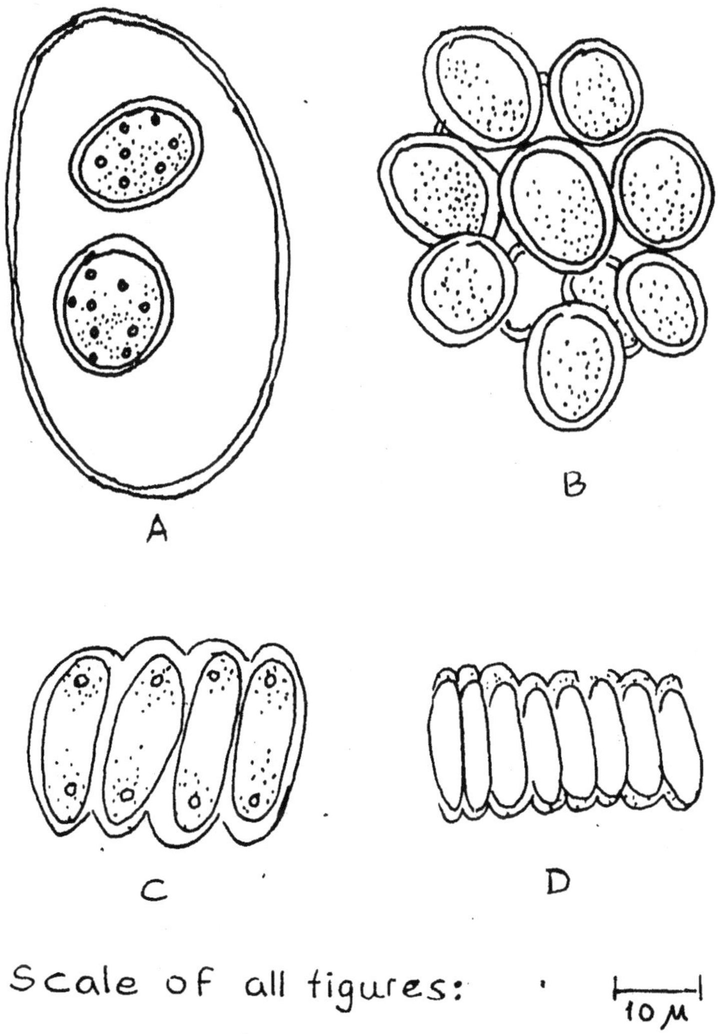

Plate 52.2: (A) *Nephrocytium agardhianum*; (B) *Coelastrum microsporum*; (C) *Scenedesmus bijugatus*; (D) *Scenedesmus bijugatus, V. altenas* forma *parvus*

Gloeotaenium loitlesbergerianum **(Hansgirg) [Plate 52.1; Figure (G)]**

Colonies ellipsoid, surrounded by two broad stratified gelatinous envelopes. Cells spherical to ovoid. Chloroplasts with prominent pyrenoids. Colonies 50–55μ long and 36–39μ wide. Cells 27–30μ in diameter.

Found in the month of August in shallow water along with *Ulothrix, Anabaena, Oscillatoria, Oocystis* and innumerable diatoms.

Oocystis borgei **Snow [Plate 52.1; Figure (H)]**

Colonies of 4 cells each, surrounded by a gelatinous envelope. Ceils in each colony arranged closely. Broadly ellipsoid with rounded ends. Diameter–13.2 to 22μ.

Collected in August from the edge of the water body along with *Ulothrix, Gloeotaenium, Anabaena, Oscillatoria* and diatoms.

Nephrocytium agardhianum **(Naegeli) [Plate 52.2; Figure (A)]**

Four reniform cells in each colony, within a thin gelatinous envelope. Chloroplast single, parietal, with a prominent pyrenoid. Cell walls thin. Cells 3.5–4.8μ broad; 8.4–11.3μ long. Colonies ranging in size from 29.4–32.8μ. In the monograph on Chlorococcales by M.T. Philipose, colony size is given as 40–60μ. Thus the present material smaller in colony size.

Collected in September, when the lake was full, along with *Chlorococcum vitiosum, Merismopedia* and diatoms.

Coelastrum microporum **Naegeli [Plate 52.2; Figure (B)]**

Colonies almost spherical. Eight to sixteen cells in each gelatinous colony. Outer boundary of colony thin, unlike that in *Oocystis* and *Gloeotaenium.* Intercellular spaces between cells small. Cells spherical; surrounded by indistinct mucilaginous investment. Chloroplast parietal, plate-like, with single pyrenoid. Diameter of cells 13–16.4μ diameter of colonies 46.2–49.4μ.

Collected in November, along with *Tetraedron trigonum, Cosmarium* and *Oscillatoria.* Sample collected from clear water at a site with no human interference.

Scenedesmus bijugatus **(Turpin) Kuetzing [Plate 52.2; Figure (C)]**

Colonies flat or slightly curved of 2, 4 or 8 cells arranged in a single linear series. Cells oblong, 3–5μ broad and 7–23μ long.

Found in the month of April. Collected along with *Chlorella, Microcystis* and diatoms entangled in floating masses of *Phormidium.*

Scenedesmus bijuga **variety *altenas* form *parvas* (Smith G.M) [Plate 52.2; Figure (D)]**

Colonies with four cells arranged in a regular subalternating series. Cell walls smooth. Cells 2.4–4μ broad and 4.8–9μ long.

Collected in January, along with *Cosmarium, Oscillatoria* and diatoms, from very shallow water at the edge of the dam.

Ecology of Chlorococcales

The observation of M.T. Philipose (1959, 1967) that the majority of Chlorococcales are free floating and planktonic, growing mostly in shallow, confined waters has been amply proved by the present work. Gonzalves and Joshi (1946) had also observed that in the North East, the order abounds in shallow waters. The taxa collected, unicellular as well as colonial, were all from shallow water at the edges of the dam and they were all planktonic while *Chlorococcum* was found free floating and also on moist soil.

The gelatinous parent membrane in *Oocystis* and *Nephrocytium,* the flat disc like or quadrate nature of the colonies in *Coelastrum* and *Scenedesmus,* presence of spines or bristles in *Tetraedron* and *Scenedesmus* etc., are all ecological adaptation for a pelagic habit.

Most of accounts available regarding Indian Chlorococcales are only morphological and taxonomical not much work has been done on their ecology. This was one of the reasons the present authors decided to correlate the occurrence of different members of Chlorococcales in different months of the year with the ambient temperature and some physico-chemical characteristics of the water in which they grow.

Ambient Temperature

Fritsch and Rich (1913) had observed that "the important genera of Chlorococcales usually occur during the warm months of the year, though high summer temperatures are as unsuitable as lower ones". Chlorococcalean members in the Sitarampur Dam have been most abundant in November, at an average of 19°C, before the ambient temperatures became too low. A secondary maximum was observed in March–April, at 27–29°C, before the ambient temperatures rose too high.

pH

It was in the months of October and November that the pH was recorded as 7.1. In all the other months, pH was either 7 or 6.9. This shows that Chlorococcales prefer slightly alkaline media for growth.

COD

Maximum recorded Chemical Oxygen Demand was in November–104 mg/l and the second and third highest recorded were 92 mg/l in April and 76 mg/l in March. In all other months, COD, were lower, ranging, in a descending order, from 71 to 17. It is thus probable that Chlorococcales prefer higher COD's.

BOD

Similarly, Biochemical Oxygen Demand recorded was also the highest in the months of maximum growth of the Chlorococcalean members, it was 21 mg/l in November, 22 mg/l in March and 26 mg/l in April in other months, BOD's were lower, ranging from 18 to 06.

TSS

Similar was the case with Total Suspended Solids. Higher values were recorded in the months of better Chrococcalean growth–57 mg/l in November and 62 mg/l in March. Values in other months ranged from 48–11.

Oil and Grease

Presence of oil and grease was nil from May to October continuously. In November, it was 6 mg/l and in December 2 mg/l. In January and February, they were in traces and in March, the value rose to 3 mg/l. In April, traces were recorded. These values tend to give an impression that oil and grease enhance Chrococcalean growth, but the values vary only slightly from month to month and hence may be ignored.

DO

Dissolved Oxygen did not vary much through the year. Values from 6.8 to 7.2 were recorded in different months, but no discernible pattern could be observed for the peak months November and March–April. So it is concluded that DO does not influence luxuriance in growth.

TDS

Similar was the case with Total Dissolved Solids as no specific correlating pattern was seen.

The fact that *Chlorella vulgaris* was observed throughout the year shows that it can grow in all ecological situations. Indeed, Tiwari, Singh and Kaur (2002) reported that *Chlorella vulgaris* grows not only in clean but also polluted waters. According to Round (1950), luxuriant growth of Chlorococcales indicates eutrophic condition of the water body. This must be the reason for higher values of COD, BOD and TSS in the months of maximum growth of Chlorococcales in the water body studied for the present work. Bairagi and Goswami (1992) had reported that higher the dissolved organic matter, more luxuriant is the growth of Chlorococcales in North Eastern India. In the present work, it has been observed that suspended solids enhance the growth of Chlorococcales in the Sitarampur Dam.

Acknowledgement

The authors are thankful to the Principal, Jamshedpur Women's College and the Head, Department of Botany for providing laboratory facilities required for algal identification. Thanks are also due Mr. Abid and Mr. Navin Singh of the laboratory of Quality Assurance (CML 2), Tata Motors for helping in water analysis.

References

APHA. 1985. *Standard Methods for Examination of Water and Wastewater.*

Kumar, N.N. Singh and M.S. Agarkar, 1986. Some rare and interesting Chlorococcales from Jabalpur, M.P. *Bionature*, 6(1): 35–36.

Kumara Hegde, B.A., G. Suresha, K. Ramdas and B. Yashovarma. 2005. Studies on limnological characteristics of Guruvayanakere pond near Belthangady of Karnataka. *Indian J. Environ. and Ecoplan.*, 10(1): 165–168.

Munawar M. 1966. Limnological studies in certain polluted and unpolluted water ponds of Hyderabad.

Pandey, U.C. and F.C. Gangwar. 1986. Chlorococcales of Bareilly. *Phykos*, 25: 144–147.

Patel, R.J. 1970. An enumeration of Chlorococcales of Gujarat. *Bombay Nat. Hist. Soc.*, 66: 665–669.

Philipose, M.T. 1967. *Chlorococcales*. Pub. ICAR, Krishi Bhawan, New Delhi.

Ramteke, D.S. and C.A. Moghe. 1988. *Manual on Water and Wastewater Analysis*. NEERI.

Smith, G.M. 1950. *The Freshwater Algae of the United States*. McGraw Hill Book Co., London.

Srivastava, P.N. and Bhoj Raj Odhwani. 1990. Additions to the Chlorococcalean flora of Jodhpur and its suburbs. *Phykos*, 29(1&2): 149–154.

Tiwari, D., S. Singh and P. Kaur, 2002. Chlorophyceae of the river Pandu. *Phykos*, 41(172): 115–120.

Zafar, A.R. 1964. On the ecology of algae in certain fish ponds of Hyderabad, India: Distribution of unicellular and colonial forms. *Hydrobiol.*, 24(4): 556–566.

Chapter 53

Detection of Trace Elements in Dahod Reservoir Water, Raisen District, M.P.

Arun Singh, H.C. Kataria and Subhash Pandey

Assistant Professors, P.G. Department of Chemistry, Government Geetanjali Girls P.G. College, Bhopal – 462 038, M.P., India

ABSTRACT

The water of 'Dahod' reservoir is polluted by man made activities, there are various reasons of pollution *e.g.*, sewage, industrial wastes and natural contaminants, Dahod reservoir has annul rainfall of 170.18 cm, catchment area 52 sq.km, live storage capacity 604 million calf and constructed on river Bagwani. To evaluate the water quality status, different trace elements Fe, Cu, Zn, Cd and Mn that contaminate in Dahod-reservoir and silt sediments has analyzed by Atomic absorption spectrophotometer model (2380) with air-acetylene flame. Water samples are acidified to pH 2 with nitric acid and then analysed on Perkin Elmer AAS. In the present study Fe, Cu, Cr, Zn, Cd and Mn has ranged from 0.010–0.50, 0.040–6.8, 0.010–0.30, 0.020–8.4, 0.01–0.06, 0.050–0.80 mg/l at different sampling stations. The concentrations in surface waters are lower than that of silt-sediments.

Introduction

Degradation of good water quality due to trace elements has received considerable attention during the recent years. Pollution in surface and groundwater system through human activities and natural process may enter into water reservoir. Solid waste from industrial units is being dumped near factories and sewage water is subjected to reaction with percolating rainwater and reaches the surface water level.

The source of metal pollution in the environment and their health hazards has briefly given by Handa (1994). Higher content of iron is causing a serious problem in water supplies from groundwater.

Pure water is never found in nature, as pure water is that contains only 2 parts of H_2 and 1 part of O_2 by volume water in nature contains a number of impurities in varying amounts. The rainwater, which is originally pure, also absorbs various gases, dust and other impurities. This water when moves on the ground further carry silts organic impurities etc.

The present study aimed to see the suitability of surface-reservoir water as a source of drinking and irrigation purpose. The study will help the authorities for planning in drinking and irrigation supply schemes in coastal catchment area of Raisen district of M.P.

Inappropriate disposal and accidental release into environment of toxic and hazardous substances from chemical and industrial activities have created environment and public health problem all over the world. Detection of traces of toxic chemicals in surface water has boasted the public awareness of the risks posed by hazardous wastes (HWs) to human health and quality of environment.

Dahod village in situated in Tehsil-Gohar Ganj, District Raisen of Madhya Pradesh at longitude 77°20′30″ and latitude 23°2′0″. Dahod dam is the main source of water to fulfil the requirement for drinking and irrigation of Raisen district, the distance of dam is 25 kms far from Bhopal, state of M.P. Dahod reservoir is under environmental stress due to Siltation (by annual rainfall of 170.18 cm), human encroachment, high macrophytic population and sewage input from various resources. Due to decreasing in water quality day-by-day, the present study becomes necessary to assess the trace elements for water quality index of this dam. The soil inflows by rain run-off water from catchment are (52 sq.km) and basin directly affects the water quality of reservoir. Sampling Stations has chosen are:

1. Near Village Side
2. 1 km Downstream
3. 2 km Downstream
4. Tank Bund
5. Waste-wear
6. Spill Channel
7. Dahod Village–Guest House
8. Dhobi Ghat

Materials and Methods

Eight sampling stations were chosen at different points of Dahod reservoir. Water samples are acidified 2 or 3 times and evaporated up to dryness and finally dissolved in HNO_3 and the samples them used to Perkin Elmer, Atomic Absorption Spectrophotometer with air-acetylene flame to detect the trace elements Fe, Cu, Zn, Cr, Cd, and Mn at different wavelength of light and width of specified slit.

The instruments are calibrated for each metal to be detected, with known standard solution and programming aspirate the pretreated samples to the flame. Set zero with double distilled water before each detection. Results are displayed in μ gm/l from which the actual concentration in the water sample can be calculated and total metal content of the water samples can be obtained in (mg/l). The method is similar to APHA (1985), NEERI (1986).

Results and Discussion

Iron

Iron may be present in groundwater in varying quantities from 0.5 to 100 ppm (Ernst, 1991), ISI 1983, and WHO (1984) have set a desirable limit of 0.3 mg/l and maximum permissible limit of 1.0 mg/l for drinking purpose. In the present study iron found in the range of 0.020–0.56 iron in silt sediments, 0.020–0.036 in sewage disposal and 0.018–0.036 ppm in domestic sewage confluence points, lower values of 0.020 ppm at SS_4, SS_8 while higher values are reported at SS_4 and SS_2, SS_7 and SS_3, SS_8, SS_6 has lower values. Higher values have found in silt sediments, sewage disposal and domestic confluence points.

Table 53.1: Detection of Heavy Metals in Dahod Reservoir Water

(A) Concentration of heavy metals in Dahod Silt–Sediments sampling metals in mg/w during 2003–04

Stations	*Fe*	*Cu*	*Cr*	*Zn*	*Cd*	*Mn*
1	0.50	6.80**	BDL	8.4**	3.8**	0.80
2	0.56**	4.2	0.08	8.0	1.0*	0.62*
3	0.48	2.6	0.06*	4.3	1.2	1.2**
4	0.02*	1.4*	BDL	2.4*	nf	BDL
5	0.026	5.6	0.09**	6.2	1.6	1.0
6	0.028	6.2	0.06	6.8	2.4	BDL
7	0.030	4.7	nf	5.4	3.0	0.90
8	0.020*	4.1	nf	5.0	2.8	0.86

(B) Heavy metals near sewages disposal point

Stations	*Fe*	*Cu*	*Cr*	*Zn*	*Cd*	*Mn*
1	0.030	3.0**	0.08**	0.52	BDL	0.14
2	0.036**	1.20	0.07*	0.70	BDL	BDL
3	0.30	0.50	nf	0.72	0.010*	0.80
4	0.026	1.10	nf	0.03*	0.014	0.84**
5	0.022	0.14	BDL	1.20	0.016	0.42
6	0.030	0.05	BDL	2.40**	0.02	0.48
7	0.020*	0.06	BDL	2.00	0.03*	nf
8	0.024	0.04*	BDL	1.10	0.010*	0.08*

(C) Heavy metals near sewages disposal point

Stations	*Fe*	*Cu*	*Cr*	*Zn*	*Cd*	*Mn*
1	0.036**	2.40	BDL	0.60	BDL	0.14
2	0.024	2.30	BDL	0.70	BDL	BDL
3	0.018*	2.50	nf	0.04*	0.010*	0.80**
4	0.026	2.80**	0.05*	0.08	0.03*	0.75
5	0.030	1.60	0.06**	2.40	0.02	0.48
6	0.021	1.80	BDL	2.80**	BDL	nf
7	0.022	0.30	BDL	1.20	BDL	0.06*
8	0.018*	0.08*	BDL	1.16*	nf	

nf: Not found; BDL: Below detection limit.

TLV mg/m^3 Cu: 0.2; Cr: 0.01; Zn: 1.0; Mn: 5.0.

Copper

Copper is ubiquitous in the environment and hence frequently present in the water. Since copper is both an essential and potentially toxic element there may be risk to living beings if there is too little or too excess of copper in the environment. Large dose of Cu irritate the Stomach (Ross, 1955). ISI has fixed a desirable limit of 0.05 mg/l for Cu. In the present study copper present in the range of 1.4–6.8, 0.10–3.0 and 0.08–2.80 ppm in Dahod dam silt sediments, near sewage disposal and in domestic confluence points respectively are well within the permissible limit. Copper is essential component of Key–metalloenzyme that maintains the vascular and nervous system. Its deficiency causes livestocks disease. Excessive accumulation of copper in tissues occurs in Wilson's disease, causes neurological dysfunction, progressive cirrhosis.

Chromium

Chromium salts are widely used in industrial processes and the products used by human beings, they enter into water bodies through discharge of waters. The normal level of Cr (V1) in drinking water is 3 to 40 ppm. Natural water contains traces of Cr (Cr^{3+}) under strong oxidizing conditions. It is converted into Cr^{6+} and occurs as CrO_4^{2-}, which is usually results pollution from industrial and domestic effluents and wastes. The maximum contaminant level of Cr is 0.05 mg/l. In the present study Cr has found in BDL or nil 0.06–0.08, BDL–0.08, BDL–0.06 ppm at SS_4, SS_7, SS_8–SS_2 in silt sediments, sewage disposal and domestic confluences respectively.

Zinc

The toxicity has been reported with dizziness, muscular incardination, dizziness, renal failure and anaemia. Most of the cases of Zinc toxicity have resulted from the irrigation of acid food kept in galvanized metal container. Galvanized pipes are contaminated by the flow of waste from industrial sources or toxic waste sites. Bio-chemically, Zinc is present in over 160 enzymes with many functions. Zinc deficiency is associated with impaired growth, parakeratosis, failure of sexual maturation and malnutrition. The. permissible limit for Zn is 5.0 mg/L by ISI: 10500. Zn conc. In the present study, lower values of Zn concentration is recorded at SS_4, SS_3 and higher at SS_1, SS_2, SS_6, SS_5 in silt sediments. Lower values of Zn near sewage and domestic effluents are at SS_4, SS_1 and at SS_3, SS_1 while the higher values are observed at SS_8, SS_5, SS_7, and SS_6 and at SS_5, SS_6 Sampling stations due to sewage.

Cadmium

Cd is a non essential non-beneficial element known to have high toxic potential. The Cd content in the study area varies from 1.0–3.8, 0.010–0.016 and 0.01–0.03 ppm at A, B, C, points. Desirable limit for Cd is 0.01 mg/l prescribed by ISI for drinking water (BIS, 1991). Beyond the limit, water becomes toxic. Cadmium can cause bronchitis, anaemia and renal stone formation in animals. The water is contaminated with Cd by industrial discharges and leaches from land filled areas. Drinking water is generally contaminated with galvanized iron pipe and plated plumbing fittings of water distribution system.

Manganese

Manganese is a vital micro-nutrient for has both plants and animals. The deficiency state of Mn has not been found in human. In other mammals, Mn deficiency is characterized by defective growth, bone abnormalities, reproductive dysfunction. The concentration of Manganese recorded a maximum level of 1.2 in silt sediments at SS_3, and 0.84 and 0.80 ppm in SS_4 and SS_3 in domestic sewage. Mn is an essential element, which does not occur naturally as a metal but is found in various salts and minerals frequently in association with iron-compounds. A concentration of 1.0 mg/l has been recommended as a desirable limit and 0.3 mg/l as the maximum permissible limit for drinking water (BIS, 1991). The presence of Mn beyond the permissible limit imparts alien taste to water. Mn also has adverse effects on domestic uses and water supply structures. Higher concentration may be attributed to the reducing conditions of the water and dissolution of Mn bearing salts from the soil strata (Jain *et al.*, 2004).

Acknowledgement

We are thankful to M.P. Pollution Control Board to help in this study and EPCO, Bhopal to facilitate the Library Consultation.

References

APHA. 1986. *Standard Methods for the Examination of Water and Wastewater*. APHA, AWWA, WPCF, New York.

BIS. 1983. *Indian Standards Specification for Drinking Water*, IS: 10500. Bureau of Indian Standards, New Delhi.

BIS. 1991. *Specifications for Drinking Water*, IS: 10500. Bureau of Indian Standards, New Delhi.

Drinking Water and Health, Vol. 1. National Academy of Sciences, Washington, DC; National Academy Press.

EPA (Environmental Protection Agency). 1980. Ambient water quality criteria for zinc, publication 440/5–80–79.

Ernst, M. 1991. *Metals and their Compounds in the Environment*. VCH, Weinheim, p. 894–1332.

Jain, C.K., Sudhir Kumar and Y.R. Rao. 2004. Trace element contamination in a coastal aquifer of Andhra Pradesh, *Poll. Res.*, 23(1): 3–23.

Kaplary, R.D. *et al.* 2003. Trace elements on groundwater of Tuppa Area New Nanded, Maharashtra. *Poll. Res.*, 22(4): 595–599.

Kataria, H.C. 1994. Heavy metals contamination and pollution in Betwa River. *IJEP*, 15(1): 34–38.

Kataria, H.C. 1995. Analysis of trace elements in lower lake water of Bhopal. *Oriental J. of Chemistry*, 11(3): 288–289.

Kataria, H.C. 1995. Heavy metal water pollution of Pipariya Township of Madhya Pradesh. *J. Freshwater Biol.*, 8(2): 105–107.

More, T.G. *et al.* 2003. Impact of heavy metals on DNA contents *Lamillidenes marginalis*. *Poll. Res.*, 22(4): 605–611.

NEERI. 1986. *Manual on Water and Wastewater Analysis*. National Environmental Engineering Research Institute, Nagpur, pp. 340.

Ross, A.I. 1995. Lancet, 11,87, e.f. Holaen, W.S. 1970.

Srikanth, R.A., Madhumohan Rao, C.H., Shravan Kumar and Anees Khanum. 1993. Pb, Calcium, Ni and Zn contamination of groundwater around Hussein Sugar lake, Hyderabad. *Bull. Env. Contam. Toxicology*, 50(1): 138–143.

Warrant-Viessman, Jr. and Mark J. Hammer. 1985. *Water Supply and Pollution Control*, 4th ed., Harper and Row, Publ. Inc., p. 220–223.

WHO (World Health Organization). 1984. *Guidelines for Drinking Water Quality*, Vol. 1.

WHO. 1972. *Geochemical Environments, Trace Elements and Cardiovascular Diseases Bull.*, pp. 47.

Chapter 54

Limnological Studies of Nambol River, Manipur with a Note on its Aquatic Bioresources

R.K. Rajeshwari Devi, Haobijam Dhamendra**, R.K. Gyaneshwari*** and Laishram Kosygin***

**Department of Zoology, Oriental College, Imphal West – 795 001, Manipur, India*
***Loktak Development Authority, Ningthoukhong – 795 126, Manipur, India*
****G.P. Women's College, Imphal West – 795 001, Manipur, India*

ABSTRACT

Water quality and aquatic bioresources of the Nambol River were studied during September 2000–June 2002. The levels of free CO_2 (12.21 mg/l), nitrite-nitrogen (0.035 mg/l), inorganic phosphorus (0.088 mg/l) and flora and fauna suggest organic pollution and nutrient enrichment in the river. Aquatic bioresources of the river are given.

Keywords: Nambol river, Water quality, Bioresources, Pollution.

Introduction

Water, the most vital resource for all kind of life is adversely affected both: qualitatively and quantitatively by all kinds of human activities on land, in air or in water (Trivedy and Goel, 1986). Today the river water which plays an important role in agriculture and domestic purposes receive untreated sewage, domestic waste, industrial and agricultural effluents containing substances varying from simple nutrients to highly toxic substances. As a result many rivers are polluted and brought a veritable water, crisis. In India, studies on water quality of rivers were reported by many workers (Sangu and Sharma, 1985; Das and Sinha, 1993; Jameson and Rana, 1996; Sharma and Agarwal, 1999; Mahanta and Patra, 2000; Baruah and Baruah, 2003 and Kumar, 2003).

Manipur is blessed with numerous rivers and streams, of which many are feeding important lakes of the state. However, limnological studies of these rivers of Manipur in general, have received little attention. Nambol River is a perennial river which arises from Kangchup hills and finally falls into the Loktak Lake near Ishok. The river while traversing through hills, before reaching plain, maintain unpolluted water quality due to less human activities, but due to human population growth, urbanization, and associated land use changes have led to water quality deterioration in the downstream of the river. In the present study an attempt has been made to study the water quality and aquatic bioresources of the Nambol river.

Materials and Methods

Surface water samples were collected between 9 and 11 h from the down stream of Nambol river near Nambol Market on quarterly basis (September /postmonsoon, December/winter, March/ premonsoon and June/monsoon) during 2000 and 2002. Physico-chemical parameters were analyzed following APHA (1989) and Trivedi and Goyal (1986). Fishes were identified following Jayaram (1981), Talwar and Jhingran (1991) and for snails following Rao (1989). Aquatic Macrophytes were identified following Adoni *et al.* (1985) and by consulting other relevant literatures.

Results and Discussion

Water Quality

The physico-chemical characteristics of the Nambol River are summarized in Table 54.1. The surface water of the river was characterized by high concentration of free CO_2 (12.21 mg/l), moderate total dissolved solids (133.87 mg/l), high concentration of nitrite nitrogen (0.035 mg/l) and inorganic phosphorus (0.088 mg/l). The river water is not favourable for fish culture, as average value of free CO_2 exceeds the tolerance limit (I.S.I., 1982). Presence of nitrite-nitrogen indicates the organic pollution in the Nambol river (Trivedy and Goel, 1986). The average concentration of inorganic phosphorus exceeds the standard permissible limit (0.050 mg/l) recommended by USEPA (1976) for any tributary discharging into a lake.

Well mark seasonal variations have been recorded in respect of physico-chemical characteristics. The lower value of pH was observed during monsoon and post-monsoon periods. It showed negative correlation with free carbon dioxide ($r = -0.555$). Highest value of Dissolved Oxygen was observed during December and lowest during June (Figure 54.1). It was negatively correlated with free carbon dioxide ($r = -0.667$). The free CO_2 concentration was highest during June and lowest during December.

The concentrations of nutrients were higher during winter (December) and pre-monsoon (March), decreasing in the monsoon (June) and lowest in post-monsoon (September) as shown in Figure 54.2. This may be attributed to the fact that during winter and pre-monsoon period the river water is more or less stagnant with less water volume thereby increasing the nutrient concentration. Moreover, the river receives domestic and municipal wastes from the Nambol area, which were deposited on the river bank during the lean season, through the surface runoff created by the pre-monsoon rainfall.

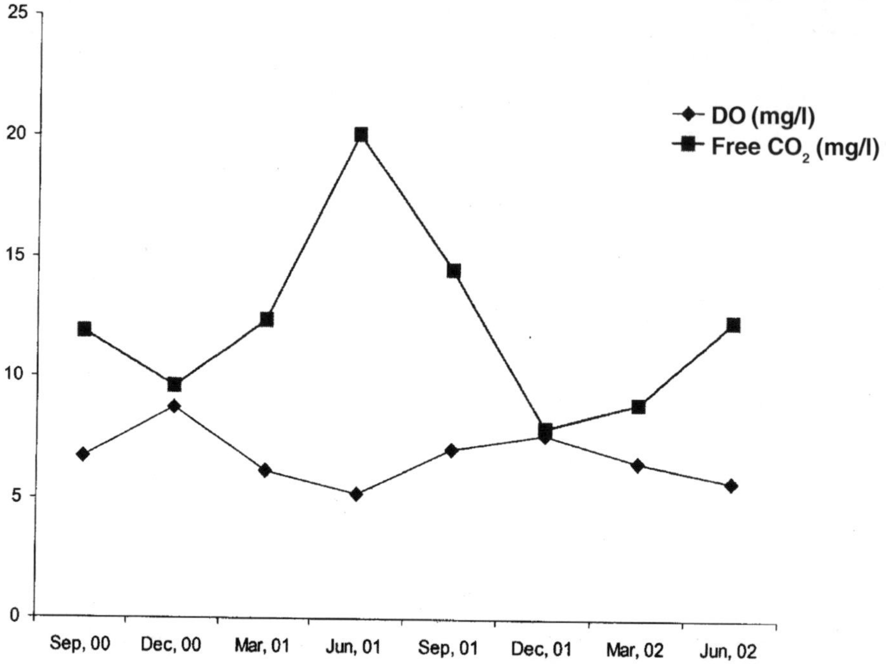

Figure 54.1: Seasonal Variations in Free CO_2 and Dissolved Oxygen (DO) of Nambol River (September 2000–June 2002)

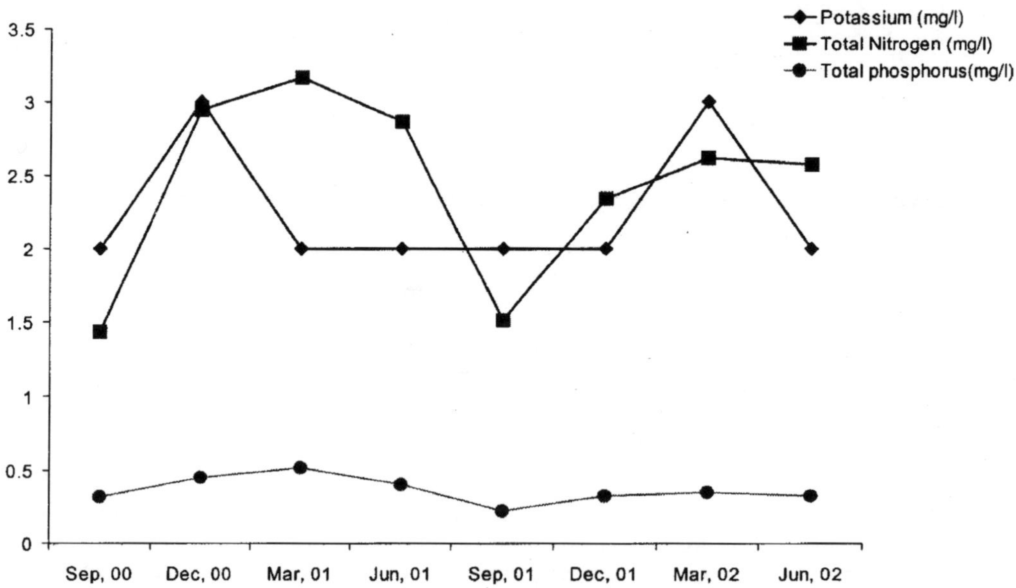

Figure 54.2: Seasonal Variations in Total Nitrogen, Total Posphorus and Potassium of Nambol River (September 2000–June 2002)

Aquatic Bioresources

List of fishes, and macrophytes of the Nambol river are given in Tables 54.2 and 54.3 respectively. The fish fauna of the river included 23 species belonging to 19 genera and 13 families. Out of which 21 species are distributed in the down stream of the river and 4 species are identified as species of high fishery potential. The malaco fauna was represented by 5 species of snails belonging to 4 genera and 3 families. They are *Angulyagra oxytropis* (Benson), *Pila globosa* Swainson, *Pila* sp., *Brotia costula* (Rafinesque) and *Lamellidens* sp. These species of snails are edible and sold in vegetable markets of Manipur. Kosygin (2001) remarked that there is a high potential of edible shellfish production in the north eastern part of India through aquaculture and this may provide additional food resources in the region.

Table 54.1: Physico-chemical Characteristics of Nambol River at Nambol during September 2000–June 2002

Sl.No.	*Parameters*	*Min.*	*Max.*	*Average*	*SD*
1.	Water Temp.(°C)	15.4	28.5	23.21	4.67
2.	pH	6.2	8.1	6.96	0.6
3.	DO	5.2	8.7	6.67	1.13
4.	Free CO_2	7.9	20.1	12.21	3.87
5.	Total Dissolved Solids	89	136	133.87	40.11
6.	Chloride	7.2	12.9	9.95	1.71
7.	Calcium	4	12.4	9.65	2.66
8.	Magnesium	4.8	8.9	7.02	1.45
9.	Sodium	16	23	19.5	2.93
10.	Potassium	2	3	2.25	0.46
11.	Nitrite-nitrogen	0.017	0.054	0.035	0.01
12.	Nitrate-nitrogen	0.215	0.881	0.398	0.24
13.	Total Nitrogen	1.427	3.164	2.431	0.64
14.	Inorganic Phosphorus	0.04	0.151	0.088	0.03
15.	Organic Phosphorus	0.134	0.420	0.272	0.09
16.	Total Phosphorus	0.218	0.517	0.36	0.09

Note: All the values are given in mg/l except water temperature and pH.

The macrophytes of the river included 11 species belonging to 10 genera and 8 families. These species are utilized as fodder, fuel, and materials for fish breeding and fencing. Singh (2005) remarked that some of the aquatic plant species are becoming rare, endangered due to accumulation of pollutants in the water bodies and hence required conservation which includes the preservation of habitats and species. However, a detail survey on the endangered plant genetic resources of the region and their habitats are yet to be taken up. In the Nambol river biodiversity of both flora and fauna was poor and most of the species collected are widely distributed forms which can thrive well in the polluted waters.

From the above discussion, it is clear that down stream of the Nambol river is polluted as it has poor water quality with high concentration of free CO_2, nitrite–nitrogen, phosphorus and flora and fauna, which can thrive well in polluted water bodies. Proper conservation measures including awareness programs to the local communities should be taken up to save the river and its aquatic bio-resources.

Table 54.2: Fish Fauna of Nambol River

Sl.No.	Fishes	Local Name	Distribution		Fishery Potential
			Upstream	Downstream	
Family: Cyprinidae					
1.	*Cyprinus carpio* (Linnaeus)	Puklaobi	–	+	High
2.	*Cirrhinus mrigala* (Ham-Buch)	Mrigal	–	+	High
3.	*Ctenopharyngodon idellus* (Val.)	Npichabi	–	+	High
4.	*Puntius sophore* (Ham)	Phabounga	+	+	Low
5.	*Ampblypharyngodon mola* (Ham)	Mukanga	+	+	Low
6.	*Esomus danricus* (Ham)	Ngasang	+	+	Low
7.	*Garra* sp.	Ngamu sangum	+	–	Low
Family: Balitoridae					
8.	*Schistura* sp.	Ngatup	+	–	Low
Family: Cobitidae					
9.	*Lepidocephalus guntea* (Ham)	Ngakijou	+	+	Medium
10.	*Lepidocephalus irrorata* (Hora)	Nganup.	–	+	Medium
Family: Bagridae					
11.	*Mystus bleekeri* (Day)	Ngasep	–	+	Medium
Family: Sisoridae					
12.	*Ompok bimaculatus* (Bloch)	Ngaten	–	+	Medium
Family: Clariidae					
13.	*Clarias batrachus* (Linnaeus)	Ngakra	–	+	Medium
Family: Heteropneustidae					
14.	*Heteropneustes fossilis* (Bloch)	Ngachik	–	+	Low
Family: Symbranchidae					
15.	*Monopterus albus* (Zuiew)	Ngaprum	–	+	Medium
Family: Chandidae					
16.	*Chanda nama* (Hamilton)	Ngamhai	–	+	Low
Family: Gobiidae					
17.	*Glossogobius giuris* (Ham)	Nailon ngamu	–	+	Low
Family: Anabantidae					
18.	*Anabas testudineus* (Bloch)	Ukabi	–	+	High
Family: Belontiidae					
19.	*Colisa fasciatus* (Schn)	Ngabemma	+	+	Low
20.	*Colisa* sp.	do	–	+	Low
Family: Channidae					
21.	*Channa gachua* (Hamilton)	Meitei ngamu	+	+	Medium
22.	*Channa marulius* (Hamilton)	Porom	–	+	Medium
23.	*Channa punctatus* (Bloch)	Ngamu bogra	–	+	Medium

Table 54.3: Macrophytes of Nambol River

Sl.No.	*Macrophytes*	*Local Name*	*Economic Utilization*
Family:	**Poaceae**		
1.	*Brachiaria mutica* (Forssk) Stapf	Paragrass	Fodder
2.	*Echinochloa stagnina* Retz.	Hoop	Fodder
3.	*Phragmites Karka*	Tou	Fodder, Fuel, Fencing
Family:	**Convolvulaceae**		
4.	*Ipomea aquatica* Forsk	Kolamni	Food
Family:	**Amaranthaceae**		
5.	*Alternanthera pheloxeroides* (Mart) Grised	Kabonapi	Fodder
Family:	**Pontederiaceae**		
6.	*Eichhornia crassipes* (Mart) Solms.	Kabokang	Fish breeding
Family:	**Araceae**		
7.	*Colocasia esculenta* (Linn.) Schott.	Lambal	Fodder, Food
Family:	**Hydrocharitaceae**		
8.	*Hydrilla verticillata* (L.F.) Royle	Charang	Fodder
9.	*Hydrilla* sp.	Charang	Fodder
Family:	**Salviniaceae**		
10.	*Salvinia natans* Hoffim	Kangkup	Fish breeding
Family:	**Lemnaceae**		
11.	*Lemna* sp.	Kangmacha	Fodder

References

Adoni, A.D., Joshi Gunwant; Ghosh Kartik; Chaurasia, S.K. Vaishya, A.K. Yadav, M. and Verma, H.G. 1985. *Work Book on Limnology*. Pratibha Publishers, Sagar, pp. 216.

APHA, AWWA, WPCF. 1989. *Standard Methods for the Examination of Waters and Wastewaters*, 17th Edition. American Publication Health Association, Inc. New York.

Baruah, B.K. and Baruah, D. 2003. Study on water quality of Subansiri river in Asssam: An EIA approach for a proposed hydroelectric power project. *Indian J. Environ. and Ecoplan.*, 7(2): 381–384.

Das, N.K. and Sinha, R.K. 1993. Assessment of water quality of Ganga River, Patna, India. *Environment and Ecology*, 11: 829–832.

I.S.I. 1982. Indian standard, tolerance limit for inland surface waters subject to pollution, 2nd revision. Indian Standards Institution publication, pp. 18.

Jameson, J. and Rana, B.C. 1996. Pollution status of the river complex Sabarmati at Kheda region of Gujarat, I: Physico-chemical characteristics. *Poll. Res.*, 15: 53–55.

Jayaram, K.C. 1981. *Freshwater fishes of India, Pakistan, Bangladesh, Burma, and Sri Lanka*. Handbook, Zoological Survey of India, Kolkata, pp. 475.

Kosygin, L. 2001. Biodiversity and utilization of Molluscs resources of north-east India. *Journal of the North Eastern Council*, 21(4): 29–33.

Kumar, A. 2003. Ecology of river Mayurakshi in Jharkhand, India. In: *Ecology and Conservation of Lakes, Reservoirs and Rivers*, (Ed.) A. Kumar. APH Publishing Corporation, New Delhi, p. 1–38.

Mahanta, B.K. and Patra, A.K. 2000. Studies on the water quality index of river Sanamachhakandana at Keonjhar Garh, Orissa, India. *Poll. Res.*, 19: 371–385.

Sangu, R.P.S. and Sharma, K.D. 1985. Studies of water pollution of Yamuna river at Agra. *Ind. J. Environ. Hlth.*, 27: 257–261.

Sharma, B.S. and Agarwal, A. 1999. Assessment of water quality of river Yamuna at Agra. *Poll. Res.*, 18: 109–110.

Singh, P.K. 2005. Aquatic bioresources of Manipur. Souvenir: *National Seminar on Fishery, Aquatic Bioresources and Sustainable Development*, p. 30–33.

Subba Rao, N.V. 1989. *Handbook: Freshwater Molluscs of India*. Zoological Survey of India, Kolkata, pp. 289.

Talwar, P.K. and Jhingran, A.G. 1991. *Inland Fishes of India and Adjacent Countries, I.* Oxford and IBH Publ. Co. Pvt. Ltd., New Delhi, Mumbai, Kolkata, pp. 541.

Trivedy, R.K. and P.K. Goyal, 1986. *Chemical and Biological Methods for Water Pollution Studies.* Environmental Publication, Karad, pp. 248.

U.S.E.P.A. 1976. Quality Criteria for water. EPA–440/9–76–023, Washington D.C.

Index